AST 101
Fundamentals of Astronautics
Version 3.0

Jason D. Reimuller, Ph.D.
editor-in-chief

Contributing Authors

Erik Seedhouse, Ph.D.	Embry-Riddle Aeronautical University, Daytona, FL.
Armin Kleinboehl, Ph.D.	NASA Jet Propulsion Laboratory, Pasadena, CA
Ken Ernandes	Integrated Spaceflight Services, LLC., Boulder, CO.
Paul Buza, M.D.	Southern Aeromedical Institute, Melbourne, FL.
Dave Fritts, Ph.D.	G.A.T.S. Inc., Boulder, CO
Ravi Komatireddy, Ph.D.	Scripps Research Institute, San Diego, CA
Ted Southern	Final Frontier Designs, Brooklyn, NY.
Gerald Lehmacher, Ph.D.	Clemson University, Clemson, SC.
Zoltan Sternovsky, Ph.D.	University of Colorado, Boulder, CO.
Van Wampler	CinemaRaven, Boulder, CO.

Integrated Spaceflight Publications Office
3360 Mitchell Lane, Suite C
Boulder, Colorado 80301

First Printing 2015

ISBN: 978-0-578-16142-6

Project PoSSUM is a 501(c)3 non-profit corporation
    International Institute for Astronautical Sciences Website: www.astronauticsinstitute.org
    Project PoSSUM Website: www.projectpossum.org

Printed and bound in the United States of America.

Preface

This is an exciting time for citizen-science in our upper atmosphere. Whether your intent is to pursue a career in astronautics through one of our professional degrees or certification programs, or your journey is simply one of curiosity of the science of aeronomy and of human spaceflight, I would like to personally welcome you to the International Institute for Astronautical Sciences.

As a graduate of AST 101, you will find many avenues to enable cutting edge research in a variety of fields. You will also be part of Project PoSSUM, IIAS's aeronomy program, which will open doors for you to both do research and to engage and educate the public about the upper atmosphere and the vital role that it plays in the study of our global climate. Even though humanity has maintained a presence in orbit for decades, the mesosphere is still largely unknown. The mesosphere is a region that we have only briefly transited in our forays to orbital space. It is a region that harbors strange 'space clouds', unusual electrical phenomena, and ionization that brings silence to vehicles re-entering through it. It is an area too high to access by balloon or aircraft yet too low to access by orbital spacecraft. It is the most unknown part of our atmosphere, and yet soon we will have the means to access this elusive region and claim our presence there. IIAS researchers working with Project PoSSUM will be among the first explorers of the mesosphere; not just travelers passing through to orbit or returning from orbit, but venturing there to understand.

As the tourist travels 'away from'; the explorer travels 'towards'. An explorer travels with an unbiased mind seeking to understand. To the explorer, the journey is the classroom. The explorer welcomes surprise. The explorer invites challenge to his assumptions and beliefs. The explorer realizes that everything and everyone that crosses the journey brings a lesson and an opportunity to grow, and welcomes the changes these influences bring. And as the explorer's environs reveal their secrets, the explorer accepts a responsibility to preserve the beauty of what is seen and experienced. The explorer becomes an ambassador and advocate of all that reveals itself during the journey, because one can never regain the ignorance of the times before the journey started.

The editor would like to dedicate this manual to Mr. Bruce Hulley (1934-2015). Bruce carried a passion of flying through life, logging over 22,000 flight hours in small aircraft, and served as the editor's aviation mentor since 2007. His work organizing and executing a noctilucent cloud flight research campaign in North Canada was essential to the completion of the editor's doctoral dissertation, from which Project PoSSUM was born.

Jason D. Reimuller, PhD,
Executive Director, *International Institute for Astronautical Sciences*

Table of Contents

Table of Figures

List of Tables

INTRODUCTION

PoSSUM OBJECTIVES

Project PoSSUM is an atmospheric science program within the International Institute for Astronautical Sciences (IIAS) that uses next-generation spacecraft to study the upper atmosphere. The objectives are twofold: conduct important science and engage the public through direct participation. In this way, atmospheric science is communicated easier to the general public.

Conducting Science in the Mesosphere

The *mesosphere* lies at an altitude that is too high to access in an aircraft or balloon, and as it lies below the minimum altitude required by orbital spacecraft, the only in-situ data that has been obtained has been through use of sounding rockets. As a result, the mesosphere is not well understood and has frequently been termed 'the Ignorosphere'.

Yet several very interesting phenomena occur in this poorly understood region. Red sprites and blue jets are a type of electrical discharge often observed in the lower part of the mesosphere, and noctilucent clouds and density shears form in the upper regions of the mesosphere. The PoSSUM imagery and tomography experiments will focus on *noctilucent clouds* (NLCs). These clouds are the highest clouds in the Earth's atmosphere, 83 km (50 miles) and are observed slightly below the mesopause in the polar summertime. These clouds are of special interest, as they are sensitive to both global climate change and to solar/terrestrial influences.

Engaging the Public and Communicating Science

The primary objective of Project PoSSUM is to realize the opportunities presented by this new generation of reusable spacecraft to access and study the mesosphere in means previously unavailable or cost prohibitive. An important secondary objective of PoSSUM is to engage the public in aeronomy research through direct public participation. The mesosphere is one of the most sensitive parts of our entire planet; small changes in our environment can reflect themselves profoundly through observables in the upper atmosphere. A key goal of PoSSUM is to inspire the public and to communicate the science to broader audiences.

WHY DO WE CARE ABOUT NOCTILUCENT CLOUDS?

Noctilucent Clouds (NLCs), also referred to as *Polar Mesospheric Clouds* (PMCs) from space-based observations, are widely studied by global climate scientists as well as developers of re-entry vehicles and comparative planetologists. Climatologists are interested in the connection between NLC presence and our changing atmosphere; NLCs have been observed in increasing frequency and geographic extent throughout the last century and increase is seen by many scientists as a probable indicator of long-term global climate change. Re-entry vehicle operators and architects are interested in the possible risks of re-entry through NLC activity. Comparative planetologists are interested in studying our upper atmosphere as a proxy of low-density atmospheres on other worlds.

Implications on Climate Change

When people are asked about climate change, they typically think of a polar bear, alone on a shrinking piece of sea ice! Yet this image, and many others, is an effect of a warming climate. Yet it is the atmosphere that is the cause of this change. The atmosphere is the medium that both natural and anthropogenic (man-made) contributors of global climate change and the result is a warming planet.

The known drivers of global climate change include both natural and anthropogenic elements. Known natural drivers include solar output, orbital variations, volcanism, and ocean processes. The net radiative input to the Earth is predominantly produced by the sun, and both long- and short-term variations in solar intensity, as well as variations in the tilt of the Earth, termed *Milankovitch cycles*, are known to affect global climate. Volcanism is also considered a natural driver of climatic variability as large eruptions can affect the radiative balance of the Earth and cause brief periods of cooling by partially blocking the transmission of incident solar radiation.

However, it is generally accepted that global climate change is principally driven by anthropogenic, or man-made, causes largely through the atmospheric forcing of carbon dioxide (CO_2), methane, nitrous oxide, and other gases. Human activities such as the burning of oil, coal and gas, and deforestation, generate more than 100 times the amount of carbon dioxide emitted by volcanoes. The increase of concentration of atmospheric CO_2 from pre-industrial concentrations has doubled since 1980 and the present level is higher than at any time during the last 800 thousand years.

The largest source of anthropogenic CO_2 emissions globally is the combustion of fossil fuels such as coal, oil and gas in power plants, automobiles, industrial facilities and other sources. In addition, a number of specialized industrial production processes such as mineral production, metal production and the use of petroleum-based products also lead to CO_2 emissions. Furthermore, since trees and plants absorb or remove CO_2 from the atmosphere, deforestation can lead to increased levels of CO_2 emissions.

In addition to CO_2, methane (CH_4) is a significant driver of global climate change. The US EPA estimates that the majority of methane emissions come from human-related activities such as fossil fuel production, animal husbandry (enteric fermentation in livestock and manure management), rice cultivation, biomass burning, and waste management. Natural sources of methane include wetlands, gas hydrates, permafrost, termites, oceans, freshwater bodies, non-wetland soils, and other sources such as wildfires.

Nitrous oxide (N_2O) is produced by both natural and human-related sources. Primary human-related sources of N_2O are agricultural soil management, animal manure management, sewage treatment, mobile and stationary combustion of fossil fuel, and nitric acid production. Nitrous oxide is also produced naturally from a wide variety of biological sources in soil and water, particularly microbial action in wet tropical forests.

Numerous gases are categorized as having a high potential to contribute to global warming due to their ability to catalyze reactions in the atmosphere. These gases are emitted from a variety of industrial processes including aluminum production, semiconductor manufacturing, electric power transmission, magnesium production and processing, and the production of HCFC-22. Chlorofluorocarbons (CFCs), hydrochlorofluorocarbons (HCFCs) and halons have long been known to deplete ozone in the atmosphere.

Observable Indicators of Atmospheric Climate Change

The remote sensing of the atmosphere has generally been focused on better characterizing the physical thermodynamical processes (atmospheric physics) and the chemical composition (atmospheric chemistry) of the atmosphere. Climate change is studied through analysis of the frequency and trends of meteorological systems over periods spanning years to millennia. In the upper layers of the atmosphere where dissociation and ionization become key factors, the atmosphere is particularly sensitive to the natural or human-induced factors that cause climates to change.

Over the past few decades NLCs have been considered an observable phenomenon of the upper atmosphere that may also be a good indicator of global climate change, due to the direct relationship that has been observed between their presence and man-made industrial products, such as atmospheric aerosols from industrial gasses, water vapor from rocket exhaust, and warmer tropospheric temperatures driven by 'greenhouse processes'.

NLCs have been observed in increasing frequency and geographic extent throughout the last century. This increase is seen by many scientists as a probable indicator of long-term global climate change as they are likely attributable to increasing levels of atmospheric 'greenhouse gasses', such as carbon dioxide and methane. Carbon dioxide warms the lower atmosphere, but radiates heat into space in the thin upper atmosphere. Thus, as carbon dioxide levels rise, the upper atmosphere cools. Further, as methane levels rise, more hydrogen is released through photo-disassociation in the middle atmosphere (~60km), which reacts with oxygen to form water vapor. Since these clouds form on condensation nuclei through cold temperatures and the presence of water vapor, and since the presence of carbon dioxide has been observed to act as a cooling agent in the mesosphere, and since the presence of methane has been observed to photo-disassociate in the mesosphere to form water vapor, the anthropogenic causes of climatic change are believed to be directly related to the presence of NLCs .

The argument for PMC research was compellingly made by Dr. James Russell III, who is the principal investigator for NASA's Aeronomy of Ice in the Mesosphere (AIM) small explorer mission:

> *"It is clear that PMCs are changing, a sign that a distant and rarefied part of our atmosphere is being altered, and we do not understand how, why or what it means...These observations suggest a connection with global change in the lower atmosphere and could represent an early warning that our Earth's environment is being altered."*

It is becoming more and more evident that the industrial revolution has introduced the presence of NLCs. We know with certainty that methane, carbon dioxide, and water vapor all trigger the presence of these clouds. We have observed directly that the water vapor produced by the Space Shuttle's main engines trigger the formation of NLCs. Aeronomers (upper atmospheric scientists) and climate change scientists are interested in determining if 1) NLCs are extending and migrating to more equatorial latitudes, and if 2) NLCs are getting thicker and thus play a role in altering the albedo of the Earth and thus the thermodynamic balance of the Earth. Thus, by better understanding NLCs, we hope to gain insight into the elements of global climate change believed to cause their expanding presence.

Implications on Crewed Re-entry Vehicles

Noctilucent clouds reside near a poorly understood yet critical region to re-entry vehicles. The ground track of space vehicles traveling into orbits inclined greater than 50 degrees (such as those to the International Space Station (ISS) which resides at an inclination of 51.6 degrees) reach latitudes where the low-latitude noctilucent clouds may be present. Since we don't know enough about the mesosphere, NASA's manned space missions have enacted conservative flight rules regulating re-entry, constraints that may be irrelevant. Through a better understanding of noctilucent clouds and the mesosphere, we may be able to design more operable space architecture.

The Space Shuttle had always avoided re-entry trajectories which fly farther north than 50°N during the noctilucent cloud season around the summer solstice. Originally it was believed that noctilucent cloud nuclei were large enough to damage the thermal protection system of the Space Shuttle, but measurements since the late 1980s have suggested the cloud nuclei are 100 to 1000 times smaller than previously believed and likely not a danger to the Thermal Protection System (TPS) of the Shuttle. Future air-breathing hypersonic vehicles may still have to consider avoiding the clouds so as not to ingest the particles, but that is a question for engine and aircraft designers to answer. The cloud particles may also increase drag substantially from the standard atmosphere which would also impact vehicle performance and navigation.

Tim Garner's 2002 paper, submitted to the International Conference on Space Planes and Hypersonic Systems and Technologies, presented concerns that NLC activity could have contributed to the destruction of Space Shuttle Columbia's STS-107 mission. Though the cause was eventually determined to be a result of ice strikes incurred during launch, Mr. Garner suggested that there was too much uncertainty regarding the behavior of airflow over the Shuttle as it streaks downward into the thinnest air at speeds of 17,000 to 14,000 mph.

Space re-entry vehicles can experience unpredictable airflow problems on the wings for a variety of reasons: the wrong trajectory, unknown atmospheric conditions or any roughness on the shuttle's surface, all of which contribute to intense heating. Understanding these airflow problems is also one of the top priorities of the Air Force, which is planning to build its own space plane and upper atmosphere aircraft. This uncertainty was one of four major reasons cited a decade ago for canceling the National Aerospace Plane (NASP), which was supposed to replace the shuttle.

In a 2002 study, NLC presence was the top concern about the impact of upper atmospheric phenomenon on the space shuttle: "the most severe effect of entry through a noctilucent cloud would probably be the erosion of the thermal protection system during the most critical heating region.". Density shears were also listed as a major uncertainty. These phenomena act as patches of thicker-than-expected air that can increase the shuttle's roll and pull on one wing. On a Columbia mission in 1992 and an Endeavor mission in 1993, hitting such patches forced them to use up its fuel for the thrusters that help keep it on course during re-entry. Also suspect was blue jets; as the space shuttle streaks through the upper atmosphere, it leaves a wake in the air just as a boat leaves a wake behind it in the water. The shuttle's wake could become electrified and draw a blue jet to the vehicle.

Implications on Comparative Planetology

Much is learned about other worlds by comparing anticipated or calculated conditions there with observed conditions here. This method is called 'Comparative Planetology'. By studying noctilucent clouds here on Earth, we can better model high-altitude, low-density clouds that may appear elsewhere, such as Mars!

PART I: SCIENCE CONCEPTS

The Mesosphere

Noctilucent Cloud Observations

Clouds in the Middle Atmosphere of Mars

THE MESOSPHERE

Module Objectives

- ➢ Understand the regions of the atmosphere
- ➢ Become familiar with the observables in the mesosphere
- ➢ Understand the density profiles in the mesosphere
- ➢ Understand the temperature distributions in the mesosphere
- ➢ Learn about Ionization and chemistry in the mesosphere
- ➢ Gain a basic understanding of the dynamics of the mesosphere
- ➢ Become familiar with the different types of instabilities in the mesosphere

Contributing Authors:

Jason Reimuller, Ph.D.
Dave Fritts, Ph.D.

REGIONS OF THE ATMOSPHERE

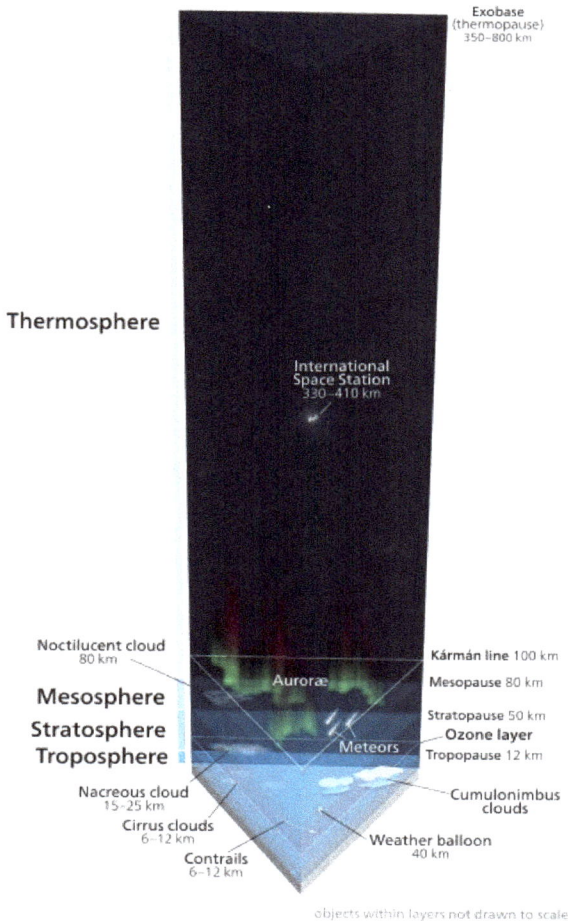

Troposphere: This is the lowest atmospheric layer and is about seven miles (11 km) thick. Most clouds and weather are found in the troposphere. The troposphere is thinner at the poles (averaging about 8 km thick) and thicker at the equator (averaging about 16 km thick). The temperature decreases with altitude.

Stratosphere: The stratosphere is found from about 7 to 30 miles (11-48 kilometers) above the Earth's surface. In this region of the atmosphere is the ozone layer, which absorbs most of the harmful ultraviolet radiation from the Sun. The temperature increases slightly with altitude in the stratosphere. The highest temperature in this region is about 32 degrees Fahrenheit or 0 degrees Celsius.

Mesosphere: The mesosphere is above the stratosphere. Here the atmosphere is very rarefied, that is, thin, and the temperature is decreasing with altitude, about −130 Fahrenheit (-90 Celsius) at the top.

Thermosphere: The thermosphere starts at about 55 kilometers. The temperature is quite hot; here temperature is not measured using a thermometer, but by looking at the motion and speed of the rarefied gases in this region, which are very energetic but would not affect a thermometer. Temperatures in this region may be as high as thousands of degrees.

Exosphere: The exosphere is the region beyond the thermosphere. In this region, lighter molecules (e.g. Hydrogen) can achieve velocities great enough to reach escape velocity, leaving the Earth's gravity permanently.

The Mesosphere

The *mesosphere* is characterized by a decrease in temperature with increasing altitude. The upper boundary of the mesosphere is the *mesopause*, which can be the coldest naturally occurring place on Earth with temperatures below 130 K (−226 °F; −143 °C). The exact upper and lower boundaries of the mesosphere vary with latitude and with season, but the lower boundary of the mesosphere is usually located at heights of about 50 kilometres (160,000 ft; 31 mi) above the Earth's surface and the mesopause is usually at heights near 100 kilometres (62 mi), except at middle and high latitudes in summer where it descends to heights of about 85 kilometres (53 mi).

OBSERVABLES IN THE MESOSPHERE

Several very interesting phenomena occur in the mesosphere. Red sprites and blue jets are a type of electrical discharge often observed in the lower part of the mesosphere, and noctilucent clouds and density shears form in the upper regions of the mesosphere.

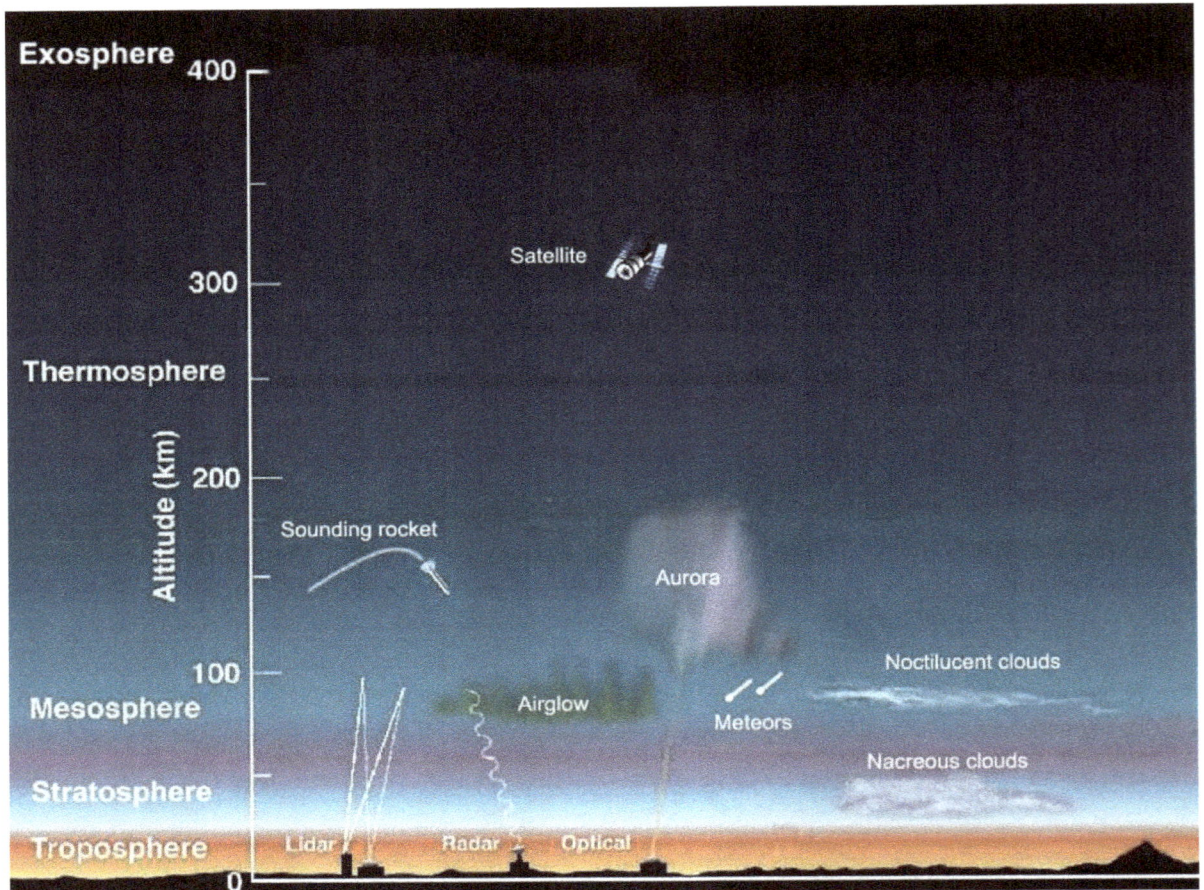

Figure 1. The layers of the atmosphere showing the observable phenomena common in each region.

Noctilucent Clouds

Noctilucent clouds are the highest clouds in the Earth's atmosphere, residing at an altitude of approximately 83 km (50 miles) and observed slightly below the mesopause in the polar summertime. These clouds are also called *Polar Mesospheric Clouds* (PMCs) when observed from space and are of special interest as they are sensitive to both global climate change and to solar/terrestrial influences, i.e. the coupling between the heliosphere and the Earth's atmosphere. NLCs have also provided a basis to understand high-altitude, low-pressure clouds that may be present on other planets and have also been seen to be a risk driver of spacecraft missions.

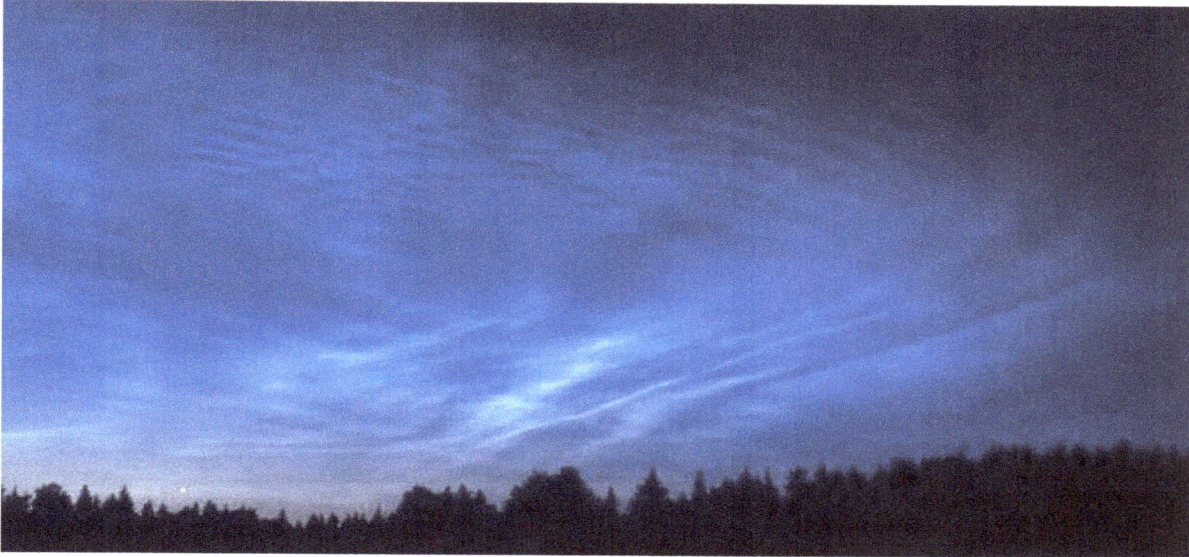

Figure 2. Noctilucent clouds observed from the ground showing band and billow structures (credit: IAP)

Red Sprites and Blue Jets

Red sprites and *blue jets* are upper atmospheric optical phenomena associated with thunderstorms that have only recently been documented using low light level television technology. Also known as upper-atmospheric discharge, these electrical phenomena occur above normal lightning; blue jets occur around 40-50 kilometres (25-30 miles) above the Earth, while red sprites are higher at 50-100 kilometres (32-64 miles).

Red Sprites are massive but weak luminous flashes that appear directly above an active thunderstorm system and are coincident with lightning strokes produced by the thunder cell. When positive lightning goes from the cloud to the ground, sprites can extend from the cloud tops and can reach altitudes of up to 95km, though the most optically intense characteristics occur at altitudes around 65-75 km. Below the bright red region, blue tendril-like filamentary structures often extend downward to as low as 40 km. Sprites rarely appear singly, usually occurring in clusters, and may extend across horizontal distances of 50 km or more. Sprites are very difficult to observe; they occur randomly with only about one percent of lightning strokes and last only about 3-10 milliseconds.

Figure 3. Red Sprites typically form in clusters above thunder cells.

Blue jets also form in thunder cells; they emerge directly from the tops of clouds and shoot upward in narrow cones through the stratosphere. However, these cone-shaped emissions form as a result of ionized nitrogen emissions; they are not related to lightning. Their upward speed has been measured to be about 100 km per second.

11

Meteors

An average of 40 tons per year of *meteors* enter the Earth's mesosphere, where most melt or vaporize as a result of collisions with the gas particles contained there. From these meteors, only iron and other refractory materials reach the surface. Most of the meteor ablates in the mesosphere, contributing to the overall chemistry of this region.

Airglow

Airglow is an illumination caused by various processes in the upper atmosphere, such as the recombination of atoms that were photoionized by sunlight during the day. This recombination of nitrogen and oxygen to form nitric oxide (NO), which causes the emission of a photon. Other species that can create air glow in the atmosphere are hydroxyl (OH), atomic oxygen (O), sodium (Na) and lithium (Li) through a process called 'chemiluminescence'. Airglow is also caused by luminescence caused by cosmic rays. Airglow is very difficult to observe from the ground, since airglow limits the sensitivity of telescopes at visible wavelengths. Observations are typically performed from aircraft or from space-based platforms.

Figure 4. Airglow observed from the International Space Station (credit: NASA)

DENSITY OF THE MESOSPHERE

The *scale height* is the relationship in which the density of the atmosphere decreases. This relationship is defined as:

$$H = \frac{RT}{mg}$$

Where: R= the Universal Gas Constant, or 8.314 (J K^{-1} mol^{-1})
T is the mean atmospheric temperature (in Kelvins),
m= the mean molecular mass of dry air (in kg/mol), and
g=acceleration due to gravity (m/s^2)

The scale height is approximately constant up to about 120 km altitude, above which the temperature begins to rise rapidly while the mean molecular mass decreases. Also note that for all altitudes, the acceleration due to gravity also decreases slowly. At NLC altitudes, the scale height is roughly 7000m.

The density of the atmosphere decreases exponentially with altitude. The atmospheric pressure near 50 km is only one thousandth of the surface pressure (1/1000 bar = 1 mbar = 100 Pa). As in the lower atmosphere, the pressure keeps decreasing exponentially with the e-folding "scale height" ($H = RT/Mg \sim 7$ km). This is equivalent to a factor of 10 every 16 km, so the pressure at the mesopause (near NLC layer altitudes) is only one thousandth the density at 50km – or only 0.001 mbar! However, despite the low density, the composition is almost identical as we find near the Earth's surface and the relationship between densities, pressure and temperature can still be represented by the ideal gas law.

The Upper Limit of the Homosphere

The mesosphere also marks the upper limit of the '*homosphere*', the region of the atmosphere where the atmospheric constituents are evenly mixed (78% Nitrogen, 21% Oxygen, 1% Argon, etc.). Above the mesopause, the atmosphere begins to stratify according to the molecular densities of these constituents; the denser constituents (e.g. diatomic oxygen) stratify at lower altitudes while the lighter constituents (e.g Helium and Hydrogen) form at the highest altitudes. This is why we see the green bands of the aurora at the lowest altitudes and red at the highest; green light is produced when oxygen molecules are photo-ionized and red light is produced when hydrogen is photo-ionized.

TEMPERATURE OF THE MESOSPHERE

Though not as pronounced as in the stratosphere, the primary source of heating in the mesosphere occurs internally as solar ultraviolet radiation is absorbed by molecular oxygen and water vapor. Temperatures in the mesosphere decrease with altitude due to less ozone heating and more infrared cooling to space. The atmospheric pressure near 50 km is only one thousandth of the surface pressure (1/1000 bar = 1 mbar = 100 Pa). Near the mesopause, this decreases to only 0.001mbar, though the relationship between densities, pressure and temperature are generally consistent with what they would be at sea level.

Ultraviolet radiation, having a wavelength of less than 240nm, is absorbed by molecular oxygen and gives off two products: atomic oxygen in both the ground O(^{3}P) and first excited state O(^{1}D). The remainder of this process is available as heat. Further, the excited oxygen atom, O(^{1}D), can transfer its energy via collision to excite another atom or molecule or by spontaneous emission of a photon to return to its ground state.

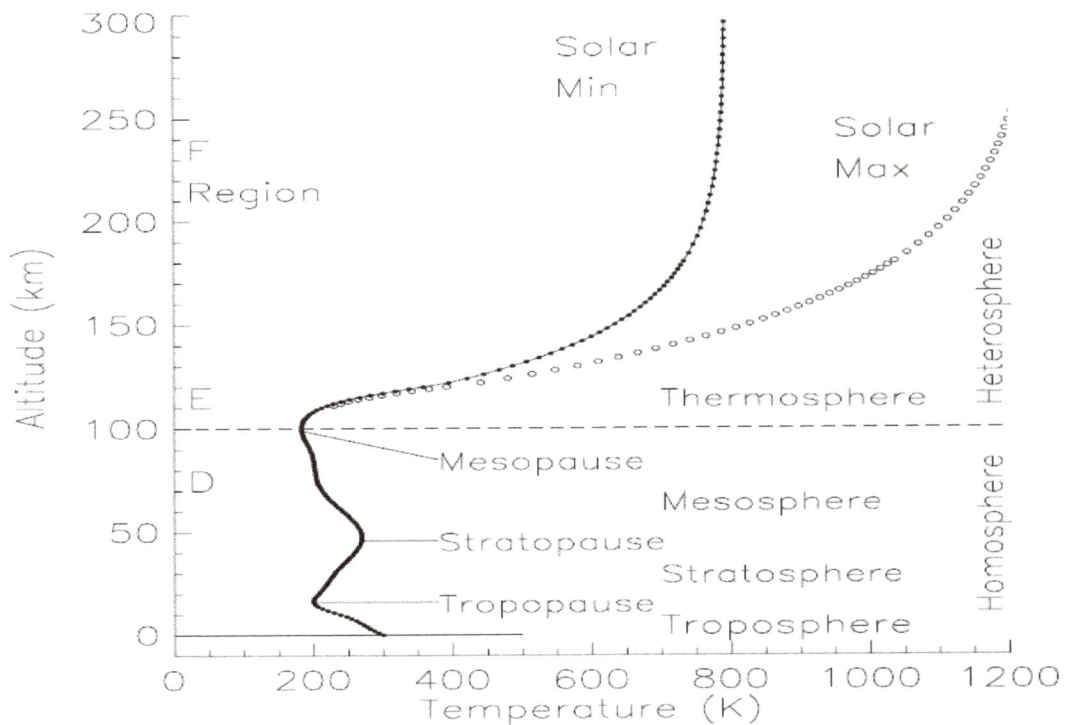

Figure 5. Temperature Distributions in the Atmosphere

The temperature of the mesosphere and its gradient vary strongly with season and latitude. The high-latitude stratopause is warmest in summer (~270 K ~ 0 °C) due to stronger ozone heating, and the resulting temperature and pressure gradients drive zonal mesospheric wind jets, easterly in the summer hemisphere and westerly in the winter hemisphere. However, the mesopause is observed to be coldest in summer (~140 K ~ -130 °C) due to a global meridional circulation from the summer hemisphere to the winter hemisphere. Air slowly ascends in the summer mesopause, thereby expanding and cooling adiabatically. This circulation is mechanically driven by a continuous flux of internal atmospheric gravity waves. They mostly originate in the troposphere from flow over mountains or deep convection also responsible for thunderstorms. Many waves will propagate upward, grow in amplitude due to the decreasing density, until they "break". They transfer their energy and momentum to the mean flow, slowing down and even reversing the zonal mesospheric jets. The result is the global meridional circulation in the mesosphere.

IONIZATION AND CHEMISTRY OF THE MESOSPHERE

Ionization

The ionosphere is a collective term that describes the regions where the atmosphere is ionized by the Sun's ultraviolet light and where the resulting ionization and recombination play important roles in the atmosphere. The mesosphere is home to the lowest layer of ionization, the "D-region" of the ionosphere. The D-region is only present during the day, when ions are produced mostly by the photo-ionization of the trace atmospheric constituent NO ([NO] $\approx$ 107 cm^{-3} (as compared with [N2] $\approx$ 1014 cm-3 (as compared with [N2] ~ 1014 cm^{-3} at 85 km) by short wavelength ultraviolet (Lyman α with $\lambda = 121.6$ nm) and x-rays, however ionization from solar x-rays is small except during a solar flare.

At night, when there is no incident radiation, the electrons quickly recombine with the molecular positive ions, so that the D-region disappears, except at latitudes greater than about 65°, where particle bombardment sustains the ionization. During the course of a day, the electron density increases from about 100 electrons/cm^3 at 60 km to about 104 electrons/cm^3 at 90 km, around noon. It is greater in the summer than in the winter, and greater at sunspot maximum. Above the D-region, the ionosphere continues upwards through the thermosphere and includes the E (Heaviside-Kennelly), and the F (Appleton) regions.

Composition and Chemistry

Like the stratosphere, the mesosphere is homogeneously mixed and the composition maintains similar proportions of gasses. One significant difference is that in the lower mesosphere, the catalytic cycles that destroy ozone are dominated by hydrogen species such as hydrogen (H), hydroxides (OH), and hydroperoxyl (HO$_2$). These are the products of the destruction of methane (CH4) and water (H2O).

Constituent gas and formula	Content, percent by volume	Molecular weight [b]
Nitrogen (N_2)	78.084	28.0134
Oxygen (O_2)	20.9476	31.9988
Argon (Ar)	0.934	39.948
Carbon Dioxide (CO_2)	0.0314 ξ	44.00995
Neon (Ne)	0.001818	20.183
Helium (He)	0.000524	4.0026
Krypton (Kr)	0.000114	83.80
Xenon (Xe)	0.0000087	131.30
Hydrogen (H_2)	0.00005	2.01594
Methane (CH_4)	0.0002 ξ	16.04303
Nitrous oxide (N_2O)	0.00005	44.0128
Ozone (O_3)	Summer: 0 to 0.000007 ξ Winter: 0 to 0.000002 ξ	47.9982
Sulfur Dioxide (SO_2)	0 to 0.0001 ξ	64.0628
Nitrogen Dioxide (NO_2)	0 to 0.000002 ξ	46.0055
Ammonia (NH_3)	0 to trace	17.0306
Carbon Monoxide (CO)	0 to trace	28.01055
Iodine (I_2)	0 to trace	253.8088

Table 1. Atmospheric constituents in the Mesosphere

The composition of the upper mesosphere begins to differentiate itself from the lower mesosphere in several ways: 1) the gasses are no longer well-mixed but begin to stratify based on their molecular masses, 2) ion chemistry begins to affect the chemical system, 3) atomic oxygen becomes more abundant as it becomes a long-lived species, 4) excited states of atoms and molecules affect the chemistry and energetics of the upper mesosphere, and 5) several important molecules, such as carbon dioxide (CO_2), are not in local thermodynamic equilibrium due to the low rate of collisions in the rare atmosphere. The last factor produces strong cooling of the upper mesosphere through emission by vibrationally excited CO_2; the low collision rate limits the exchange of energy from kinetic to vibrational levels.

The ablation of meteors, combined with photo-disassociated atmospheric constituents, allow for some exotic chemistry in the mesosphere. The presence of several airglow layers (e.g. OH, O2, and O) introduce a variety of molecular constituents at different regions of the atmosphere. Prominently, a 5 km (3.1 mi) deep sodium layer is located between 80–105 km (50–65 mi) and is made of unbound, non-ionized atoms of sodium introduced through meteoric ablation.

DYNAMICS OF THE MESOSPHERE

Dynamically, the mesosphere is characterized by strong *zonal* (East - West) winds. There is also a *meridional* (North - South) component of wind that is seasonal. Zonal means "along a latitude circle" or "in the west–east direction"; while meridional means "along a meridian" or "in the north–south direction". The meridional winds are strongest during the winter and summer months, where the summertime pole is oriented towards the sun. The heating received at the summertime pole causes an upwelling which drives meridional flow (a 'source') to the wintertime pole, which acts as a 'sink'. For vector fields (such as *wind* velocity), the *zonal* component (or x-coordinate) is denoted as *u*, while the *meridional* component (or y-coordinate) is denoted as *v*.

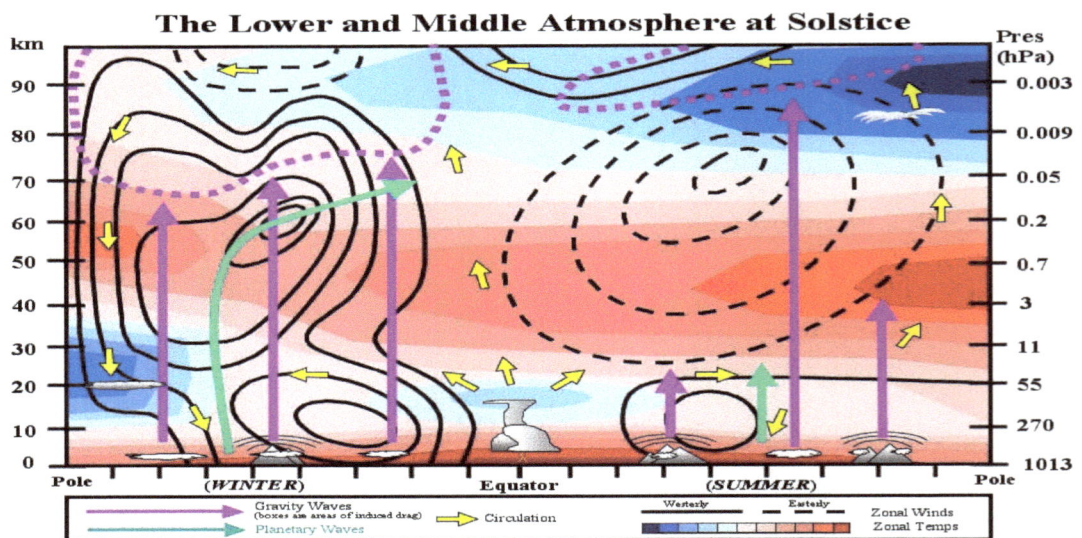

Figure 6. Dynamics of the Mesosphere at Solstice

The mesosphere is a highly dynamic region as it is a region where gravity waves break and turbulence structures are created. Gravity waves are waves that are formed at the surface or in the lower atmosphere and propagate upwards. As they rise, they expand until the air becomes too thin to carry the wave and it breaks, imparting its energy and momentum into the upper atmosphere. Figure 6 shows the circulation of the mesosphere and the contributions of gravity waves and planetary waves. Note how air is transported from the summertime pole to the wintertime pole during times near the solstice.

Shear

Mesosphere altitudes of 100-105 km are characterized by high values of shear, driven by high winds that are commonly seen in the 95-100 km region. Instabilities are driven by this sheer and also gravity wave breaking that is believed to impart momentum and increase temperature in the upper mesosphere, causing an increase in the topside lapse rate and making the atmosphere more unstable.

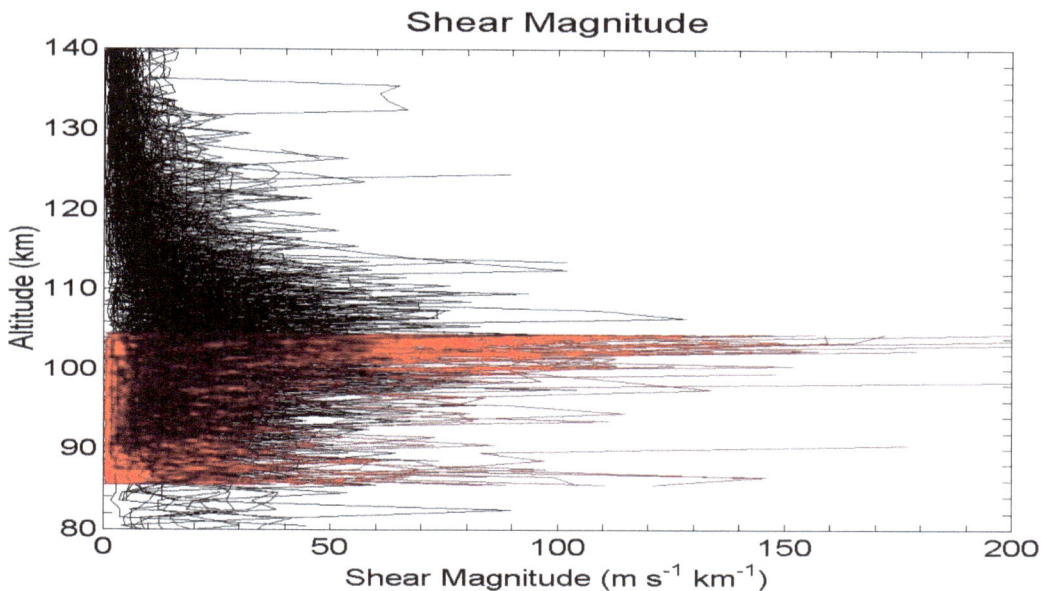

Figure 7. Shear Profile in the Mesosphere

INSTABILITIES

Complex Circulation

The mesosphere exhibits a complex circulation that indicates that the region is far from radiative equilibrium. Observations of the mesosphere are very contrary to physical intuition: the region records the lowest temperatures on Earth in the summer polar region, and thus achieves temperatures cold enough to form noctilucent clouds. However, the mesosphere warms in the winter where there is no solar irradiation. One would think this would not be possible; how could the pole that is not receiving radiation from the sun actually be warmer? The answer has to do with the type of radiation that is being imparted into the polar regions. Since most of the absorbing radiation of the sun gets absorbed in the stratosphere (UV) and troposphere (visible and UV), heating occurs BELOW the mesosphere. As the stratosphere is warmed, the air rises. As the air rises, it expands adiabatically and, as a result, cools the mesosphere.

Gravity Waves

Gravity waves are waves that are set into motion by disturbances in the lower atmosphere. For example, gravity waves may be initiated by atmospheric flow over mountains; as the flow is abruptly compressed and expanded, it excites gravity waves that propagate upward. But orographic sources (caused by topographic changes) are not the only means by which gravity waves are created. These waves may also be created by air flow over a convective cloud or the presence of instability around the jet stream.

Seasonal changes affect gravity wave forcing. Gravity waves are filtered by the stratospheric wind system and will be reflected or absorbed; in winter, the stratospheric jet is eastward and gravity waves in the mesosphere tend to be dominant westward. However, during equinox, the wind in the stratosphere reverses and the dominant direction of gravity wave in the mesosphere also reverses due to filtering.

Ultimately, gravity wave propagation, interaction with other waves and the background flow, wave instabilities and dissipation, and the generation of turbulence play crucial roles for the accurate physical description of the mesosphere, as well as short-term predictions of "space weather" and long-term trends of temperatures and densities.

Gravity Wave Effects on Temperatures

Temperature modulations due to gravity waves are often 20 K or more over a depth of a few kilometers, giving rise to stable mesospheric inversion layers (where the temperature gradient is positive) and adiabatically unstable layers (when the temperature decreases more than ~10 K/km). Such temperature variations correspond to density variations of 10%, and can persist for hours or days.

Since gravity wave amplitudes increase with altitude, the upper mesosphere is the region of the strongest effects of turbulence, such as mixing of heat and trace gases. The PoSSUM MCAT instrument will be able to characterize these very largest scales of mesospheric "clear air turbulence", however, turbulence is also and more often evident in density and temperature fluctuations at spatial scales much smaller than gravity wave scales. With greater altitude and diminishing atmospheric density, turbulence will eventually become less important in favor of viscous fluid motion, which dominates near the turbopause, the altitude where the homosphere ends and the heterosphere begins (~100km). Therefore, depending on altitude and turbulence strength, turbulent density fluctuations are observed at scales of 100 meters down to meters and centimeters.

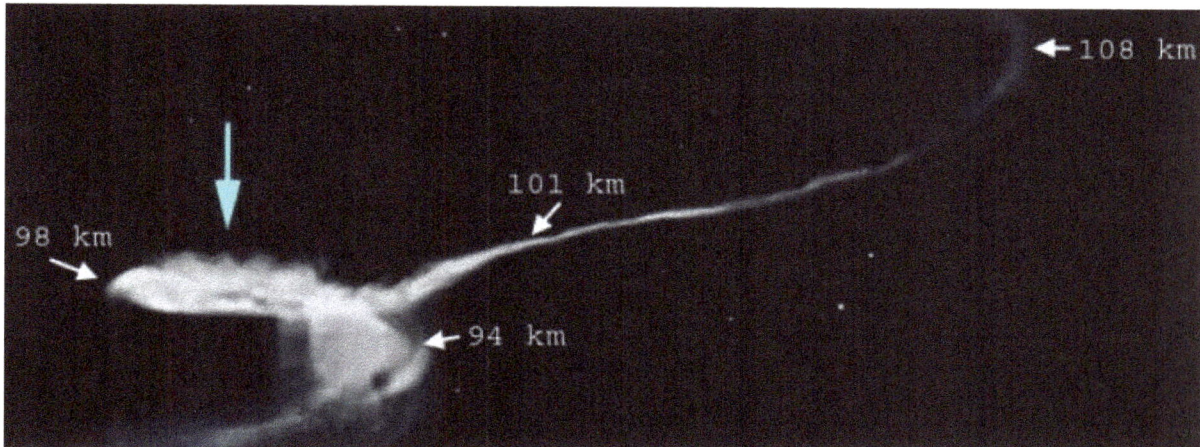

Figure 8. A photograph of chemical tracers (TMA) descending through the upper mesosphere. The arrow points to the Kelvin-Helmholtz billows which occur between 96 and 99 km. (credit: M. Larsen)

The PoSSUM sorties will allow for immediate and local observation ("*in-situ*") of the density and temperature structure. This is the only way to observe the distribution of small-scale turbulent density fluctuations.

Mesosphere Inversion Layer (MIL)

The *Mesosphere Inversion Layer* (MIL) phenomenon is a region of enhanced temperatures ($\Delta T \sim 15$–50K) that spans a distance of about 10 km. The MIL is observed with great regularity in both the upper mesosphere (60–70 km) and the mesopause (90–100 km) at low and mid-latitudes. Gravity wave activity is believed to play an important role in the development of a linkage between the MIL and the tidal structure through gravity wave coupling that results in an amplification of the tidal thermal structure.

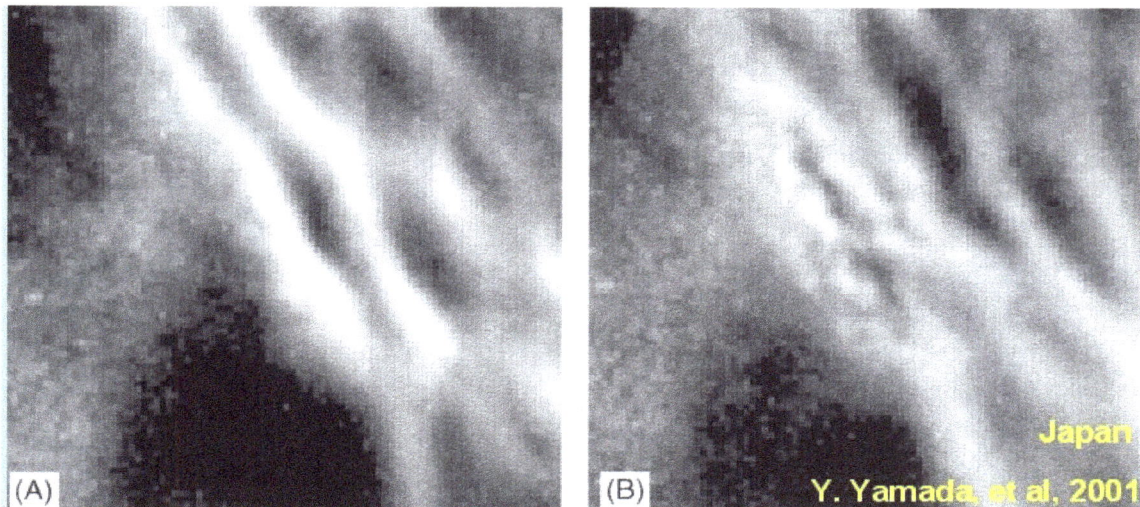

Figure 9. Gravity Wave breaking a) GW bands propagating upward, b) same wave after breaking near the mesopause (credit: Y. Yamada)

There are three major sources for the production of MILs: 1) breaking of long-period gravity waves, 2) planetary wave activity which occurs when enough gravity wave events make a zero-wind line and absorption of planetary waves produce a MIL at lower altitude, and 3) gravity wave events combined with tidal (planetary) waves decrease stability and contribute to the formation of a critical layer that causes the gravity wave to break, creating a MIL at higher altitudes.

Kelvin Helmholtz Instabilities (KHI)

Kelvin-Helmholtz Instabilities (KHI) occur through a velocity shear in a continuous fluid. As we have noted the shear that is intrinsic in the upper mesosphere region, KHI is common. Figure 10 illustrates the formation of KHI in neutral shear and the onset of instability and transition to turbulent flow.

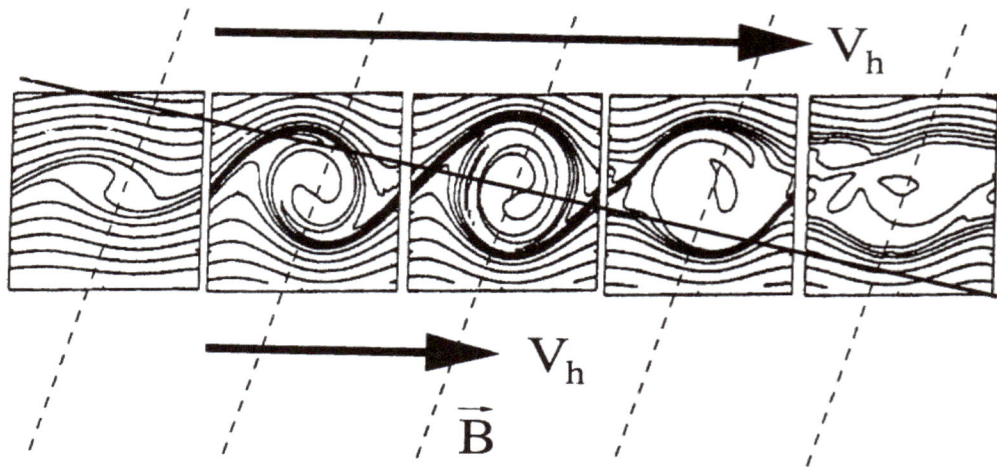

Figure 10. Kelvin-Helmholtz billows generated by a neutral shear. The heavy arrows show the neutral flow above and below the layer.

KHI are observed in tropospheric clouds as well. Figure 11 shows several formations of KHI that form under shear. KHI have been observed in the atmospheres of other planets as well.

Figure 11. Kelvin Helmholtz instabilities in Tropospheric Clouds

NOCTILUCENT CLOUD OBSERVATIONS

Timo Leponiemi 2001

Module Objectives

- ➤ Become familiar with Observational Trends of Noctilucent Clouds from the Ground
- ➤ Become familiar with Remote Sensing Techniques for Observing Noctilucent Clouds

Contributing Author:

Jason Reimuller, Ph.D.

NOCTILUCENT CLOUDS

The first recorded sightings of NLCs were reported in 1885 at high latitudes and were recognizable because they were of such altitude as to still reflect sunlight after the sun had set, thus giving the appearance of glowing at night. Observations of these clouds were never recorded before 1885, even in the clearly documented histories of indigenous Arctic cultures. Since these initial sightings, observations from the ground have recorded NLCs with increasing frequency and in increasing area and albedo. In addition, satellite observations over the past four decades have confirmed earlier observations by indicating that the presence of these clouds has been increasing in frequency and extending to lower latitudes.

Figure 12. Noctilucent clouds

OBSERVATIONAL TRENDS OF NOCTILUCENT CLOUDS FROM THE GROUND

NLCs are generally observable by a ground observer situated between a latitude range of 50-65 degrees and visible to the naked eye during the hours of twilight when the observer and the lower atmosphere lie in the Earth's shadow, while the clouds themselves are still exposed to direct sunlight. Thus, there is a limited viewing opportunity corresponding to a range of solar depression angles between 6 and 12 degrees that occur within the latitude range of 50–65°. The strong forward scattering of sunlight from the ice particles characterize the extensive cloud layers observable within NLC structures. The mesosphere is no longer illuminated south of 50°, due to the sun being more than 12° below the horizon (astronomical twilight). Conversely, the sky is too light north of 65° as the sun is less than 6° below the horizon (civil twilight). Since NLCs are so optically thin that they scatter less than 1 part in 1000 of light incident upon them, this makes them invisible to a ground observer against the residual brightness present during civil twilight. At the end of civil twilight, the clouds become visible as sky brightness has decreased by a factor of several hundred. Here, the low angle of the sun below the horizon also allows the high altitudes at which the NLC resides to be in direct sunlight.

NLC formations typically span large areas of the polar upper atmosphere, extending at times up to 4×10^7 km^2 for time periods lasting from several minutes to more than 5 hours. Dynamically they tend to migrate to the southwest at an average velocity of 40 m/s, though individual bands often move in different directions and at speeds differing from the NLC display as a whole. NLCs are also highly polarized in the same sense as, but less sharply than the twilight sky.

Structure and Color

Ground-based photographs of NLCs often exhibit distinct wave-like structures with alternate dark and bright bands spaced from 10–100 km. Larger scale structures are commonly described as 'bands' while smaller-scale features often occur orthogonal to the bands, with a spacing of 3–10 km and are commonly termed 'billows'. These features are generally seen as the manifestation of internal gravity waves (GWs). As such, GWs play an important role in the thermal structure of the polar upper atmosphere, and hence the structure and evolution of NLCs.

Each PMC is composed of different forms, of which there are four major groups, veil (Type I), bands (Type II), billows (Type III) and whirls (Type IV). There are further sub divisions of these four major classes. Generally, NLCs are an electric blue color, but red, gold and white are not uncommon and are due in part to the angle of the sun below the horizon. The general blue coloring is caused by absorption of incident sunlight by ozone in the Chappuis bands which reside in the yellow portion of the spectra. The standard reference for description of cloud forms is the "International Noctilucent Cloud Observation Manual" (WMO 1970), which breaks down NLCs into five types:

Type I NLCs: Veils

Veils are the simplest of NLC formations and are very tenuous with no well-defined structure, and are often present as a background for other categories or forms. They are somewhat like cirrus clouds of uncertain shape; however, occasionally they exhibit a faintly visible fibrous structure. They often flicker.

Type II NLCs: Bands

Band formations are characterized by long streaks with diffuse edges (type IIa) or sharply defined edges (type IIb). They are sometimes hundreds of kilometers long and often occur in groups arranged roughly parallel to each other or interwoven at small angles (perhaps visible evidence of the gravity waves propagating through the region). Occasionally an isolated band is observed. Bands change very little with time and blurred bands with little movement are often the predominant structure in the noctilucent cloud field. When they do move, it is often in a direction and with a speed that is different than that of the display as a whole. Very closely spaced thin streaks, called serrations, are occasionally seen in the veil background. They look like a continuous cloud mass as serrations are separated by only a few kilometers.

Type III NLCs: Billows

Billow formations are characterized as groups of closely spaced short bands which sometimes consist of straight, narrow, and sharply outlined parallel short bands (type IIIa). Sometimes they exhibit a wave-like structure (type IIIb). The distance separating pairs of billows is about 10 km. Billows sometimes lie orthogonal to the long bands and their alignment usually differs noticeably in close portions of the sky. Unlike the long bands, billows may change form and arrangement or even appear and disappear within a few minutes.

Type IV NLCs: Whirls

Whirls of varying degrees of curvature are also observed in veils, bands, and billows; infrequently, complete rings with dark centers are formed. Whirls of small curvature are classified as type IVa while whirls having a single simple band or several bands are classified as type IVb. Larger scale whirls are classified as type IVc.

Type V NLCs: Amorphous

Amorphous formations are similar to veils in that they have no well-defined structure but they are brighter and more readily visible than the veil type NLC.

Figure 13. Type I NLC with veil structures (credit: M. Gadsden)

Figure 14. Type II NLC with band structure (credit: M. Gadsden)

Figure 15. Type III NLC structure with billow structure (credit: M. Gadsden)

Figure 16. NLC Formation with whirl structure (credit: M. Gadsden)

Figure 17. Complex NLC formation (credit: M. Gadsden)

Altitude

The visibility of NLCs is due to their great altitude. Their altitudes have been accurately determined by ground-based parallactic photography, LiDAR observations, and in-situ rocket measurements. The clouds reside at an altitude of 80-85 km with an average altitude of 83 km. This places the clouds just below the temperature minima of the mesopause which is approximately 150K (-123°C). It is theorized that the clouds nucleate at the mesopause over several hours during nucleation and growth and then descend by a few kilometers to the altitude at which they are observed.

Formation

A number of models have been constructed to model the microphysics involved with the evolution of NLC particles. These models assume that nucleation occurs under the supersaturated conditions in the cold summer mesopause region, when the ambient temperature can drop to 135K. The particles grow through condensation of water vapor until sedimenting out of the supersaturated region. Sublimation of the particles once out of the supersaturated region, however, is slow compared to the fall speed of the particles so the particles retain their size over most of their downward trajectory, leading to an accumulation of particles near their peak particle size, generally around 50nm. NLCs are very stratified, possessing a thickness in the vertical of only 0.5 to 2.0 km. The visible band structures result from vertical waves that reach an amplitude of approximately 1.5 to 3.0 km.

The mechanisms of NLC formation are still highly contested. Some maintain that the ice crystals nucleate around terrestrial dust particles and there are even some hypotheses that suggest deposition of dust particles from libration clouds provide the nuclei upon which NLCs grow. The Cosmic Dust Experiment (CDE) onboard the AIM satellite is an instrument designed to monitor the variability of the cosmic dust influx into Earth's mesosphere in order to address its role in the formation of NLCs, testing the viability that the nucleation particles may be of extraterrestrial origin. Meteoric smoke has been detected just above the sulfate layer (~35 km) to ~85+ km by the Solar Occultation for Ice Experiment (SOFIE) onboard the AIM satellite. It is currently estimated that between 10 to 100 tons of meteoric material enter Earth's atmosphere per day, where about 70% of the incoming meteoroids ablate at altitudes between 70 and 110 km.

The mesosphere is extremely dry and cold so it is unusual that NLCs form at all. Nucleation occurs due to super-saturation in the mesosphere. This is possible due to the very low temperatures and pressures at and near the mesopause. Temperatures as low as 111K (-162 degrees C) have been measured just a few kilometers above a NLC. These temperatures are set by mixing in the atmosphere due to hemisphere-to-hemisphere circulation from the movement of upward cooling air in the summer polar regions and the downward, warming movement of air over the winter polar regions. These temperatures are further maintained or adjusted by the attenuation of upwardly propagating gravity waves (buoyancy waves).

Gravity waves also supply the water molecules from which NLCs form by upward diffusion from lower down in the atmosphere. Water vapor is also supplied by the photo-dissociation of methane molecules in the mesosphere by ultraviolet light. Even with these sources of water the levels of water in the mesosphere available for NLC formation are extremely low - a few water molecules are present for every million atmospheric molecules. Mixing ratios for water at 80 km have been estimated to be approximately 3ppm, explaining why NLCs have been observed to be very thin. Indeed, measurements have suggested the optical thicknesses of NLC layers to be in the region of 10^{-4} with cloud particle diameters in the range of 50nm.

It is generally believed that upper atmospheric water vapor is increasing because of increases in methane, leading to observable effects on NLC brightness, which has been recorded since 1885. At present, methane oxidization accounts for about one half of the total water content above the troposphere and it is believed that the increase in water vapor in the upper atmosphere from methane oxidization will continue and increase the brightness and occurrence frequency of NLCs. The relationship between water vapor concentration and solar cycle is less determined. While relationships between the solar cycle and upper atmospheric temperature and humidity have been documented at low to middle latitudes, much less information is available for high latitudes, particularly in summer.

REMOTE SENSING TECHNIQUES FOR OBSERVING NOCTILUCENT CLOUDS

Following the initial observations of NLCs in 1885, NLCs have been observed by a range of methods including ground-based photography including stereography and globally imagery campaigns. Data have also been obtained from ground-based LiDAR, airborne, and space-borne methods through imagery and spectrometry as well as from in-situ methods using sounding rockets.

Noctilucent Cloud Observations using LiDAR

Light Detection and Ranging (LiDAR) systems have more recently verified the altitudes and some vertical structure of noctilucent clouds, but they provide only point observations with limited time resolution and background removal.

Nevertheless, new techniques in LiDAR technology are being employed to study particle size distributions and study the distributions of particle shapes through techniques of depolarization.

Figure 18. Ground Lidar Observation of the Upper Atmosphere.

History of Noctilucent Cloud Research using Sounding Rockets

The first successful sounding rocket flights were conducted in 1962 from Northern Sweden. They employed particle collectors that opened at altitudes between 75 and 98 km that revealed an abundance of particles at least 1000 times the size of particles that were detected following a flight that did not penetrate NLC formations. These rocket experiments measured airglow, nitric oxide ionization, and solar irradiance, showing that a large percentage of collected particles contained high atomic number elements (e.g. Hf, Ni, Co, and Ce).

Several rocket launchings from Kiruna in 1982 during the *Cold Arctic Mesopause Project* (CAMP) proved that in summer mesopause, temperatures at high latitudes are as low as 110 K and demonstrate a graduated temperature profile. In 1993, The NLC-93 rocket campaign at Esrange, Sweden investigated the vertical structure of a noctilucent cloud layer in-situ, showing little vertical variation of the population of NLC particles through the layer, though the lower part of the cloud was observed to have an increase in particle size and a decrease in particle density near the cloud base.

The *MIDAS-DROPPS* campaign was a rocket-borne campaign conducted in Norway in 1999 that synchronized rocket and ground-based observations to determine particle sizes through analysis of the scattering phase functions of NLC particles. Optical photometers were flown onboard two rocket payloads through different mesospheric conditions. Using Mie calculations, the NLC observed in the first flight was determined to have small particles (r<20nm) and the relatively stronger NLC observed in the second flight was dominated by particles in the size range of 40-50 nm.

In 2007, a rocket-borne experiment was conducted to quantify the smoke number density and size distribution as a function of altitude as well as the fraction of charged particles of NLCs. This experiment showed how the size-dependent, altitude-dependent and charge-dependent detection efficiency can be determined for a given instrument design, related the measured particle population to the atmospheric particle population, and found that rocket-borne smoke detection with conventional detectors is largely limited to altitudes above 75 km. Further, they found that there is no general difference between neutral and charged particles for particle sizes down to 1 nm and that the use of a Brownian motion model is mandatory to correctly describe the statistical motion near the Faraday cup detector in the rocket payload [Hedin et al, 2007].

A rocket-borne mass-analysis of charged aerosol particles in the NLC layer was performed on 3 August 2007 as the first of two "MASS" (Mesospheric Aerosol Sampling Spectrometer) rockets was launched from the Andøya Rocket Range into an NLC layer, verified by the ALOMAR LiDAR site, approximately 26 minutes after an AIM satellite overpass. The rocket carried an electrostatic mass analyzer for the charged fraction of the aerosol particles. The mass-analyzer detected both positively and negatively charged particles that exist in about equal numbers for the 1 – 2 nm size range where the largest particles (>3 nm) are mostly negatively charged and the 0.5 – 1 nm particles are mostly positive. The transition of the sign of charge from positive at the smaller sizes to negative at the larger sizes is proposed to have been caused by positive ions or cluster ions being the condensation nuclei and subsequent collection of electrons after growth to a radius of 1–2 nm.

History of Noctilucent Cloud Research using Space-Based Remote Sensing Techniques

Satellite observations over nearly three decades have provided a consistent measure of NLC occurrence, latitudinal extent, and brightness and indicate that the presence of these clouds has been increasing in frequency and extending to lower latitudes. American and Soviet astronauts have observed the phenomenon from space as early as 1970 and subsequent observations have been made from spaceborne platforms such as the Halogen Occultation Experiment (HALOE) and the Student Nitric Oxide Explorer (SNOE). The most comprehensive advances to the understanding of NLCs have come from spaceborne platforms such as the Orbiting Geophysical Observatory (OGO-6), the Solar Mesospheric Explorer (SME), and most recently the AIM satellite, described briefly below. Noctilucent clouds are generally imaged in the UV spectra so that the ozone, having a concentration that peaks in the upper stratosphere, may be used as a background. At ultraviolet wavelengths, the albedo of the Earth is very small so that particulate (Mie) scattering observed above the altitude of the background can then be assumed to be associated with noctilucent clouds.

Orbiting Geophysical Observatory

The *OGO-6* satellite was launched on 5 June 1969 into a low polar orbit as the sixth of the OGO series of satellites tasked to study the Earth's magnetosphere. OGO-6 carried onboard visible airglow photometers that first scanned the atmospheric horizon throughout the summer polar mesopause region and was the first to trace the motions of NLCs across the polar cap.

Solar Mesospheric Explorer

The general seasonal characteristics of NLCs were observed continuously from the Solar Mesospheric Explorer (SME) satellite over the time period 1981 to 1986. Onboard SME was an ultraviolet spectrometer, which mapped the distributions of NLCs and measured the altitude profile of scattering from clouds at two spectral channels of 265 nm and 296 nm.

Student Nitric Oxide Explorer

The *Student Nitric Oxide Explorer* (SNOE), a 254 lb spacecraft built by the Laboratory for Atmospheric and Space Physics (LASP), carried three instruments: an ultraviolet spectrometer to measure nitric oxide altitude profiles in the terrestrial lower thermosphere (100-200 km altitude), a two-channel auroral photometer to measure auroral emissions beneath the spacecraft, and a five-channel solar soft X-ray photometer. In addition to its primary Mission Objective, SNOE successfully observed NLCs globally each day through seven PMC seasons and produced the first observations of 5-day period variations in NLC albedo.

Aeronomy of Ice in the Mesosphere

The AIM Satellite, launched in April 2007, is the first satellite with a primary mission dedicated to the study of NLCs. It has aboard it three payloads: 1) the *Cloud Imaging and Particle Size* (CIPS) instrument, providing a 2-D panoramic look at polar mesospheric clouds by collecting

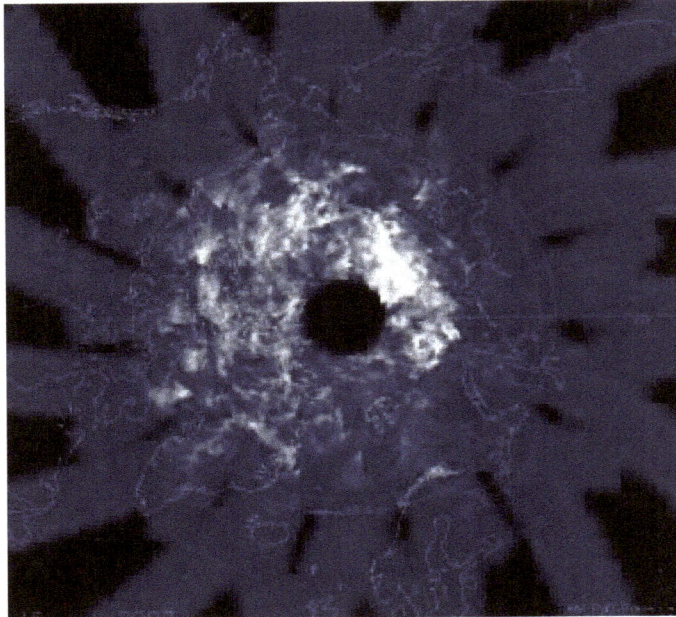

Figure 19. Summation of CIPS overpasses (credit: NASA)

360 degrees of multiple images, 2) the *Solar Occultation for Ice Experiment* (SOFIE) payload, measuring the variability of cloud particles with respect to their altitude and their chemical composition, and 3) the *Cosmic Dust Explorer* (CDE), recording the amount of space dust entering Earth's atmosphere in order to assess whether space dust provides the foundation for the cloud condensation nuclei in the formation of noctilucent clouds.

CIPS is a four-camera, near-nadir ultraviolet (UV) imager flying on the AIM satellite. AIM was launched on April 25, 2007 into a 600 km, sun-synchronous orbit with a noon/midnight equator crossing time. CIPS provides wide-angle UV images of the atmosphere over a broad range of scattering angles in order to determine the presence of NLCs, measure their spatial morphology, and constrain the parameters of cloud particle size distribution. The four identical CIPS cameras have a 15 nm passband centered at 265nm, the wavelength chosen to maximize NLC contrast due to the relative weakness of the Rayleigh-scattered sky background caused by the absorption of solar radiation in the ozone Hartley bands, and are oriented near-nadir in a configuration that allows partial overlap between cameras. Each camera uses a detector with 2048 x 2048 pixels that is electronically binned in 4x8 combinations to generate a 340 (along-orbit track) x 170 (cross-orbit track) array of science pixels. Simultaneous Level 1A images from the four cameras are merged and binned to form a single display called a scene, which has a combined field-of-view (FOV) of 120 degrees along-track by 80 degrees cross-track (approximately 2000 by 1000 km when projected to PMC cloud altitude). This merged Level 1B data is mapped to a uniform spatial grid with a resolution of 25 km^2 per pixel. The large CIPS FOV, combined with a temporal sampling rate of 43 seconds between scenes, results in a significant overlap between adjacent scenes, so that a NLC in the FOV is viewed seven times at a large range of scattering angles (typically from approximately 30° to 170°) as the satellite moves along the orbit. These multiple observations at each location facilitate the removal of the Rayleigh scattered background.

Early data acquired in 2007 indicate that NLCs form early in the spring season at high polar latitudes and travel down to lower latitudes--where they can be seen sweeping Europe and scattering over North America--as the spring and summer NLC season progresses. The Northern Hemisphere NLC season is mid-May to mid-August, the Southern Hemisphere NLC season is the southern spring and summer that falls between mid-November and mid-March.

Airborne Imagery of Noctilucent Clouds

When studying noctilucent clouds, aircraft can provide improved spatial resolution, time evolution imagery, and the ability to validate spaceborne imagers. An aircraft may be positioned in a way to obtain imagery of improved resolution and contrast and be synchronized with spaceborne imagery overpasses. An aircraft is also able to track a prominent cloud that is detected by the AIM spacecraft, a capability that does not exist with ground-based or space-based imagery, and produce imagery of the noctilucent cloud from formation to dissipation.

The first dedicated airborne campaign to study noctilucent clouds was led by Jason Reimuller. A synchronized observation of a noctilucent cloud by the airborne platform and polar mesospheric cloud by the CIPS instrument on the AIM satellite was recorded on 6 July 2009 at 06:49:24 UT from an altitude of 18000 ft (5486m). The coordinated observations enabled investigation of CIPS performance at the day-to-night terminator and made it possible to relate the different resolutions of the two types of imagery to identify noctilucent cloud features, improving the ability of the AIM satellite to extract data near the day-night terminator. This represented the first such coincident measurement of noctilucent clouds from airborne and spaceborne instruments and verified that synchronous imagery of noctilucent cloud structures could be obtained through use of airborne and spaceborne imagers.

Figure 20. Dr. Jason Reimuller images NLCs from a modified Mooney M20K. On the right, geo-located airborne and spaceborne imagery show coincident NLC feature

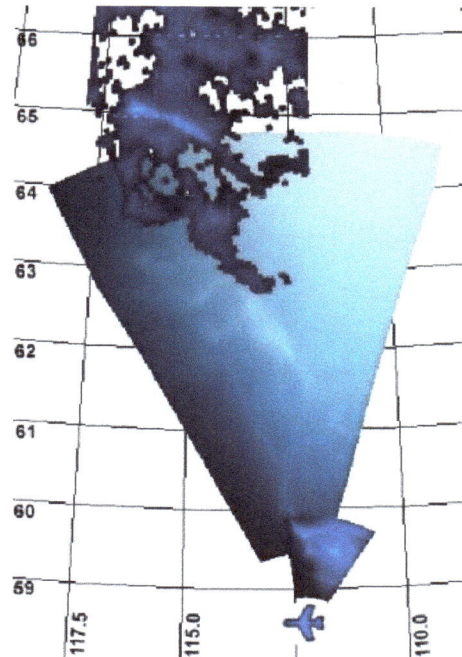

CLOUDS IN THE MIDDLE ATMOSPHERE OF MARS

Photo credit: NASA

Module Objectives

➢ Understand the differences between the atmosphere of Earth and Mars
➢ Understand the physics of ice clouds that form in the Martian atmosphere
➢ Understand the temperature structure of the Martian Atmosphere

Contributing Author:

Armin Kleinboehl, Ph.D.

The atmospheres of Earth and Mars in comparison

Mars is our neighboring planet in the solar system. Mars is a desert planet that is much colder and drier than the Earth. It is well known that frequent dust storms occur on Mars and some of them grow to a global scale, enshrouding the planet in dust for months at a time. It is less well known that clouds are also omnipresent on Mars. While lower atmospheric clouds on Mars are typically thinner than their equivalents on Earth, they still have significant effects on the temperature structure and circulation of the Martian atmosphere. Clouds in the middle atmosphere of Mars have quite a few similarities – and notable differences – to mesospheric clouds of the Earth's polar regions. In the following I want to give a brief overview of our current knowledge of Martian middle atmospheric clouds and discuss what we might be able to learn about them by studying polar mesospheric clouds on Earth. Comparing planetary and atmospheric features between our planet and other planets in the solar system – or even outside of our solar system – can help reveal fundamental processes that determine how planets work. These kinds of studies are summarized under the term 'comparative planetology'.

	Earth	Mars
Orbit		
Aphelion	152,098,000 km	249,209,000 km
Perihelion	147,098,000 km	206,669,000 km
Eccentricity	0.017	0.093
Orbital period	365.256 days	686.971 days / 668.599 sols
Planet		
Radius	6371 km	3389 km
Surface gravity	9.81 m/s²	3.71 m/s²
Rotation period	23h 56m 4s	24h 37m 22s
Axial tilt	23.44°	25.19°
Atmosphere		
Surface pressure	1013 mbar	6.1 mbar
CO_2	400 ppm	95.3%
N_2	78.1%	2.7%
Ar	0.9%	1.6%
O_2	20.9%	0.14%
H_2O	0.01% - 4.24%	15 - 1500 ppm

Table 2. Key planetary and atmospheric parameters of Earth and Mars in comparison.

Table 2 compares key parameters of Mars with our planet. Mars is a much smaller planet than the Earth with its planetary radius being only about half the Earth's radius. This leads to a surface gravity that is only about a third of the gravity on Earth. However, the rotation periods of both planets are very similar, such that a Mars day (typically called a 'sol') is only about 40 minutes longer than an Earth day. Also, the tilt of its rotation axis against the plane in which the planet orbits around the sun is quite comparable. This leads to seasons on Mars in the same way it does on Earth. Seasons on Mars are typically defined by the direction of the line between the Sun and Mars with respect to the sky noted as L_s (spoken 'L sub s'). $L_s=0°$

is defined as the beginning of northern spring, $L_s=90°$ is the beginning of northern summer, $L_s=180°$ is the beginning of northern fall, and $L_s=270°$ is the beginning of northern winter.

Mars is on average about 1.5 times farther away from the Sun than the Earth and hence receives significantly less sunlight. Its orbital period is about twice as long as Earth's so that a Mars year lasts close to two Earth years. An important difference in the orbital parameters between Earth and Mars is the orbital eccentricity. Earth's orbit is comparatively round, with the distance between the Earth and the Sun varying only a few million km over an Earth year. In contrast, Mars at its closest point (Perihelion) is about 50 million km closer to the Sun than at its farthest point (Aphelion). The seasonality of these events is such that Mars goes through Perihelion close to the beginning of southern summer and through Aphelion close to the beginning of northern summer. This leads to surface and atmospheric temperatures being considerably warmer during southern summer compared to northern summer temperatures. The increased heating also enables more lifting of dust such that the most prominent dust storms on Mars occur in the southern summer season.

Today's atmosphere of Mars is much thinner than the atmosphere of the Earth. Geologic evidence like outflow channels and canyons suggests that the Martian atmosphere was considerably more massive and probably also warmer and wetter than today. However, a considerable part of this original atmosphere has been lost to space over geologic time scales. Today the average surface pressure on Mars is only about 6 mbar. This is less than one percent of the Earth's surface pressure, and roughly corresponds to the atmospheric pressure found at 35 km altitude on Earth.

Large differences between Earth's and Mars' atmospheres exist in their compositions. While Earth's atmosphere consists of 78% of molecular nitrogen, in Mars' atmosphere nitrogen mixing ratios are only 2-3%. The main constituent of the Martian atmosphere is CO_2, over 95% of the atmosphere is made up of this gas. On Earth, the CO_2 volume mixing ratio is about 400 ppm (parts per million) and has been slowly rising over the years, largely due to fossil fuel consumption of humankind. 21% of Earth's atmosphere consists of oxygen, which is largely produced by biologic activity as a by-product of photosynthesis. To date there are no known biologic processes on Mars, and oxygen in the Martian atmosphere is only a trace gas, largely produced by the photolysis of CO_2 in the upper atmosphere.

An important constituent when it comes to cloud formation in the atmospheres of both planets is water vapor. Water vapor in Earth's atmosphere is abundant, at least in the lower atmospheric layer, the troposphere. Water vapor mixing ratios in Earth's atmosphere vary strongly depending on latitude, temperature, and weather. In tropical regions, mixing ratios up to 4% can be found. Due to the temperature range found in Earth's troposphere, clouds formed in the lower troposphere typically consist of liquid droplets, while clouds in the upper troposphere consist of ice particles. In the middle troposphere mixtures of liquid and solid cloud particles can exist. The Martian atmosphere is on average a lot drier than Earth's atmosphere. Water vapor mixing ratios are typically in the ppm range, although values up to 1.5% (1500 ppm) have been measured. Liquid water is not stable in today's Martian climate so water-based clouds in the atmosphere of Mars are made of water ice.

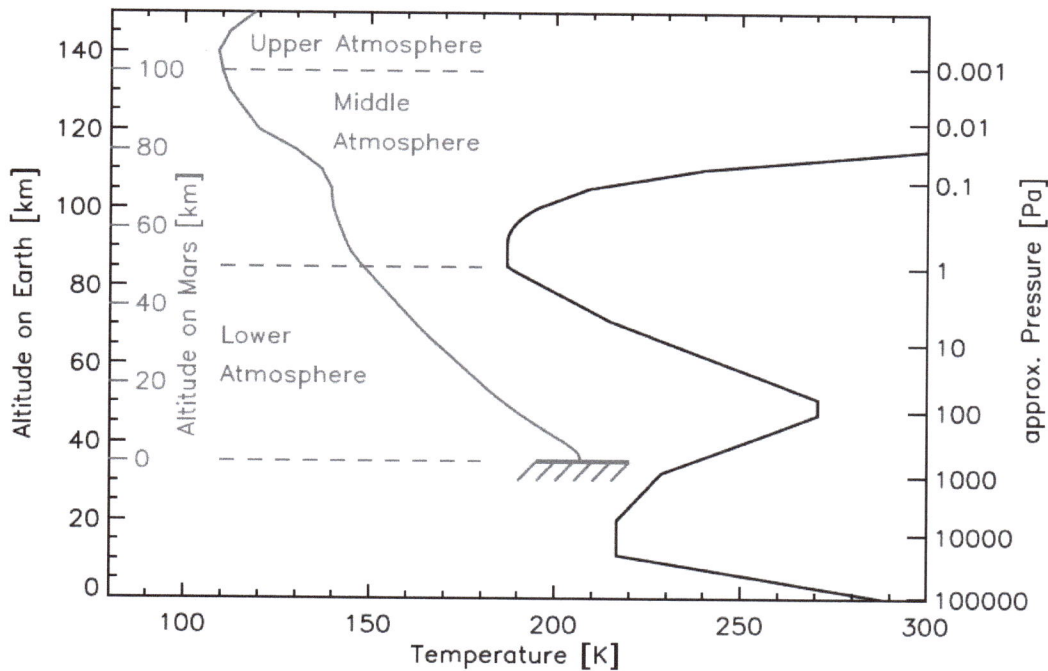

Figure 21. *Standard temperature profiles vs. altitude for Earth (black, based on the US standard atmosphere) and Mars (gray, based on the MCS temperature climatology in the non-dusty season in Haberle et al., 2017). The surface pressure on Mars roughly corresponds to the atmospheric pressure on Earth found at 35 km altitude, hence the altitude scale for Mars was offset by 35 km with respect to the altitude scale for Earth. The pressure scale is only approximate due to the differences in atmospheric temperatures.*

Figure 21 shows an average temperature profile of Mars in comparison to a standard atmospheric temperature profile of the Earth. Temperatures are plotted vs. an altitude scale. The altitude scale for Mars' atmosphere has been offset by 35 km in comparison to the altitude for Earth's atmosphere as the surface pressure on Mars corresponds roughly to the Earth's atmospheric pressure found at 35 km altitude. The pressure scale on the right hand side of Figure 1 is approximate but valid for both atmospheres. Figure 1 indicates that atmospheric temperatures on Mars are lower than on the Earth at every altitude level. Atmospheric temperatures on Earth close to the surface are about 280-290 K on average. On Mars the average temperature in the lower atmosphere is about 240-250 K. While surface temperatures can exceed the freezing point of water at 273 K in localized areas in the middle of the day, Martian atmospheric temperatures rarely reach this value. Another feature easily observed in Figure 1 is that layers in Mars' atmosphere cannot be as easily defined as in Earth's atmosphere. On Earth, atmospheric layers are defined due to changes in the sign of the lapse rate. A dominant feature is the Earth's stratosphere, which has increasing temperatures with height due to atmospheric heating through solar absorption by the ozone layer. Due to the much lower concentration of oxygen, Mars lacks an ozone layer, hence the Martian atmosphere lacks the strong temperature changes with altitude found in Earth's atmosphere. Outside of the winter polar regions, Mars' atmospheric temperature typically decreases with height up to an altitude of 40-50 km. Above this altitude, temperatures tend to become more isothermal. At altitudes above ~100 km, temperatures may increase again due to solar absorption, which leads to a strong day/night temperature contrast. Hence the lowest 50 km of the Martian atmosphere have been defined as lower atmosphere, the region between 50 and 100 km has been defined as middle atmosphere, and the region above 100 km is considered the upper atmosphere.

Martian middle atmospheric water ice clouds

Clouds in the Martian atmosphere can be made of water ice or carbon dioxide ice. The latter do not have an equivalent on Earth as temperatures are never low enough for carbon dioxide to freeze. In addition, carbon dioxide is the main constituent of the Martian atmosphere so freezing it out (either due to cloud formation or direct deposition to the surface at the winter poles) has a profound impact on the atmospheric pressure cycle and leads to significant pressure variations over the seasons.

Water ice clouds are omnipresent on Mars. Temperatures are typically suitable for the formation of ice clouds and cloud formation mainly depends on the availability of water vapor. On Earth the tropopause provides an effective cold trap for water vapor, leading to very dry conditions in the stratosphere and mesosphere. On Mars the smaller lapse rate in the atmosphere does not provide an effective cold trap for water vapor, and water ice clouds can form in the lower atmosphere as well as in the middle atmosphere. The most prominent cloud feature in the lower atmosphere is the aphelion cloud belt, which was already observed from Earth-based measurements. It is a band of clouds that appears in the equatorial region in the northern spring and summer season. It typically extends from about 10°S to 30°N in latitude and the clouds reach up to altitudes of roughly 40 km. The aphelion cloud belt is fed by water vapor coming off the north polar cap in spring and summer. The cooler global temperatures during the aphelion season cause clouds to form in the equatorial region. As the perihelion season approaches, the aphelion cloud belt starts to dissipate due to the rising global temperatures. This corresponds to northern fall. Temperatures drop in the northern high latitudes, causing condensation of water vapor in the lower atmosphere of the polar region that leads to the formation of the northern polar hood cloud. With an extent from the north pole down to 50°N latitude it covers not only the polar region but reaches well into the mid-latitudes. It starts forming in late northern summer around $L_s=160°$ and dissipates again in early northern spring around $L_s=20°$. The southern polar region also develops a polar hood cloud in southern fall and winter. However, the southern polar water ice clouds are not nearly as dense or as extended as their counterpart in the north. They are present between about $L_s=20°$ and $L_s=180°$, with a notable gap in occurrence between $L_s=70°$ and $L_s=110°$. Like the northern polar hood, also the southern polar hood is constrained to the lower atmosphere. However, in contrast to the north, the southern polar hood cloud is shaped like an annulus, mostly covering the latitudes between 60°S and 80°S.

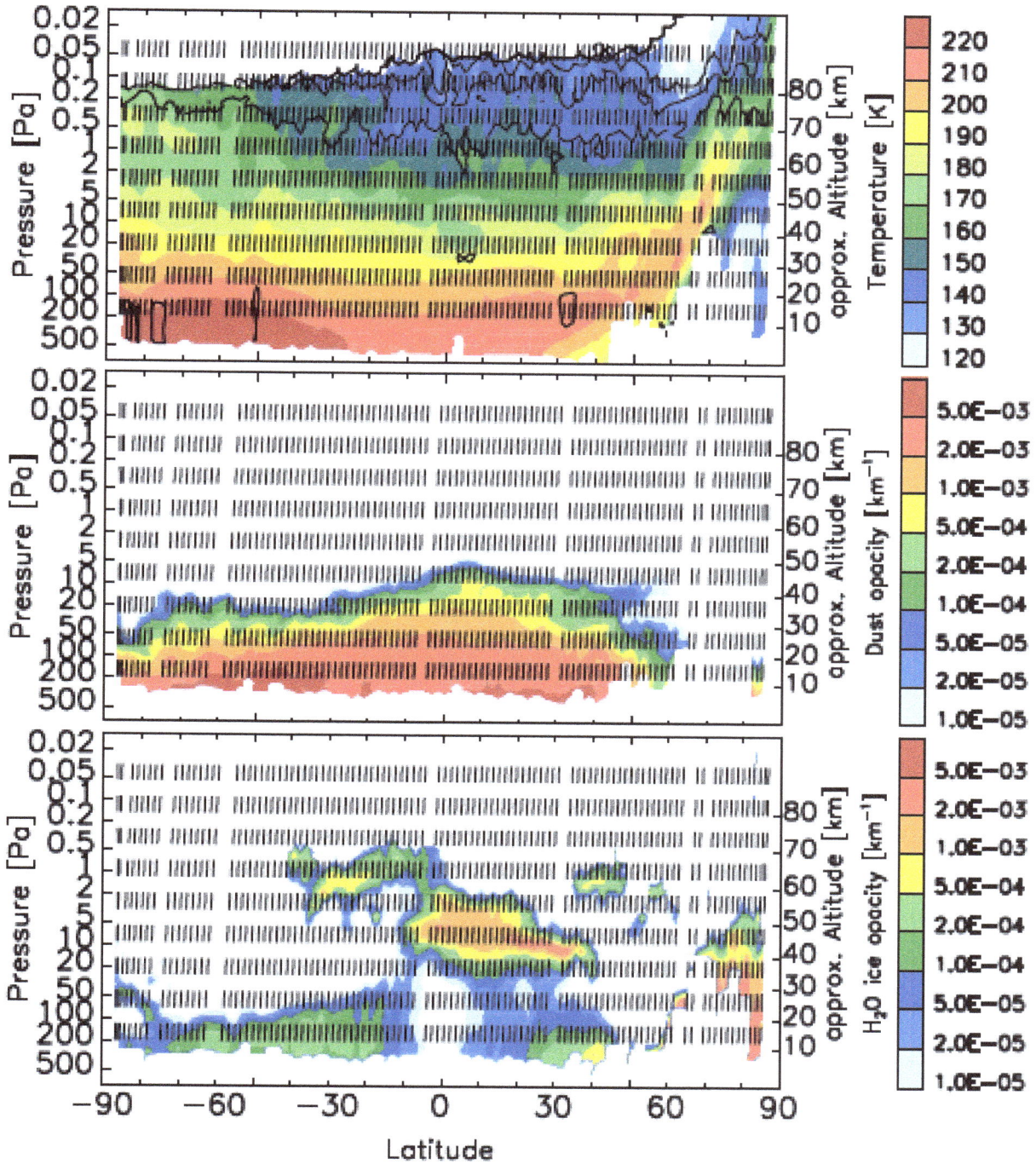

Figure 22. Transect of atmospheric temperature (top), dust opacity (center, at 463 cm⁻¹) and water ice opacity (bottom, at 843 cm⁻¹) as measured by MCS in September 2009 (Lₛ=247°). The nearly vertical dashed lines indicate the locations of individual measurements. Water ice clouds reach altitudes of 60-70 km at this season.

Figure 22 shows a pole-to-pole transect of temperature, dust and water ice opacity at $L_s=247°$ during northern fall as measured by the Mars Climate Sounder (MCS). MCS is a thermal emission radiometer on board the Mars Reconnaissance Orbiter (MRO). It has eight channels in the mid- and far-infrared and one channel covering a broad band in the visible and near-infrared wavelength region. It predominantly measures in limb geometry, which means that it looks at the horizon of the planet. The limb geometry provides a long optical path, which increases the sensitivity of the measurement, and it allows the observation of the vertical structure of dust and clouds. From the radiance measurements, profiles of atmospheric temperature, dust and water ice are retrieved from the surface to ~80 km altitude with a vertical resolution of ~5 km. The most prominent feature of Figure 2 is a cloud in the equatorial region. This is not during the season of the aphelion cloud belt. However, a temperature minimum at 30-50 km altitude between about 10°S and 40°N causes water vapor to condense and form a cloud. In addition, temperature minima at higher altitudes to the south (40°S-10°S) and to the north (40°N-60°N) lead to cloud formation. Water ice clouds in these regions are found at altitudes of 60-70 km in the middle atmosphere. This shows that at least in this season, water vapor can penetrate high enough into the atmosphere to allow the formation of water ice clouds in the middle atmosphere. The measurements of these clouds were made with an MCS infrared channel at 12 µm, suggesting that these clouds consist of moderate-size ice particles, likely of order 1 µm radius, a size typical for clouds in the lower atmosphere. This is significantly larger than the particle sizes of polar mesospheric clouds on Earth. However, other measurements in the ultraviolet or near-infrared wavelength ranges, like solar occultation observations from Mars Express or observations by the CRISM instrument on MRO have also suggested clouds with particle sizes down to ~0.1 µm radius, which would be more in line with polar mesospheric cloud particle sizes on Earth. Other features that can be identified in Figure 2 include the northern polar hood cloud, which was discussed previously. It extends from the pole to about 60°N in the lower atmosphere. The extent of the polar hood cloud at this season is determined by the polar vortex, a region of confined air formed by the descent of air masses over the northern polar region.

Martian middle atmospheric carbon dioxide ice clouds

In the previous section it was shown that water ice clouds on Mars can reach mesospheric altitudes. This happens predominantly in the perihelion season ($L_s=180°-360°$) when the atmosphere is dustier and lower atmospheric temperatures are higher, allowing the transport of water vapor to higher altitudes of the atmosphere.

However, parts of the Martian atmosphere can become cold enough to allow carbon dioxide to condense, leading to the formation of CO_2 ice clouds. A feature that makes this process particularly intriguing is that CO_2 is the main constituent of the Martian atmosphere. In the lower atmosphere of the polar regions, temperatures in winter regularly drop to values at which CO_2 condenses. The condensation of CO_2 is the main driver of the seasonally varying surface pressure on Mars. In Figure 2, temperatures drop below the frost point of CO_2 (~145 K at Martian pressures in the lower atmosphere) in the center of the vortex close to the pole. Hence the conditions in these regions are favorable for the formation of lower atmospheric CO_2 ice clouds. Due to the high abundance of CO_2 in the Martian atmosphere, these clouds are expected to grow to large particle sizes rather quickly, causing CO_2 snowfall in the winter polar region.

Observations of high clouds or detached aerosol layers in the aphelion season raised the question whether the Martian mesosphere could also become cold enough to allow atmospheric CO_2 to condense and form mesospheric clouds. In 1997 the atmospheric temperature profile reconstructed from the entry trajectory of the Mars Pathfinder lander revealed temperatures below the CO_2 frost point in the middle atmosphere, suggesting that CO_2 cloud formation may be possible. Shortly thereafter, these measurements were used together with early observations by the Mariner 6 and 7 as well as the Viking missions, ground-based submillimeter observations, and pre-dawn cloud observations by Pathfinder from the Martian surface to provide evidence for the existence of middle atmospheric carbon dioxide clouds on Mars.

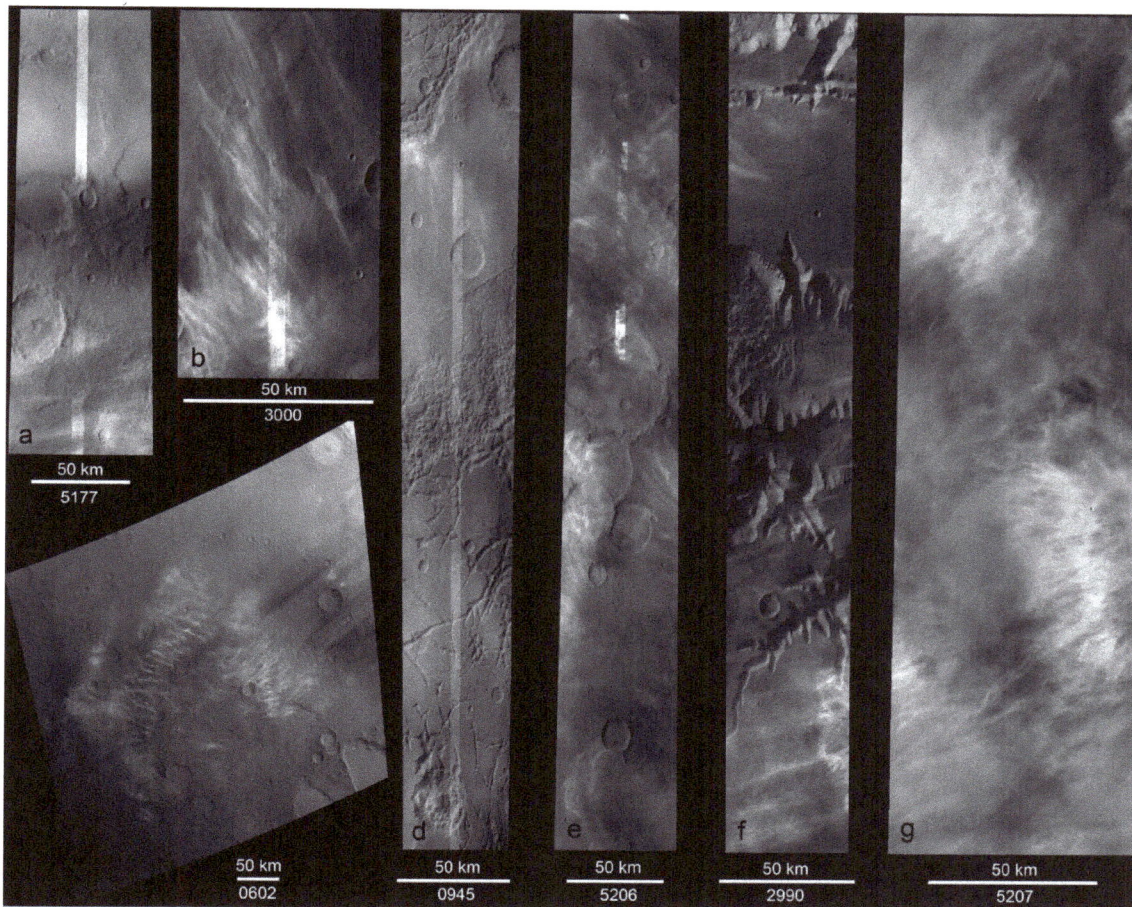

Figure 23. Image examples of martian ice clouds taken by the HSRC camera onboard Mars Express. Vertical stripes in the center of images (a), (b), (d), and (e) indicate simultaneous measurements with the OMEGA imaging spectrometer onboard the same spacecraft. From Määttänen et al. (2010), used with permission.

Since the 2000s, observations by instruments on several Mars orbiters including Mars Global Surveyor (MGS), Mars Express (MEx), Mars Odyssey (ODY), and the Mars Reconnaissance Orbiter (MRO) have been providing a multitude of evidence of martian middle atmospheric CO_2 clouds. Limb observations by the camera and thermal emission spectrometer instruments on MGS allowed the detection of mesospheric clouds and provided estimates of their altitudes. Stellar occultation observations from MEx, by which the absorption of starlight is observed as the star sets or rises through the atmosphere, also provided cloud extinctions and their vertical distribution. In addition, atmospheric density measurements by the same instrument could be used to derive temperature structure and to show that cold pockets with atmospheric temperatures below the CO_2 frost point existed in the vicinity of the detected clouds. Nadir-viewing imagers and imaging spectrometers also provided evidence for mesospheric clouds. Figure 3 shows a set of mesospheric cloud images by the HSRC camera onboard Mars Express. These images can be used to identify clouds and evaluate cloud structures. The images show striking similarities to images of Polar Mesospheric Clouds (PMCs) on Earth taken with camera instrumentation flown on Antarctic circumpolar balloon flights or with the CIPS instrument onboard the AIM satellite from Earth orbit. While the lateral extent of mesospheric clouds can be easily determined from images, it is much harder to derive the cloud altitude. Knowing the solar illumination angle it is possible to derive cloud altitude from the shadow that is cast by the cloud onto the surface. However, this works only for clouds that are sufficiently optically thick to cast a discernible shadow on the surface. Another possibility for altitude determination is the use of the parallax in two or more subsequent images taken at varying off-nadir angles. In an image taken straight down in nadir geometry a cloud will obscure a different part of the surface than in an image taken at an angle in off-nadir geometry. With the knowledge of the spacecraft positions at the times the two images were taken and the observation angles, the altitude of a cloud can be calculated. This method was used extensively to characterize clouds imaged with the visible subsystem of the THEMIS instrument on Mars Odyssey.

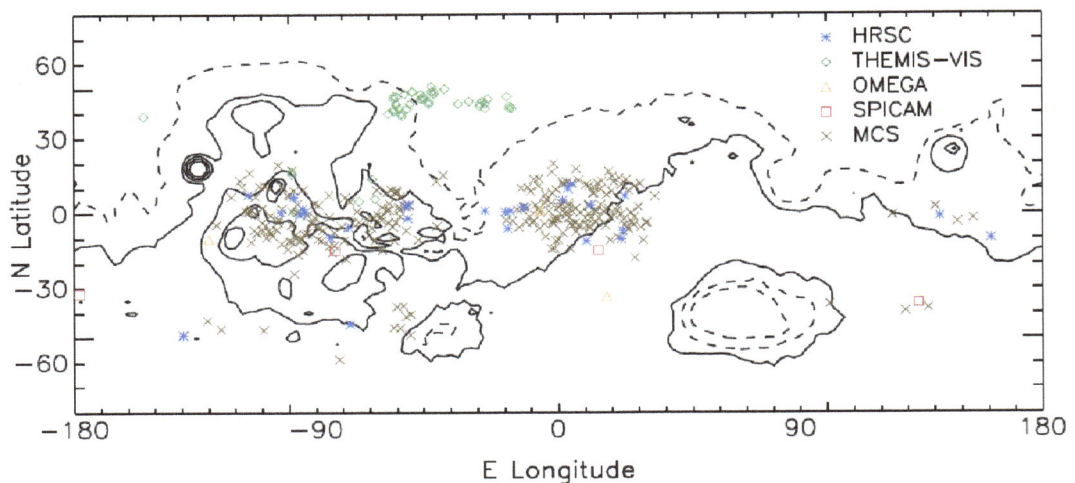

Figure 24. Latitude/Longitude distribution of mesospheric clouds on Mars as measured by various instruments from Mars orbit. MCS measurements have been restricted to be between $L_s=0°-90°$ (northern spring) and above 60 km altitude. Contours show Mars topography (solid: positive, dashed: negative) in intervals of 3 km.

None of these methods are equally suited to determine that a cloud actually is composed of CO_2 ice rather than water ice so for many of the identified mesospheric cloud examples the actual composition is not necessarily known. Limb observations of cloud extinction can be combined with limb observations of temperature, either based on the density structure as derived from solar or stellar occultation measurements or directly measured by means of thermal emission radiometry. If the temperatures in the vicinity of a cloud are close to the CO_2 frost point the cloud is likely composed of CO_2 ice. Direct evidence for cloud composition requires the spectroscopic identification of CO_2 ice in the clouds. This has been achieved by identifying the CO_2 ice scattering peak at 4.24 μm in near-infrared observations by the OMEGA instrument, an imaging spectrometer onboard Mars Express. The CRISM imaging spectrometer has also been used to identify cloud composition and estimate particle sizes. Its observations suggest typical particle sizes of order 0.5 μm radius for mesospheric CO_2 clouds.

Figure 24 shows the latitude/longitude distribution of mesospheric clouds identified in measurements from various instruments. A significant fraction of the observed mesospheric clouds occur in a band of about 20° latitude around the equator. They cluster around longitudes of 90°W and 10°E, a few clouds were also observed around 120°-150°E. Most of these clouds are made of CO_2 ice. The clustering is related to regional temperature minima in the Martian mesosphere. In addition, the clustering occurs in regions close to large changes in topography, suggesting that gravity waves driven by wind over high topography may play a role in the formation of the clouds. Equatorial mesospheric clouds typically form in the aphelion season around $L_s=0°-150°$. Figure 4 also shows the occurrence of mesospheric clouds in mid-latitudes. Most of these mid-latitude clouds tend to appear in the fall of their hemisphere (around $L_s=50°$ in the southern hemisphere and around $L_s=250°$ in the northern hemisphere). It is thought that most of these mid-latitude clouds also consist of CO_2 ice although the occurrence of water ice clouds cannot be excluded (compare Figure 22).

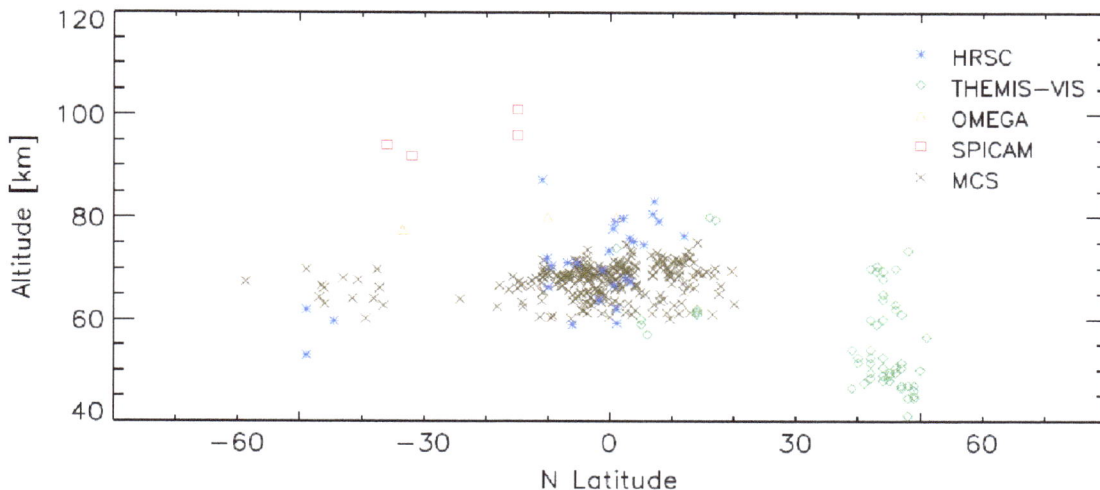

Figure 25. Latitude/Altitude distribution for the same set of mesospheric clouds on Mars as in Figure 4. Again, MCS measurements have been restricted to be between $L_s=0°-90°$ (northern spring) and above 60 km altitude.

Figure 25 shows the vertical distribution of the clouds in Figure 4. The bulk of the equatorial clouds in the aphelion season occur at altitudes between 60 and 90 km. A few clouds have even been observed to occur as high as 100 km. Clouds found at mid-latitudes in both hemispheres tend to occur at lower altitudes and occupy a typical altitude range of 40-70 km.

Temperature structures controlling cloud formation on Mars

Deviation of nearest available retrieved temperature from CO_2 frost point calculated at pressure level (K)

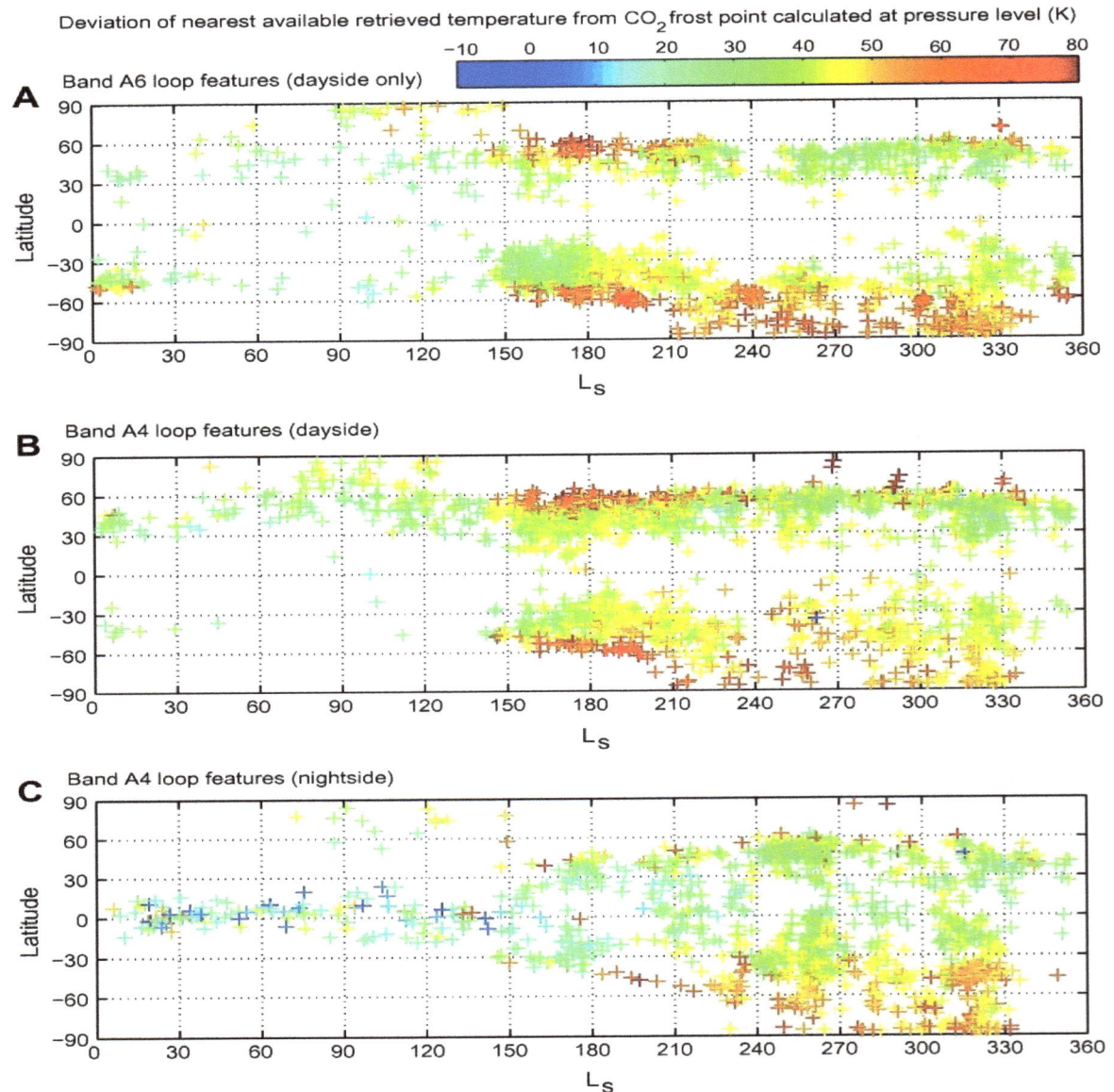

Figure 26. Latitude/L_s distribution of mesospheric clouds from MCS. Band A6 corresponds to the MCS visible/near-IR channel while Band A4 corresponds to the MCS infrared channel centered at 12 μm. The color coding indicates atmospheric temperature deviation from the CO_2 frost point from the temperature measurement closest the cloud measurement. From Sefton-Nash et al. (2013), used with permission.

What causes clouds to form in the Martian mesosphere? A clue can be found by studying cloud occurrence in relation to the temperature structure of the Martian atmosphere. Figure 26 shows the distribution of mesospheric clouds observed in various channels of the MCS instrument over the course of a Martian year. The color coding gives the temperature that was measured by MCS in the thermal infrared at the closest location and the same pressure level as the cloud observation. Note that the temperature is given as the deviation to the CO_2 frost point, which is of order 100-120 K at pressures typical for the Martian mesosphere. Figure 6 shows that temperatures close to clouds observed during the aphelion or northern spring and summer season tend to be within 10-20 K of the local CO_2 frost point. This suggests that most of the mesospheric clouds during this season are made of CO_2 ice. Starting around the northern fall equinox ($L_s=150°-180°$) an increasing number of clouds is found for which nearby temperatures are much higher, in many cases 40-50 K and in some cases even 70-80 K above the CO_2 frost point. These temperatures indicate that the clouds observed in these locations are water ice clouds. Note that at equatorial latitudes around $L_s=180°-210°$ and at northern mid-latitudes around $L_s=270°-330°$ there are cloud populations with nearby temperatures around 20-30 K above the CO_2 frost point. It is possible that local temperature perturbations could bring these temperatures close to the frost point, allowing CO_2 clouds to form even at this season.

In order to better understand the conditions under which mesospheric clouds form we should have a closer look at the processes controlling the temperature structure. One of the main drivers of temperature variations in the Martian atmosphere are atmospheric tides. Tides are periodic changes in atmospheric parameters like temperature, pressure, and wind that have periods of a fraction of a solar day. They are driven by the changes of solar energy input to the Mars surface and atmosphere over the course of a day. This is in contrast to ocean tides on Earth that are driven by the gravitational pull of the moon and the sun. Due to the thin atmosphere on Mars, most of the solar radiation reaches the surface, where it causes strong differences in temperature between day and night. Surface temperature maxima are typically reached at local noon or slightly later, while surface temperature minima are reached in the early morning. The heat flux from the surface causes changes in pressure and temperature in the lowermost atmosphere. The propagation of these changes gives rise to global oscillations in atmospheric pressure and temperature, and subsequently also wind. The most prominent oscillation is the diurnal tide, which has a period of one solar day, meaning that for example temperature will exhibit one minimum as well as one maximum over the course of a day. While thermal tides also exist on Earth, their temperature perturbations typically start to become significant only at altitudes of the upper mesosphere and above. On Mars, thermal tides cause significant temperature variations throughout the atmosphere.

Figure 7 illustrates the diurnal tide in the temperature field as observed by MCS. Due to the sun-synchronous orbit of MRO, MCS collects temperature data predominantly around 3 AM and 3 PM local time. In Figure 7 these temperature measurements were combined to zonal averages separately for day and night and then used to create diurnally averaged temperatures and temperature differences between day and night. The average temperature field shows a temperature structure that is typical for atmospheric conditions close to equinox ($L_s=135°-165°$), with cold temperatures in the lower middle atmosphere of both polar regions, overlaid by layers of warmer temperatures, and cold temperatures in the equatorial middle atmosphere. The temperature differences show a complex pattern of minima and maxima, which is highlighted in the schematic in the bottom left panel of Figure 26. A temperature maximum is found above the equatorial surface. Above this region, the temperature deviations show a nodal pattern consistent with the expectation of a vertically propagating tide with a vertical wavelength of order 40 km. Around 30° N and S the phase of the tide reverses, such that in northern and southern mid-latitudes temperature minima are located at altitudes of temperature maxima in the equatorial region and vice versa.

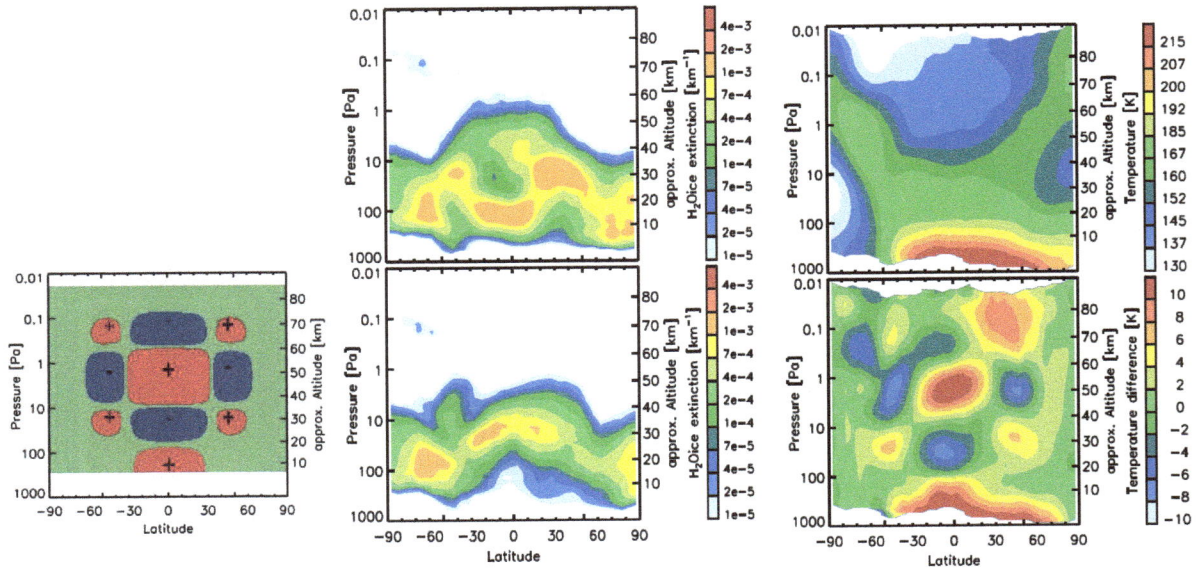

Figure 27. Diurnally-averaged temperature (top center panel) and day/night temperature difference (bottom center panel, defined as (T$_{PM}$-T$_{AM}$)/2) as a function of latitude and pressure for the period from L$_s$=135°-165° derived from MCS data. The far left panel shows a schematic of the temperature minima and maxima. The right panels show water ice clouds from MCS data at different local times (top: 3 AM, bottom: 3 PM) and indicate that cloud coverage is strongly correlated with temperature minima.

The right panels of Figure 27 shows the occurrence of water ice clouds as observed by MCS, separated for day and night. Few water ice clouds are observed above 50 km altitude in this season. The only notable cloud occurrence in the upper middle atmosphere is observed at 60°-70°S around 70 km altitude, coincident with very low temperatures found in this region. Elsewhere water ice clouds mainly form in locations consistent with temperature minima driven by the diurnal tide. At nighttime, clouds tend to form close to the surface and around 40 km altitude in the equatorial region. At daytime the pattern reverses and clouds tend to form predominantly between 20 and 30 km altitude, where the equatorial daytime temperatures are lower than at night. The formation of water ice clouds is very common at temperatures found in the martian atmosphere and limited largely by the availability of water vapor. During aphelion season, water vapor in the middle atmosphere is limited by the extensive cloud formation below 40-50 km such that water ice clouds at mesospheric altitudes are rare. During perihelion season (southern spring and summer) the warmer and dustier lower atmosphere allows water vapor to be transported to higher altitudes, enabling the frequent formation of water ice clouds. Small dust particles, advected together with water vapor to mesospheric altitudes, could serve as nuclei for cloud condensation. The presence of clouds at altitudes of 50-70 km in perihelion season (Figure 2) indicates localized water vapor mixing ratios of tens of ppm, which would be about an order of magnitude higher than in Earth's stratosphere and mesosphere.

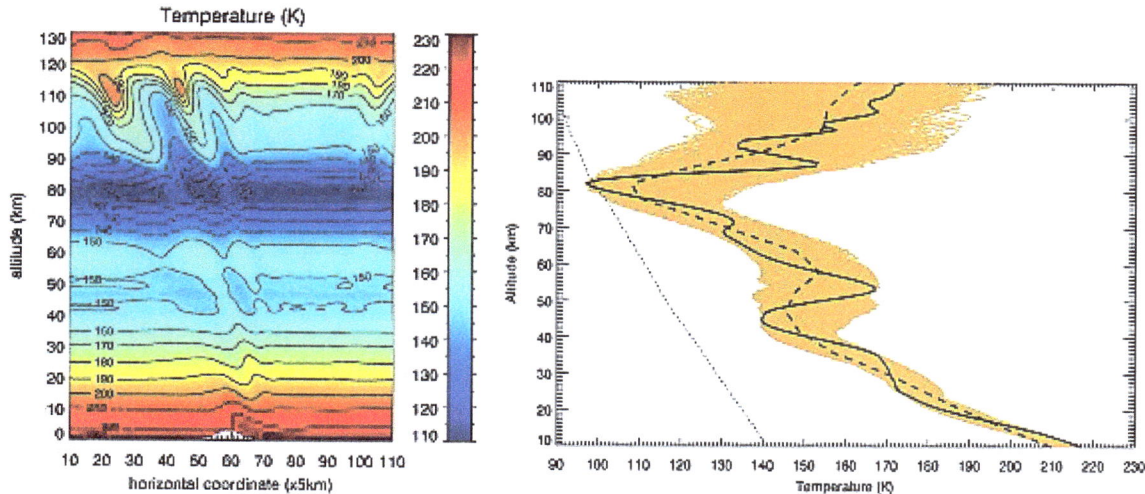

Figure 28. Mars atmospheric temperature structure derived from a regional-scale numerical computer model for conditions of wind passing over a mountain range and causing gravity waves (left). Cross-section through the temperature structure of the left panel showing the undisturbed background temperature (dashed) and the perturbation range due to gravity waves (yellow-shaded area). The solid line indicates a temperature profile within this range that reaches the CO_2 frost point (right). From Spiga et al. (2012), used with permission.

The formation of CO2 ice clouds is obviously not limited by vapor supply as CO2 is the main constituent in the Martian atmosphere. However, measurements show that temperatures in the Martian middle atmosphere rarely reach the CO2 frost point. In addition, computer models that are used to simulate Mars atmospheric circulation on a global scale do not tend to simulate temperatures at the frost point outside of the winter polar regions.

Figure 28 gives a clue about the processes that are likely required for the formation of mesospheric CO_2 ice clouds. It shows temperature simulations by a regional-scale computer model. The banded temperature structure in the left panel of Figure 8 shows the temperature variation with altitude as it would be simulated with a global model. This temperature structure is controlled by radiative processes, the global atmospheric circulation, and atmospheric tides. The perturbations in the banded structure are caused by gravity waves due to the wind in the lower atmosphere passing over a mountain range (depicted as a white spot at the bottom of the left panel of Figure 8). As the wind passes over the mountain, it induces gravity waves that propagate into the upper atmosphere. As they propagate to higher altitude their amplitudes grow due to the decrease in atmospheric density. The right panel of Figure 8 shows a cross-section through the temperature structure. The dashed line indicates the background temperature due to the general circulation and the tides. The yellow-shaded area gives the temperature perturbations caused by gravity waves. If the phase of a gravity wave coincides with the tidal structure such that the gravity wave deepens a temperature minimum caused by the tide (black line) the perturbation can be large enough to reach the CO_2 frost point (dotted line). This process is thought to be essential for the formation of CO_2 ice clouds in the Martian mesosphere. The temperatures are still not low enough to allow homogeneous nucleation, such that the formation of CO_2 ice requires some form of condensation nuclei. It is suggested that particles from meteoric infall could provide these condensation nuclei, similar to what is suggested for the formation of PMCs in Earth's mesosphere.

Continued operations of limb sounding instruments like MCS on MRO and the deployment of new limb sounding instruments would complete and further refine the climatology of mesospheric clouds in the Martian atmosphere. Characterization of the atmospheric environment will help separating water ice and CO_2 ice clouds and further constrain the conditions required for cloud formation. Imaging of Martian mesospheric clouds by orbiters and landed assets and the analysis of small-scale cloud structures could help to characterize gravity waves in the Martian atmosphere and their influence on mesospheric clouds. Analysis of these images could be done analogously to image analyses of spaceborne or suborbital images of PMCs on Earth. This would provide additional details to further our understanding of the formation and development of mesospheric clouds on Mars.

Figure references:

Määttänen A., Montmessin F., Gondet B., Scholten F., Hoffmann H., González-Galindo F., Spiga A., Forget F., Hauber E., Neukum G., Bibring J.-P., and Bertaux J.-L. (2010) Mapping the mesospheric CO_2 clouds on Mars: MEx/OMEGA and MEx/HRSC observations and challenges for atmospheric models. *Icarus, 209,* 452-469.

Sefton-Nash E., Teanby N. A., Montabone L., Irwin P. G. J., Hurley J., and Calcutt S. B. (2013) Climatology and first order composition estimates of mesospheric clouds from Mars Climate Sounder limb spectra. *Icarus, 222,* 342-356.

Spiga A., González-Galindo F., López-Valverde M.-A., and Forget F. (2012) Gravity waves, cold pockets and CO_2 clouds in the martian mesosphere. *Geophys. Res. Lett., 39,* L02201, doi:10.1029/2011GL050343.

Further reading:

Haberle, R. M, Clancy, R. T., Forget, F., Smith, M. D., and Zurek, R. W. (eds.) (2017) The Atmosphere and Climate of Mars. *Cambridge University Press.*

Read, P. L. and Lewis, S. R. (2004) The Martian Climate Revisited – Atmosphere and Environment of a Desert Planet. *Springer.*

Kleinböhl, A., Wilson, R. J., Kass, D. M., Schofield, J. T., and McCleese, D. J. (2013) The Semi-diurnal Tide in the Middle Atmosphere of Mars. *Geophys. Res. Lett., 40,* 1952-1959, doi:10.1002/grl.50497.

Määttänen A., Pérot K., Montmessin F., and Hauchecorne A. (2013) Mesospheric clouds on Mars and on Earth. In Comparative Climatology of Terrestrial Planets (S. J. Mackwell et al., eds.), 393-413. *Univ. of Arizona Press,* doi:10.2458/azu_uapress_9780816530595-ch16.

PART II: FUNDAMENTAL CONCEPTS OF SPACE FLIGHT OPERATIONS

Space Physiology

Space Life Support Systems

Spacesuit Operations and Biometric Monitoring

Hypoxia

SPACE PHYSIOLOGY

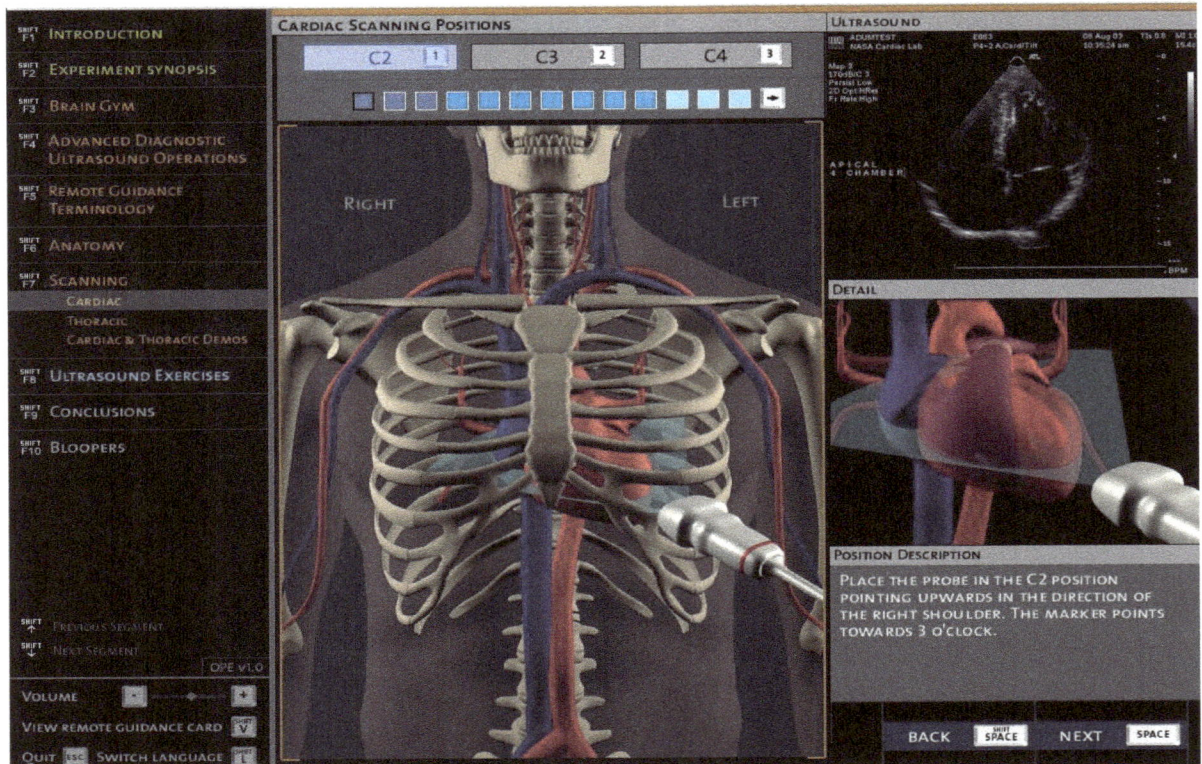

Module Objectives

- ➢ Describe the control of blood flow and blood pressure
- ➢ Explain what is meant by orthostatic hypotension
- ➢ Describe some of the countermeasures to muscle atrophy
- ➢ Describe the effects of spaceflight on the fluid regulating system
- ➢ Describe the effects of spaceflight on the neurovestibular system
- ➢ Explain the space motion sickness syndrome.
- ➢ Explain the effects of noise and vibration
- ➢ Describe the etiology of space motion sickness
- ➢ List three symptoms of space motion sickness

Contributing Author:

Erik Seedhouse, Ph.D.

The human body has, thanks to its terrestrial development, adopted potent physiological mechanisms that enable man's upright posture to be compatible with Earth's 1-G environment. Microgravity adversely affects many of these mechanisms, the most seriously affected being those associated with the cardiovascular and the musculoskeletal systems. Since the effects upon these physiological systems may compromise your performance as a crewmember returning from orbit, it's important to be familiar with the fundamentals of space physiology. This section introduces you to the physiological consequences of exposure to microgravity and to the countermeasures required to maintain physiological conditioning.

CARDIOVASCULAR SYSTEM

The cardiovascular system consists of a heart that functions as a pump and blood vessels that function as a high and low-pressure distribution circuit. The heart can be divided into two pumps as shown in Figure 29, the chambers on the right side receive blood returning from the body and pumps blood to the lungs for aeration, and the left side receives oxygenated blood from the lungs and pumps blood into the aorta for distribution throughout the body.

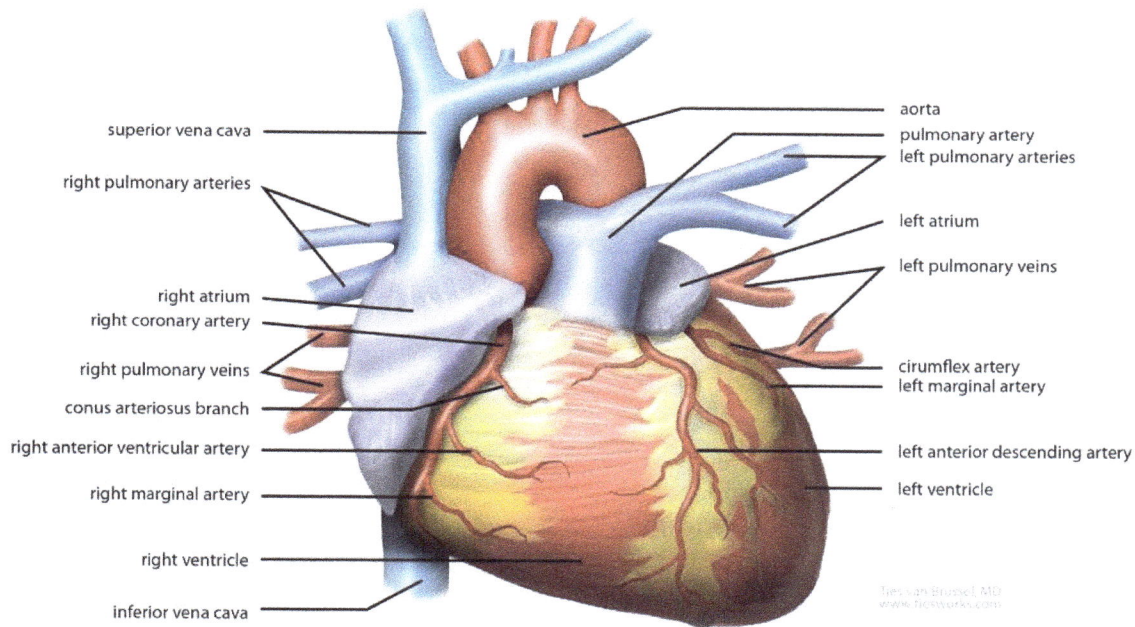

Figure 29. The Cardiovascular System

The low-pressure system - the *pulmonary circulation* - is the pathway of blood from the right ventricle to the lungs and back to the left atrium. The high-pressure system, also termed the *systemic circulation*, is the pathway of blood from the left ventricle to the capillaries and back to the right atrium. To prevent the backflow of blood, atrioventricular valves provide a one-way flow of blood from the right atrium to the right ventricle. Similarly, the semilunar valves prevent backflow into the heart between contractions.

Blood Pressure

When your left ventricle contracts, blood is forced through the aorta, creating pressure throughout the arterial system and causing a pressure wave, or *pulse*, to travel down the aorta and throughout the arterial tree. The highest pressure generated by your heart is termed *systolic blood pressure*. Between beats, your heart pauses to allow the atria to refill with blood for the next contraction, a period of lower pressure termed *diastolic blood pressure*. The difference between systolic and diastolic pressure is termed pulse pressure (PP). Normally, blood pressure is expressed as Mean Arterial Pressure (MAP), calculated as MAP = DP + 1/3 PP

Blood Volume

Your cardiovascular system contains 7% of the body's water in the form of plasma (about 3 liters for a 70kg male) and serves as a major fluid transportation system, a function that has implications discussed later for all astronauts before, during and following orbital flight. What is important to understand at this stage is that blood volume changes may occur due to changes in the water content of blood plasma, a process caused by the dynamic interaction with body tissues and blood. Blood is not passively trapped within the circulatory system because there is a constant exchange of fluid between blood plasma in the capillaries and the fluid between cells of the tissues. This exchange is governed by physical forces and physiological laws that explain why areas of the body can undergo dehydration or swell with excess fluid, a state termed 'edema'. Edema may occur due to an increase in blood pressure, which results in a concomitant rise in capillary pressure, which in turn causes fluid to filter out of the capillary and edema of the tissues.

Control of Blood Flow and Blood Pressure

Pressure drop, fluid flow, and resistance to that flow are the principle components of fluid mechanics. For those mathematically inclined, the control of blood flow can be described by the following equation which describes the flow of blood through a given tissue bed as being directly proportional to the pressure gradient flow across the bed and inversely proportional to the resistance encountered during transit:

$$F = (PA - PV) / R \qquad R = 8\,\eta\,L / \pi\,r4$$

where **F** = flow, **PA** = arterial resistance, **PV** = venous pressure, **R** = resistance to flow, η = viscosity, **L** = length of tube, and **r** = radius of tube

Now that you understand how blood flow is regulated it is important to also understand how blood pressure is controlled. Before putting all this together it is necessary to understand a little more about the systemic circulation. Firstly, the central force for driving blood around the systemic circulation and through the capillaries of the organs of the body is mean arterial pressure. Secondly, the total flow through the systemic circulation is equal to the cardiac output (CO), and thirdly, the total vascular resistance to flow is the sum resistance offered by the entire systemic circulation, which is referred to as total peripheral resistance (TPR). Thus:

$$MAP = CO \times TPR$$

Once again, this relationship can be expressed as:

$$Resistance = 8l\eta/r^4$$

Basically, resistance in any vessel will be dependent on the length of the vessel (**l**), the viscosity of the blood as it flows (**η**), and the radius to the fourth power of the vessel (**r⁴**). The importance of this relationship is that the ability to control blood vessel radius is a very powerful tool for the body to divert flow from one area to another and vary blood pressure. As we shall see later, this process is useful when dealing with the effects of microgravity.

Feedback control and Hydrostatic Pressure

Blood pressure regulation is achieved by a feedback control system consisting of *baroreceptors* located in the arterial and venous side of the circulation. These pressure receptors provide information about blood pressure to the cardiovascular control center in the brain, which integrates the information and initiates actions to ensure blood pressure doesn't deviate too far from normal. However, as you will experience in microgravity, the response of the baroreceptors to pressure changes is not immediate. For example, upon orbital insertion, due to the lack of gravity, your blood will pool in the cephalothoracic region, leaving you with thin legs and a stuffy head. When you return to Earth however, this blood will translocate from the cephalothoracic region into your legs leaving you with a light-headed feeling that physiologist's term *orthostatic hypotension*. Fortunately, your body can counteract this translocation up to a point by increasing heart contractility and heart rate. The reason these mechanisms are compromised upon return to Earth is because there is less blood volume, and as we shall see, blood volume regulation is an important aspect of cardiovascular function in microgravity.

Blood Volume Regulation

One of the problems of being in microgravity is losing blood volume. This is because sensors in the body detect an excess blood volume due to the translocation of blood to the cephalothoracic region and because hormones are released resulting in an excretion of urine, thereby reducing blood volume.

WHAT HAPPENS IN SPACEFLIGHT?

Pre-launch

When you are lying in your seat waiting for the countdown, you will experience the first effects that microgravity will impose upon your body, even though you haven't left the ground! Since you may be in the reclined position for several hours (depending on which vehicle is ferrying you to orbit), gravity will cause fluids to shift from your legs and settle in the cephalothoracic region resulting in a reflexive increase in kidney output and urine volume in the bladder. This is the reason Shuttle astronauts used to wear undergarments to absorb urine in case the urge to urinate became excessive! You may reason that purposely dehydrating yourself prior to launch may solve the problem, but this is a tactic that has been tried by astronauts and although it is sometimes successful, the disadvantage is that these astronauts are often severely dehydrated on orbit.

'Chicken Leg' Syndrome

You will experience the more pronounced effects of cephalothoracic fluid shift that occur immediately upon arrival on orbit as blood that was normally pooled in your legs moves to the point of least resistance - the large vessels in the chest. Because sensors located in your chest perceive the circulation is "overfilled", the body reacts by adjusting arterial pressure and getting rid of the "extra" fluid by the kidneys in the form of urine, a process termed *diuresis*. In addition to losing blood volume and becoming dehydrated you will also experience sensations of sinus congestion and headache. Looking in a mirror you will see more visible signs such as puffy faces and 'bird legs' caused by more than a liter of fluid translocating from your legs to your chest. The cardiovascular changes you can expect during your orbital vacation are summarized in Table 3.

Physiologic Measure	Change in Microgravity
Resting Heart Rate	Increased after flight. Peaks during launch and re-entry.
Resting Blood Pressure	Normal during flight but decreased after flight.
Orthostatic tolerance	Decreased after flights longer than 5 hours. Exaggerated cardiovascular response to Tilt Test, Stand Test, and LBNP after flight. RPB 3 – 14 days.
Total Peripheral Resistance	Decreased in flight. No increase following landing despite drop in stroke volume and increase in HR.
Cardiac Size	Normal or slightly decreased C/T ratio post-flight.
Stroke Volume	Increased inflight by as much as 60% but compensated by a decrease HR.
Cardiac Output	Elevated 30 to 40%. Reduced immediately post-flight.
Central Venous Pressure	Elevated above resting supine level pre-launch. Transient increase followed by levels below preflight upon attaining orbit.
Cardiac electrical activity (ECG/VCG)	Moderate rightward shift in QRS and T waves post-flight.
Arrhythmia	Usually PABs and PVB's. Isolated cases of nodal tachycardia, ectopi beats, and supraventricular bigeminy inflight.
Exercise Capacity	No change or decreased ≤ 12% post-flight. Increased HR for same VO_2. No change in efficiency. RPB 3-days.

Table 3. Cardiovascular changes associated with short-duration spaceflight

Acronyms

RPB:	Return to Preflight Baseline	**C/T**:	Cardiothoracic
LBNP:	Lower Body Negative Pressure	**ECG**:	Electrocardiogram
PAB:	Premature atrial beat	**VCG**:	Vectorcardiograph
PVB:	Premature ventricular beat	**HR**:	Heart rate

Except for occasional HR irregularities, the changes above are adaptive mechanisms to microgravity, and are not usually associated with any ill effects. One of the ways your body adapts to the perceived extra blood is to decrease ADH secretion, which in turn causes less thirst, which means you will drink less in space and after a while some of the excess fluid will disappear. After a few days, however, you will also become dehydrated such that after 3 days in orbit your total body water will have decreased by 3%.

Fluid changes on re-entry

During re-entry much of the fluid in your cephalothoracic region will translocate to your legs resulting in you recovering most of your leg-volume. However, you may experience a slight swelling in your legs since there is tendency for leg fluid content to be slightly greater than pre-flight values as fluid pools more easily in veins which have become more compliant during spaceflight.

ORTHOSTATIC HYPOTENSION

If you have spent a week in orbit - and perhaps even if you are returning from a suborbital ride - you need to be prepared for the phenomenon of *orthostatic hypotension*, a condition in which a person is unable to maintain blood pressure while standing. Orthostatic hypotension is experienced by all returning astronauts, and usually leaves most feeling light-headed, and dizzy - some actually pass out! Orthostatic hypotension is due to the cardiovascular deconditioning that has occurred during your flight. The deconditioning means that once the body is exposed to a gravitational force, the blood that was in the cephalothoracic region is translocated to the lower body resulting in a relative decrease in blood pressure in the upper body. Unfortunately, this decrease (hypotension) usually exceeds the ability of the baroreceptor responses to maintain pressure to the brain, which is why you will feel dizzy upon landing. Your situation will be compounded by the fact that, due to the deconditioning of your leg muscles, the ability of the muscle pump is reduced, and therefore cannot effectively aid in forcing blood from your legs to the heart and the brain. Since orthostatic hypotension occurs despite astronauts taking countermeasures, such a problem has the potential to be life threatening if an emergency egress is required.

What causes Orthostatic Hypotension?

We know one of the reasons astronauts become orthostatically hypotensive is due to the reduced plasma volume but what hasn't been explained is the effect this has upon other cardiovascular variables. Decreased plasma volume will also cause an increase in heart rate, a decreased venous return, and a reduction in stroke volume observed when standing post-flight. There are other mechanisms that compound the problem but a discussion of these is beyond the scope of this manual. What is important to understand is how the effect can be reduced and what countermeasures can be used to decrease the incidence of orthostatic hypotension. Unsurprisingly, both the US and Russian space programs have used countermeasures to mitigate the effects of orthostatic hypotension. Often, these countermeasures involve the form of a saline loading protocol, G-suit inflation, and inflight exercise during the mission, but before discussing these it is necessary to understand the musculoskeletal system since it is directly implicated in these countermeasures.

MUSCLE STRUCTURE AND FUNCTION

There are three types of muscle in the body. *Cardiac Muscle* is found only in the heart, *Smooth Muscle* is found in the organs and *Skeletal Muscle* comprises the working muscles. The first two are under autonomic (automatic) control, whereas the latter is under conscious control.

The characteristics of skeletal muscle fibers allow them to adapt to the under-loading that occurs in microgravity, but unfortunately the process of this adaptation is also associated with the process of atrophy that occurs within a few days on-orbit. One of the results of this atrophy is that muscles that have an antigravity function, such as the calf and quadriceps, hip, back, and neck, all rapidly shrink!

Physiologic Measure	Change in Microgravity
Stature	Slight increase during first week (~ 1.3cm) RPB 1 day
Body Mass	Post-flight weight losses average 3.4%. 2/3 due to water loss, remainder due loss of lean body mass and fat.
Body composition	Fat replacing muscle towards end of short-duration mission.
Total body volume	Decreased post-flight.
Limb volume	Inflight leg volume decreases exponentially during first flight day. Thereafter, rate declines and plateaus within 3-5 days. Post-flight decrements in leg volume up to 3%. Rapid increase immediately post-flight.
Muscle strength	Decreased during and post-flight. RPB 1-2 weeks.
Reflexes	Reflex duration decreased post-flight.
Bone density	Os calcis density decreased post-flight. Radius and ulna show variable changes.
Calcium balance	Increasing negative calcium balance inflight.

Table 4. Musculoskeletal changes associated with short-duration spaceflight

Since there is no "biological need" to activate large parts of the musculoskeletal system in microgravity, you will need to work against the adaptation process otherwise the loss of muscle mass may compromise your performance in the event of an emergency egress.

COUNTERMEASURE STRATEGIES

Countermeasures are designed to systematically neutralize spaceflight's potentially harmful deconditioning effects on crew physiologic function, performance, and overall health. In this section we look at the countermeasure strategies you will be expected to perform. **Table 5** summarizes the effects of six typical countermeasures employed by astronauts to alleviate the effects of orthostatic hypotension, the least effective of which is saline loading, although this method does have the advantage over other pharmacological interventions since there are no side effects. Current procedure among NASA astronauts is to consume a maximum of eight 1-g salt tablets with approximately 900 ml of fluid 2 hours prior to re-entry in an effort to restore blood volume. For missions of seven days or less, this procedure tends to work well. Pharmacological intervention such as Fludrocortisone has demonstrated some promise in ground studies, but alterations in dose implementation for operational use in space missions has shown few positive effects. Perhaps the most promising pharmacological agent is Midodrine, which has been successful in improving orthostatic hypotension in ground experiments and in application to spaceflight. Although pharmacological intervention offers an alternative to treatment of post-flight OH, the most effective means of reducing OH symptoms is exercise.

	Microgravity	Fluid Loading	Florine	Midodrine	Maximal Exercise	*ITD*
Cardiac Baroreflex Function	☐	↔	↔	?	☐	☐
Blood Volume	☐	↔	☐	☐	☐	☐
Stroke Volume	☐	↔	☐	☐	☐	☐
Cardiac Output	☐	↔	☐	☐	☐	☐
Orthostatic Response	☐	↔	☐	☐	☐	☐
Aerobic Capacity	☐	↔	?	?	☐	?

Table 5. Efficacy of countermeasures on cardiovascular functions

Exercise Countermeasures

Maintaining an exercise regime during spaceflight has been proven to be an effective countermeasure to the mechanisms contributing to orthostatic hypotension and also to counteracting the atrophying forces of microgravity. However, exercising in space differs from exercising on Earth, since any work you perform will be less effective due to the absence of the resistive force of gravity.

Since your mission will be several days in length, your operator will require you to perform inflight exercise to protect your emergency egress abilities. To be prepared for the exercise regime during your flight you should start an exercise-training program a few weeks prior to launch.

Treadmill

When you run on the treadmill (Figure 30) you will need to tether yourself using a subject load device to restrain you on the treadmill surface and also a subject position device to keep you in an area of the treadmill where a pitch oscillation of the treadmill cannot be initiated.

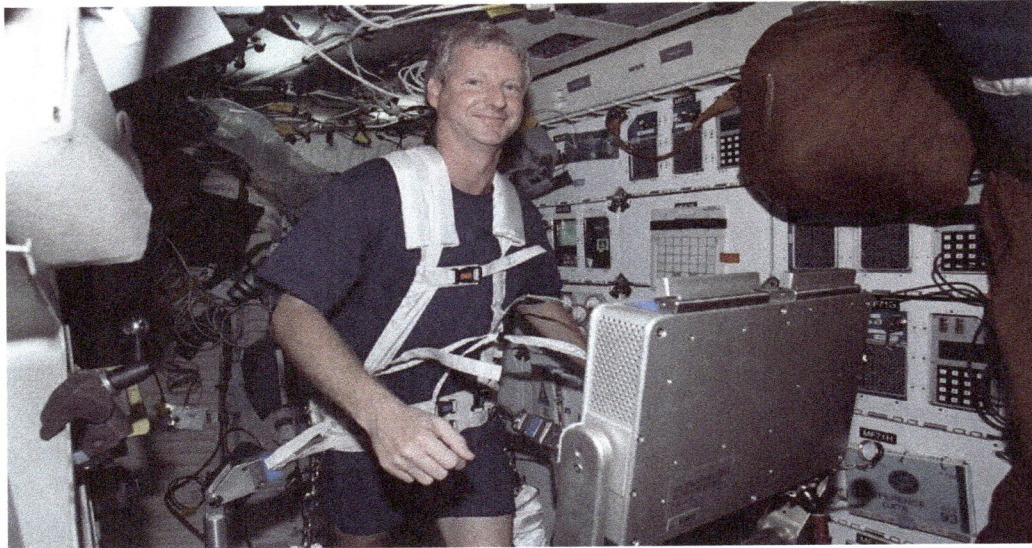

Figure 30. Treadmills used to combat atrophy during prolonged exposure to microgravity (credit: NASA)

Operational Countermeasure Procedures

In addition to the procedures described your operator may implement a program to protect you against orthostatic hypotension. A typical program may involve the following:

a. 5 weeks prior to launch you undergo a drug tolerance test for Midodrine.
b. 5 weeks prior to launch you start an exercise-training regime.
c. 10-15 days prior to launch you perform a 10-minute stand test, preceded by 6 minutes of supine rest. During this test your heart rate and blood pressure will be monitored. This test will be repeated post-flight to evaluate the efficacy of inflight countermeasures and will follow the protocol similar to the one outlined below:

Stand-test protocol
  i.   Crew members supine for 30 minutes.
  ii. Crew members assisted to freestanding position with feet 15 cm apart.
  iii. Crewmember remains in freestanding position for 10 minutes or until signs or symptoms of presyncope appear.
       Presyncope defined by any of the following:
       ▪ Sudden drop in systolic BP (SBP) > 25 mm Hg/min.
       ▪ Sudden drop in diastolic BP (DBP) > 15 mm Hg.
       ▪ Sudden drop in HR > 15 bpm.
       ▪ An absolute SBP < 70 mm Hg.
       ▪ Dizziness, light-headedness, or nausea.

    iv.   Heart rate recorded from 3-lead electrocardiogram during last 15 seconds of each minute.

    v.   Systolic blood pressure and diastolic blood pressure measured by auscultatory method during the last 30 seconds of each minute.

    vi.   Heart rhythm and change in blood pressure in the finger is monitored continuously for signs of presyncope.

d. 10 days prior to launch you conduct a tilt test.

e. Each mission day. 30 minutes resistive and 30 minutes dynamic aerobic exercise at intensity => 70% of age-predicted maximum heart rate to prevent muscle atrophy and maintain reflexes associated with autonomic regulation of blood pressure. Every other mission day. 20 minutes exercise within lower-body negative pressure (LBNP) device to maintain orthostatic function.

f. Each mission day you ingest pharmacological countermeasures to enhance autonomic responses to orthostatic challenge post-flight.

g. Each second day. Cardiac rate and rhythm monitoring will be downlinked for evaluation by the flight surgeon.

h. 3 hours pre-entry you ingest 10 mg dose of Midodrine.

i. 2 hours pre-entry you consume 15 ml per Kg of preflight body mass of an isotonic fluid or potassium citrate.

j. 1 hour pre-entry you don a Liquid Cooling Garment. This suit, worn under your landing and re-entry suit, contains a network of tubing that circulates water across the body surface, thereby minimizing water loss and reducing the severity of orthostatic symptoms upon landing.

k. During re-entry you inflate your anti-G suit to 1 psi.

l. During transport from landing to the medical clinic you consume fluid, the amount of which you report to medical personnel conducting the stand test.

m. 2 hours post-landing you perform the 10-minute stand test, preceded by 6 minutes of supine rest. During this test your heart rate and blood pressure is monitored.

n. 1-3 days post-landing a post-flight medical evaluation is conducted at your operators training center. Post-flight tests include the following:

    i.   Cardiovascular assessment to determine fluid loss, electrolyte changes, electrical activity disturbances, and neuro-reflex adjustments.

    ii.   Musculoskeletal assessment to determine muscle mass loss, muscle strength, and EMG analysis.

EFFECTS OF SPACEFLIGHT ON THE NEUROVESTIBULAR SYSTEM

In common with fluid shifts, the neurovestibular effects won't be a significant factor for most passengers in suborbital flight because these effects take time to manifest themselves. After orbital flight, astronauts often suffered an altered ability to sense tilt and roll, defects in postural stability, impaired gaze control, and changes in sensory integration; basically, they were discombobulated! While suborbital passengers probably won't be affected to the degree Shuttle astronauts were, simply because the changes are dependent on the duration of weightlessness, there have been neurovestibular alterations observed in even short exposures to zero-G in susceptible individuals. For example, illusions were reported on several of the high-altitude X-15 flights. Now, you may think why not test this in a simulator, but the problem is that the flight profile of rapid launch acceleration followed by zero-G followed by re-entry deceleration can't be tested in continuity.

Space Motion Sickness

More than 70% of first-time astronauts flying on orbital space flights suffer from space motion sickness (SMS). The syndrome, thought to be due to a sensory conflict between visual, vestibular, and proprioceptive stimuli, has been a problem as long as there have been astronauts. As well as being uncomfortable, SMS is a major headache for commercial space operators because it's impossible to predict who will be affected or when they will be affected. Symptoms typically occur within the first 24 hrs, but some astronauts have reported symptoms - dizziness, pallor, sweating, severe nausea and vomiting are the most common - immediately after main engine cut off. Vomiting, which can be especially messy in zero-G, can crescendo suddenly without any warning symptoms. In a multi-passenger vehicle, one passenger becoming nauseated can potentially trigger nausea in the other vehicle occupants – just imagine trying to take photos of the Earth while barfing into a vomit bag.

Anti-motion sickness medications could be used, but then performance would be affected (Space Shuttle flight crew members were not allowed to take medications for SMS for precisely this reason). The risk of nausea in zero-G gravity can be reduced if provocative motions, especially of the head, are avoided because head movements generate conflicts between the semicircular canals and the otoliths, with pitch movements being the most provocative.

Unfortunately, there is just no way to protect you against the possibility of being sick. Parabolic flight adaptation and experience in high performance jet aircraft don't work. Neither do rotating chairs nor centrifuge training. The only pre-flight training that has shown to be effective is the use of training aids that duplicate the sensory conflict that occurs in parabolic flight; this type of pre-flight adaptation training helps passengers become 'dual-adapted'. An example of the training aids used in this effort to duplicate sensory conflict is the device for orientation and motion environment (DOME) which is a spherical virtual reality simulator.

Effects of Acceleration on the Human Body

Seen through the eyes of your flight surgeon, the most worrying aspects of a suborbital flight profile are the launch acceleration and entry deceleration, especially when the acceleration exposure is in the head-to-foot ("eyeballs down" or +Gz) direction. That's because Gz acceleration can cause all sorts of neurovestibular, cardiovascular and musculoskeletal problems. Exposure to Gz can also have an impact on pulmonary function proportional to its applied force magnitude – at the lower end of the G scale, say 2 to 3 G's, you might experience difficulty breathing, while at the other end of the G load spectrum, say 5 to 6 G's, you may suffer airway closure.

To avoid these problems, spacecraft designers try to limit the launch and re-entryacceleration forces by ensuring most of the acceleration is in the +Gx direction (eyeballs in). That's because people are more tolerant to +Gx acceleration, and with the heart and brain located at approximately the same level within the acceleration field there is less risk for gravity-induced loss of consciousness (**G-LOC**) or almost loss of consciousness (**A-LOC**). Acceleration stress is perhaps one of the issues that most worry the flight surgeon because it is *dysrhythmogenic*, which means the heart's rate, rhythm, and conduction can be affected. In fact, high G forces or particularly long exposures to acceleration could potentially increase the frequency of a heart problem known as *dysrhythmia*. It is for this reason that spaceflight accelerations have mostly been designed to be in the +Gx axis; that is until the Shuttle came along. In the early days of manned spaceflight, the direction of acceleration was even more important than it is today because of the sheer magnitude of the acceleration. For example, the Mercury, Gemini and Apollo flights had launch accelerations of 4.5 to 6.5 +Gx for six minutes and anywhere from 6 to 11 +Gx during entry (which was why astronauts received 45 hours of +Gx centrifuge training, with some runs going up to 18 +Gx!).

By comparison, the now-retired Shuttle had a maximum of 3.0 +Gx during the 8.5-minute launch and 1.2 +Gz (briefly 2.0 +Gz during turns) for 17 minutes during entry. Fortunately for spaceflight participants the acceleration forces imposed by most of the current crop of space vehicles should be reasonably comfortable for most people, although there will be some who will do better than others.

That's because your tolerance to +Gz acceleration is dependent on your height and weight, certain physiological characteristics and the type of acceleration profile; physical conditioning, hydration, previous and recent exposure to +Gz forces, and recent centrifuge training also affect your response. This is important because the maximum +Gz level, exposure duration and the rate of onset of the +Gz determine the risk of injury to your heart and musculoskeletal system. The most problematic type of acceleration is rapid-onset rate (**ROR**) which is defined as increases greater than 0.33 G/sec. ROR tolerance limits are approximately 1 +Gz lower than gradual-onset rate (GOR) tolerances because they exceed the ability of the cardiovascular system to fully respond to preserve adequate central nervous system (CNS) blood flow; basically, if your brain doesn't get enough blood it will shut down. RORs can also result in the dreaded G-LOC without any of the usual visual warning symptoms such as tunnel vision, gray-out or black-out. To prevent this happening when they're performing aerobatic manoeuvres, fighter pilots wear anti-G suits which increase their G-tolerance to +Gz by up to 1.5 +Gz. Another way fighter pilots can increase their G-tolerance is to practice the anti-G straining maneuver (AGSM), which can increase tolerance to +Gz by as much as 3 +Gz. However, performing the AGSM is tiring and is generally used only for a short period of time. Over the years, centrifuge data has allowed scientists to develop a model of +Gz tolerance limits which incorporate +Gz magnitude, duration, and rate of onset (generally, with no protection, most healthy people can tolerate up to 4 +Gz acceleration for ROR profiles and up to +4.5Gz with GOR profiles).

Okay, so we've talked about +Gz tolerance, but what about –Gz and the transition from one type of G to another? Well, this is perhaps where most of the problems occur because transition to +Gz can cause a profound drop in cerebral blood pressure and that's bad news for the cardiovascular system because it can take quite a while before the body compensates. In fact, when there is prior exposure to –Gz, a transition to +Gz – a situation that creates what is known as the 'push-pull effect' - can be deadly. Usually, this 'push-pull effect' occurs in combat engagements and has been implicated in several combat training fatalities (it's also been identified as a possible cause of 30% of G-LOC events). Even now, with a wealth of G data available, there exists a knowledge gap in the complete understanding of this issue and no known countermeasures have been developed. It's unclear if a 'push-pull effect' will occur in transition from microgravity to entry deceleration, but it has been described in parabolic flight and there are some who are concerned that it could occur in suborbital flights. That's because the 'push-pull effect' is prolonged by increasing the duration of the prior -Gz exposure. Normally, the -Gz exposure is only several seconds in combat flight, whereas in parabolic flight profiles the exposure is 20-30 seconds. But what about after 4 minutes of suborbital flight? The truth is we just don't know if 4 minutes of microgravity would provoke the same response or a further deterioration in the +Gz tolerance.

If you're planning to fly as a passenger on board one of the new crop of space vehicles and you're worried about how you might be affected by G, you should know that the acceleration envelope recommended by the IAA for commercial aerospace vehicles should not exceed +3Gz (-2Gz), ±6Gx and ±1Gy. These levels, if experienced as GORs, should be well tolerated by unprotected, healthy individuals. Let's take Virgin Galactic's vehicle as an example. During SS2's rocket engine boost, acceleration may be as high as +3.8Gx followed by a brief spike up to +4.0Gz as the vehicle rotates to a nose high attitude. On re-entry, 6 g's will be felt mainly in the +Gz-axis by the pilots but, thanks to SS2's tilt-back seating, most of the acceleration during entry will be in the Gx axis for the passengers. Duration of these G forces is expected to be about 70 seconds during launch and about 30 sec during re-entry. Although SS2's acceleration onset rate has yet to be defined, it isn't expected to be an ROR.

What will this acceleration feel like?

During launch, your more-dense tissues will tend to be driven downwards. As a result, your liver will sink deeper into your abdomen, and your heart and great vessels will descend in your chest. The net effect of this process is to displace your diaphragm downward, which makes breathing progressively more difficult as $+ G_z$ acceleration increases. In addition, any useful activities performed by the arms, such as reaching for switches, etc., becomes progressively more difficult. At $+2 G_z$, you will experience a distinct feeling of heaviness and by $+3$ to $+4$ G_z, you will notice a marked dragging sensation in your chest and abdomen, and it will require great effort to move. By $+6 G_z$, it will be extremely difficult to reach overhead and, depending on your physical condition and stature, consciousness is generally lost at between $+3$ and $+5$ G_z in a sitting position. Another interesting effect of acceleration is degradation of visual acuity because the acceleration forces will distort the globe of your eye and reduce acuity.

Anti-G Straining Maneuver (AGSM)

In this sub-module you will learn that there are two components to the AGSM:

1. A continuous and maximum contraction of the big muscle groups including the arms, legs, chest, and abdominal muscles. Tensing these muscles reduces blood in the G-dependent areas of the body and assists in returning the blood to the chest, the heart, and the brain.

2. The respiratory element of the AGSM is repeated at 2.5 - 3.0 second intervals. The purpose of the respiratory element is to counter the G force by increasing chest pressure by expanding the lungs. This increased pressure forces blood to flow from the heart to the brain. The respiratory tract can be completely closed off at several different points, the most effective point being the glottis. Closing the glottis (located behind the Adam's Apple) results in the greatest increase in chest pressure. Note: The exhalation and inhalation phase should last no longer than 0.5 to 1.0 second.

Figure 31. Anti-G Training Device (credit: DLR)

One may also learn to anticipate the G exposure whenever possible by spending time in the AGSM trainer (Figure), a state-of-the art system proven to increase G-tolerance. The AGSM trainer is a device that helps you learn via biofeedback technology:

> ➤ how to perform appropriate breathing technique in combination with straining
> ➤ how to perform positive pressure breathing technique (PPB) under simulated +Gz using an anti-G-suit with reduced pressure via a software-controlled anti-G-valve
> ➤ how to communicate under PPB

To help you understand the effects of the AGSM you will be hooked up to a bioinstrumentation unit that measures your ECG, pulse, EMG, pressure (Thorax), and blood flow. Once you're wired up, you will sit in the trainer, a generic fighter cockpit including a generic seat, a spring-loaded center stick, a throttle and generic instrument panel with a display. Sensors are installed in the rudder panel and on the centre stick to measure forces and also in the trousers and vest to monitor pressure in those locations.

Noise and Vibration

Launching a vehicle into suborbital space requires powerful thrust which happens to be extremely noisy. This noise is transmitted through the whole spacecraft and because the vehicle is an enclosed space, the noise is reflected multiple times off the walls, bulkheads, floor and ceiling. Although the noise levels are relatively short, the magnitude can be quite intense and physiological effects such as reduced visual acuity, vertigo, nausea, disorientation, and ear pain may be experienced. Loud noise can also interfere with normal speech, making it difficult to communicate. Noise levels in the crew compartment during a Shuttle launch reached almost 120 dB (equivalent to an amplified rock concert in front of the speakers). Because of this assault on your hearing, auditory protection will definitely be required during a suborbital launch.

As well as all that noise, the power being unleashed to power the vehicle will also generate awful vibration (check out the in-cabin videos of the SS1 flights during both ascent and entry and you'll see what I mean); think about the vibration you feel when an aircraft takes off and multiply that by about ten orders of magnitude and you have some idea of what to expect. While vibration won't be more than a temporary inconvenience for the tourists, for commercial astronauts tasked with flying payloads, it could be a problem. That's because vibration can cause manual tracking errors and can interfere with your ability to visually track displays; this might be a problem for someone tasked with keeping an eye on an experiment from launch through to re-entry.

SPACE MOTION SICKNESS

Two-thirds of first-time space tourists will experience SMS symptoms, which may include headache, stomach awareness, nausea and vomiting. You may notice these symptoms shortly after orbital insertion and may find them triggered by viewing an unusual scene such as an inverted crewmember, although symptoms may also be provoked by head movements. The good news is that symptoms normally abate within 48 to 72 hours inflight, although the rate of recovery, degree of adaptation, and specific symptoms vary between individuals.

The development of SMS typically follows an orderly sequence, the time scale largely being determined by factors such as the individuals' susceptibility and the intensity of the motion stimulus. Although the incidence and severity will depend upon the particular environment involved, SMS is always unpleasant and may in certain situations compromise performance during an emergency egress.

Space Motion Sickness Symptoms

A common feeling upon orbital insertion is one of disorientation, a sensation that will probably be addressed on your vehicle by careful location of lighting and color schemes to give you a definite 'up' and 'down' feeling. The disorientation is likely to precipitate classic SMS symptoms such as GI awareness, perhaps as early as the first few minutes on reaching orbit. These symptoms may range from nausea to vomiting and retching. Unfortunately, these episodes will not be preceded by prodromal symptoms as they are on Earth. Most likely the first indication you're suffering from SMS will be a forceful expulsion of stomach contents, which you will need to capture and stow as soon as possible to avoid the ire of your fellow crew members! If you are among the lucky few who do not experience SMS you may experience milder symptoms such as malaise, lack of initiative and general irritability. One way for you to reduce the incidence of SMS symptoms is to minimize head movements since hypersensitivity to head motion is perhaps one of the more commonly suffered symptoms and may provoke stronger sensations for reasons that are explained in the following sections.

Essential Neurovestibular Physiology

The vestibular system consists of the inner ear, in which are located three semicircular canals for detecting angular acceleration and the saccule and utricle which detect linear acceleration. The semicircular canals correspond to the three dimensions in which movement occurs, each canal being responsible for detecting motion in a single plane. Flowing through each canal is a fluid (endolymph), which deflects small hair-like cells (cupula) as the head experiences angular acceleration, which in turn sends messages to the vestibular receiving areas of the brain. One vestibular component is located on each side of the head, their function being to mirror each other and act in a push (excited) pull (inhibited) manner depending on the direction the cupula moves.

When you move forward, when accelerating in a car for example (linear acceleration), this information is communicated to the brain via the utricle and saccule, structures that have a sheet of hair-like cells (macula) embedded in a gelatinous mass. This mass has areas of small crystals (otolith), which provide the inertia required to drag the 'hairs' from side to side, thereby providing the perception of motion. Once you have decided on a speed to drive, a steady velocity is detected, the otoliths stabilize and the perceived motion dissipates. The arrangement of the utricle and saccule determine motion detection, the utricle being responsible for motion in the horizontal plane since it lies horizontally in the ear and the saccule able to detect down, up, forward, and backward motion by virtue of its vertical orientation.

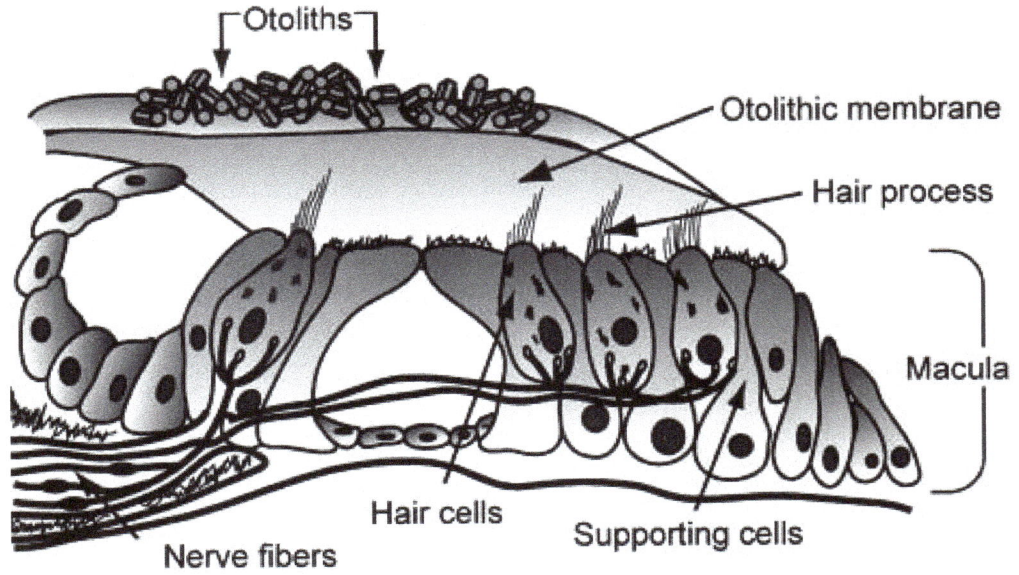

Figure 32. The Vestibular System

Etiology

One of the problems faced by space medicine experts is that despite the high incidence of SMS, there is no reliable ground-based test to predict which crew members will be affected. Also, despite 50 years of manned spaceflight, little knowledge exists of how to prevent SMS and there are no fully acceptable means of treating symptoms once they appear. Although the underlying causes of the SMS syndrome are reasonably well understood, the mechanisms are not clearly defined and no satisfactory methods have been identified for the prediction, prevention or treatment.

Unsurprisingly, considerable research has been directed at elucidating consistent and objective physical signs that correlate with the onset of the constellation of symptoms associated with SMS, but much of this research has been largely unsuccessful. A part of the reason for the lack of success in accurately identifying methods for its prediction, prevention and treatment is due to the inflight use of anti-MS drugs, a procedure that tends to obscure meaningful clues. Also, while astronauts serve as an elite subject group, their numbers are small.

SPACEFLIGHT LIFE SUPPORT

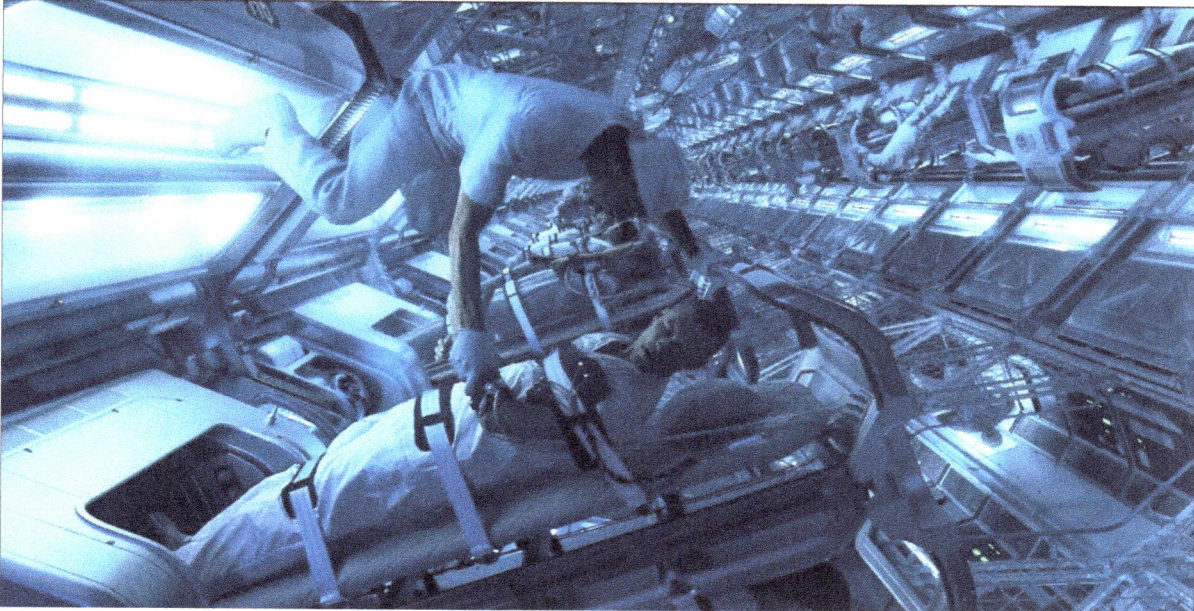

Module Objectives

- Describe the Environmental Control Life Support System, the Air Revitalization System, Atmosphere Revitalization Pressure Control System, Active Thermal Control System, and Fire Suppression System
- Explain the factors affecting cabin pressure
- Describe the factors affecting cabin temperature
- Explain how low and high humidity affect performance
- Explain the consequences of high and low oxygen concentrations
- Describe the consequences of high and low carbon dioxide levels.
- Describe the effects of hazardous gases on health
- Explain how particulate contaminants are controlled in the cabin.
- Explain how ventilation is controlled in the cabin

Contributing Author:
Erik Seedhouse, Ph.D.

In this module we take a look at the factors affecting the monitoring and control of atmospheric conditions inside a suborbital launch vehicle. Monitoring is important because it provides the crew with an idea of the atmospheric conditions inside the cabin so adjustments can be made to maintain conditions to sustain life and consciousness. The measured values can be continuously refreshed or periodically updated, depending on the hazard that an unmonitored atmospheric condition would present to the passengers. Monitoring may be the responsibility of the crew, an on-board computer system, or of a ground-based remote operator who can alert the on-board crew of an unsafe condition.

When the engineers were designing your spacecraft's life support system, they were interested in the following requirements:

1. What is the danger for passengers if the cabin's atmospheric condition is uncontrolled during normal or emergency operating conditions within the vehicle?

2. Does the uncontrolled atmospheric condition create a physiologic effect upon the flight crew at the onset of exposure under plausible flight conditions, such that a flight crew could identify a flight hazard at the onset of exposure before flight safety is compromised?

3. Is the uncontrolled atmospheric condition unlikely to change rapidly or in large magnitude, such that a flight crew could identify a flight hazard at the onset of exposure before flight safety is compromised?

4. Following the onset of exposure to uncontrolled atmospheric conditions stemming from a failed component, what corrective actions are possible?

5. What is the maximum period of time between onset of exposure to the uncontrolled atmospheric condition and the completion of corrective actions?

For each of these parameters, cabin conditions will be described with regard to hazards and the potential for rapid changes or for each atmospheric condition.

ONBOARD SYSTEMS ORIENTATION

Professional astronauts must have a comprehensive understanding of all systems and subsystems on board. They must also understand the relationships between these systems and be able to identify major hardware components, state their function, identify intra-system and inter-system interfaces, and describe the capabilities of each. Such a rigorous level of knowledge is not a requirement for space tourists, although a familiarity with the vehicle's primary systems is necessary as you may be required to assist with routine maintenance tasks and must be proficient in reacting to emergencies such as fire and decompression.

Environmental Control and Life Support System

The Environmental Control and Life Support System (ECLSS) performs vital functions such as supplying air, water and food. It also maintains temperature and pressure as well as shielding you from radiation. The subject matter expert for this system will be the flight engineer who will be familiar with every component, including the air revitalization system (ARS), the atmosphere revitalization pressure control system (ARPC) and the active thermal control system (ATCS), each of which interact to provide a habitable environment.

Figure 33. Astronauts pose with elements of Mir's life support system (credit: NASA)

Air Revitalization System

Your habitat contains several independent air loops that circulate the habitat pressure atmosphere. These loops constitute the ARS, which is responsible for:

➢ Circulating air
➢ Ensuring humidity remains between 30% and 75%
➢ Ensuring carbon dioxide and carbon monoxide levels remain non-toxic
➢ Ensuring temperature and ventilation is regulated
➢ Ensuring the habitat's avionics and electronics are cooled.

As air circulates it collects heat, moisture, carbon dioxide and debris before being drawn through a cabin loop and filter by a cabin fan which ducts the air to lithium hydroxide canisters that remove carbon dioxide and trace contaminants. Once air has passed through the lithium hydroxide canisters it passes through a heat exchanger and is cooled. At this stage, any water that is in the air is separated by a humidity separator fan which routes water to a waste water tank and the air is returned to the cabin. The frequency with which air is renewed depends on the size of the vehicle/habitat but usually a cabin air change will occur between six and eight times an hour.

Atmosphere Revitalization Pressure Control System

The Atmosphere Revitalization Pressure Control System (ARPCS) ensures habitat air pressure is maintained at 14.7 psia and that the partial pressures of oxygen and nitrogen are maintained within nominal levels. Since oxygen constitutes 20 percent of the air mixture, the partial pressure of oxygen must be maintained between 2.95 psia and 3.45 psia, whereas the partial pressure of nitrogen, which makes up 80 percent of the air, must be maintained at about 11.5 psia. Several specialized cryogenic oxygen tanks contain the source of the habitat's oxygen whereas nitrogen is contained in several nitrogen cylinders. An average of 1.76 pounds of oxygen is used per crewmember per day but these numbers do not take into account the normal loss of habitat gas to space and the amount lost to metabolic usage.

The oxygen and nitrogen supply systems are controlled by the atmosphere pressure control system that regulates the release of the gases by a system of check valves, inlet valves, relief valves, supply valves, sensors, control switches and talkback systems. A description of the function of this system is beyond the scope of this section and you will not be expected to have an understanding of how the system works. However, you will need to know how to react if the system detects pressures outside nominal parameters.

If habitat pressure falls below 14.0 psia or rises above 15.4 psia, or if oxygen partial pressure falls below 2.8 psia or rises above 3.6 psia then the Master Alarm will sound in and caution lights will illuminate on the ARPCS panel. Whichever crewmember is closest to the ARPCS panel will be responsible for dealing with the emergency, meaning you will need to know where the overpressure and negative relief valves are located. For example, if the emergency is an over pressurization you will need to activate the relief switch which will in turn activate a valve designed to relieve pressure through venting. If the emergency is a low-pressure alarm you will activate the negative pressure relief valves releasing a flow of ambient pressure into the habitat.

Active Thermal Control System

The ATCS is responsible for heat rejection, achieved by the use of cold plate networks, coolant loops, liquid heat exchangers and various other heat sink systems that reject heat outside the habitat. The habitat has a large number of electronic units and systems that generate heat and it is important the heat sink systems are not overloaded, although if the capacity of the heat sink units is exceeded the habitat can activate a flash evaporator designed to meet excess heat rejection requirements for short periods.

Smoke and fire detection and suppression

A description of each item of firefighting equipment and the fire response procedures employed by your operator is beyond the scope of this manual, but given the potentially grave implications of such an event it is appropriate to review some generic fire-fighting capabilities and systems. Smoke and fire detection and suppression capabilities are provided throughout the vehicle by means of ionization detection elements that provide information of smoke concentration levels to the general purpose computer (GPC). If the GPC detects an abnormal concentration of smoke the master alarm lights will activate and the general alarm will sound.

Fire suppression in the vehicle and habitat is dealt with by release of Freon-1301 (bromotrifluoromethane), which may be released automatically by the GPC or manually by a push button. In addition to the fitted systems, the crew can fight fires using portable, Halon-1301 (monobromotrifluoromethane) fire extinguishers (PFE's) - watch the film 'Gravity' to get a sense of the challenges operating one of these. PFE operation comprises a simple sequence of events requiring you to remove the locking pin and depress the trigger.

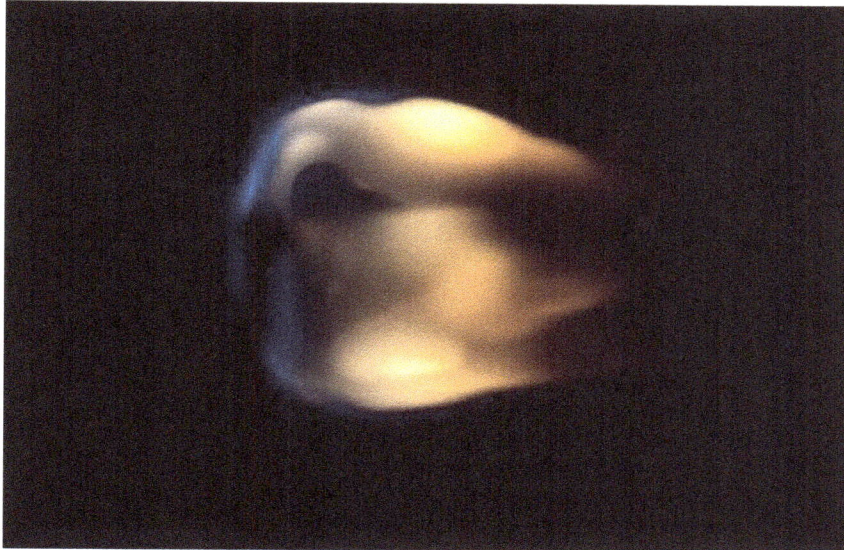

Figure 34. Combustion in a microgravity environment

During the drills you will use the same PFE's installed onboard the vehicle and habitat and also have the opportunity to become familiar with the Portable Breathing Apparatus (PBA), a system you will need to fight fires. The PBA is a space-modified breathing mask similar to the ones terrestrial fire-fighters use and comprises a mask attached to a 1.8m hose that has a quick disconnect attached at one end. The disconnect end of the hose is plugged into a small oxygen cylinder that supplies oxygen for 15 minutes.

Cabin Pressure

Although the probability may be low during suborbital flight, a puncture of the vehicle's pressure shell by micrometeoroids or failure in the pressure shell could result in a loss of cabin air. An uncontrolled decrease in cabin total pressure could be rapid, depending upon the volume of the cabin and the size of the breach in the shell. In the event of total cabin pressure loss, the pressure would decay below levels necessary for human life. Operationally, cabin depressurization can be one of the most rapidly developing, human performance-compromising emergency conditions faced by a crew and its passengers. It was the cause of the deaths of three cosmonauts during there-entry of Soyuz 11 and rapid decompression has also been a cause or contributing factor of numerous fatalities aboard commercial aircraft, notably Qantas Flight 30 on 25th July 2008.

The aircraft, carrying 346 passengers and 19 crewmembers, suffered an explosive decompression over the South China Sea while cruising at 29,000 feet on a flight from Hong Kong, to Melbourne, Australia. The event occurred 55 minutes into the flight while the aircraft was over the Pacific Ocean, 200 miles from Manila. The crew descended to 10,000 feet and diverted to Manila. None of the passengers or crew were injured. A portion of the fuselage just forward of the wing root was found missing after landing. Damage included a rupture in the lower right side of the fuselage, just in front of the area where the right wing joins the fuselage. One cylinder associated with the emergency oxygen system, had sustained a sudden failure and forceful discharge of its pressurized contents, rupturing the fuselage and propelling the cylinder upward, puncturing the cabin floor and entering the cabin adjacent to the second main cabin door.

As you can see in the photo, rapid decompression is a serious business and the reaction time of the crew or automated system is critical. While commercial aircraft such as the Qantas 747-400 are able to descend to lower altitudes in the event of depressurization, most suborbital vehicles are committed to a ballistic trajectory after a rocket burn is terminated, with little or no recourse for shortening the time to return to lower altitudes. In addition to the systems designed to replenish lost atmospheric gases within the vehicle, the design of the cabin pressure containment components are also relevant design considerations of the total cabin pressurization system. Dual pressure containment components (i.e., dual pane windows, dual seals at mated surfaces, dual hull shells, or isolation bulkheads) may decrease hazards associated with depressurization events in exchange for a small increase of mass and complexity of the vehicle, depending on vehicle design. Another danger is the fact that depressurization of small cabins occurs much more quickly than large cabins with equal puncture size, equal make-up air input, and pressure difference between the cabin and the exterior. Rapid decompression may be accompanied by a sudden drop in cabin temperature, fogging in the cabin, windblast and noise. In addition to the threat of hypoxia, these factors may lead to confusion, impairment of situational awareness and increased response times. Unless the environmental control system can compensate for the decreased temperature, passengers could suffer frostbite and other cold related problems. Now you might think that one way to reduce the danger would be to design cabins with lower total pressure because this would reduce the leak rate. The problem with this approach is you need a higher partial pressure of oxygen, thereby increasing the risk of cabin fire. So, an alternative is to use pressure suits (Figure 2). After all, if total loss of cabin pressure occurs above 40,000 feet altitude without the protection of a pressure suit, then the outcome will be fatal. Wearing a pressure suit means the cabin pressure can be reduced, but it doesn't solve the entire problem because there is still the issue of decompression sickness to consider.

Figure 35. Ted Southern demonstrating a Final Frontier Spacesuit (credit: Final Frontier Designs)

A survey of more than 400 U-2 pilots found that 75% reported in-flight symptoms of decompression sickness throughout their careers that resolved upon descent to lower altitudes, and about 13% of them reported that they altered or aborted their missions as a result. Those statistics don't bode well for a spacecraft operator and its passengers. The use of pressure suits is compounded by the need to maintain and service the suits. Then there is the problem that these suits may adversely affect the ability of flight crew to perform certain safety-critical functions by limiting range of motion, response time, communications, visibility, reach, tactile sensitivity, or hand-eye coordination. Another problem is heat dissipation: these suits are bulky and heavy and you sweat when you wear them. To reduce the risks associated with the cabin environment, chances are your spacecraft will be fitted with pressure monitoring devices such as a warning signal that is triggered in the event of rapidly decaying pressure so the crew can take corrective action. The vehicle may also be fitted with an autonomous, compressed gas release system that activates when pressure drops below a nominal pressure value would be an acceptable means of preventing cabin

Cabin Temperature

Although humans can survive in a relatively wide range of temperatures, your spaceflight will obviously be more enjoyable if the cabin temperature is properly controlled. For its spacecraft, NASA developed a comfort *box*, which is bounded by 25 to 70 percent relative humidity and by 18 to 27°C. Maintaining this cabin temperature is important because there are so many sources of heat in the cabin. First there is the heat generated by avionics and other electrical equipment located in the habitable areas of the vehicle. Then there are the temperature changes that occur during the various phases of the flight: space is cold, which means that heat needs to be added to the cabin, but when you're taxiing on the ground, there is an addition of thermal energy which means that heat must be removed from the cabin - this is because there are so many vehicle systems interfaced with the cabin such as the life support system and temperature management systems. As with cabin pressure, maintaining a comfortable cabin temperature is achieved thanks to monitoring devices and control devices. Typically, temperature control is achieved by removing heat from the circulating cabin air, with forced continuous circulation of the cabin air through one or more heat exchangers. Chilled water, ethylene glycol/water, or Freon serves as the coolant in these heat exchangers.

Cabin Humidity

Excessive humidity or a lack of humidity isn't as serious as loss of cabin pressure but high humidity and very low humidity can impact your physical comfort. High temperature and high humidity decrease your body's natural body temperature regulation processes (i.e., sweating) and low humidity has a 'drying effect' on your body and is quickly noticed in areas such as eyes, lips, nose, and mouth causing discomfort. Thus, humidity may be interrelated with a flight crew's ability to successfully perform safety critical functions. Spacecraft cabin air receives moisture as exhaled water vapor and evaporated perspiration from the humans on board the vehicle. The average metabolic rate (normal activity) is 5.02 pounds of respiration and perspiration water generated per person per day (0.21 pounds per hour). Chances are you will be producing a lot more water than 0.21 pounds per hour when you take your ride into space because stressed or excited individuals produce water vapor at higher-than-average rates: you will also be performing zero-G acrobatics, which will increase your production of water vapor. Relative humidity in commercial aircraft cabins is typically below 20 percent because air is continuously compressed from the engine, conditioned by the air-cycle machine for the cabin and then dumped overboard via the outflow valves thus preventing any significant accumulation of humidity in the cabin. To make sure your cabin stays within the comfort box of 25% to 70% relative humidity the vehicle will probably use silica gel, activated alumina, or molecular sieve materials. This approach may be combined by removing heat from the circulating cabin air with forced continuous circulation of the cabin air through condensing heat exchanger(s) (chilled water, ethylene glycol/water, or Freon serves as the coolant in these condensing heat exchangers).

Oxygen Concentration

Hypoxia is the greatest single threat to anyone who flies.
Richard M. Harding and F. John Mills, *British Medical Journal* April 30, 1983

Very low oxygen partial pressure constitutes a severe hazard because it results in impaired judgment, ability to concentrate, shortness of breath, and fatigue. In short, you don't function well without oxygen! Rapid decreases in oxygen partial pressure, which may be experienced during a rapid decompression event, result in loss of consciousness within a few seconds. The effects of gradually falling oxygen partial pressure (a slow decompression event) are insidious, as it dulls the brain and prevents realization of danger. The total atmospheric pressure and the duration of exposure affect the minimum allowable oxygen partial pressure, as some detrimental effects of hypoxia are time dependent.

Ghost Plane: Flight N47BA

A ghost aircraft flying across country with a crew disabled or dead. It reads like a script for a Hollywood disaster movie, but this actually happened. Payne Stewart's ill-fated Learjet took off on October 25, 1999. Captain Michael Kling was flying the airplane and co-pilot Stephanie Bellegarrigue was handling the radios and coordinating the ascent. Accompanying them was their famous passenger, golfer Payne Stewart, and his agents Van Ardan and Robert Fraley and Bruce Borland, a golf course designer. It was a clear and sunny Monday in Orlando, with light winds. Bellegarrigue turned around in the cockpit to face the passengers seated behind her to give them the safety briefing. She instructed them on the proper use of the drop-down oxygen masks, used in case of a loss of cabin pressure. Most air travelers ignore these safety briefings because they can't imagine the inhospitable environment outside an airplane in flight. Yet just a few miles above even the warmest places on earth, the temperature is way below zero. The cabin pressure in the Lear was not like being on the ground in Florida, because cabin pressure was maintained at five thousand feet. When the pilots program their planned cruise level and interior altitude into the cockpit controls, everything is accomplished automatically. If something goes wrong, display lights or illuminated messages alert the crew. If the cabin altitude exceeds fourteen thousand feet, the altitude at which people can quickly become affected by a lack of oxygen, a loud horn sounds and oxygen masks drop to give passengers an emergency supply of air. Flight N47BA received permission to ascend to thirty-nine-thousand feet and began its climb. During the climb it made a six-degree change to the north, a turn so slight that at first it was not even noticeable to the controllers. But with each mile, the plane was flying farther from its destination. Traffic control radioed N47BA but received no response. The jet was flying fast and still climbing and with enough fuel for four and a half hours of flight, no one could say where it might come down. Captain Christopher Hamilton, an Air Force F-16 fighter pilot was dispatched at the request of the FAA, and was the first person to get a glimpse of the runaway Learjet. "I expected just to look in and make eye contact with the pilot and get a thumbs up that everything was okay," he said. "I figured it was just a radio malfunction or something." But what he saw as he maneuvered around the Learjet was spectral: a windscreen dense with frost, a dark cockpit beyond and no sign that the airplane was under a pilot's control. Hamilton flew around the plane for eighteen minutes, his fighter jet closer than any pilot ever wants to be to a passenger plane. On autopilot and at forty-five thousand feet, the Lear was fourteen-hundred feet above the manufacturer's recommended maximum altitude and it seemed to be flying fine. But there's no pushing the design limits on the human body. As the plane ascended, the volume of air in the passengers' lungs, ears and sinuses would have expanded by about 30%. In the decompression, air would have raced out as if in a vacuum. With loss of the pressurization system, there would be no heat in the airplane. So the temperature in the small cabin dropped quickly as the arctic air seeped in. In the cockpit, the pilots' color vision would have been reduced adding to their initial difficulty seeing through the decompression fog. If they had tried to put on their emergency oxygen masks, they would have to feel their way to them and do it before they became too uncoordinated by the spasmodic contraction of the arms that typifies severe hypoxia. Restrained by lap and shoulder harnesses, the captain and first officer may have had some control over their wracking bodies. But the passengers could have been thrown free of their seats as the convulsions in their extremities increased. At 1:13 p.m. the Lear finally ran out of fuel and plunged into a grassy field near Aberdeen, South Dakota. Hitting the ground at 400 miles per hour, the plane pulverized completely. Little was identifiable beyond the wings, a fuel tank and a bag of golf clubs.

Just as low oxygen partial pressures can be lethal, high oxygen partial pressures are also a hazard because they can result in lung irritation and oxygen toxicity (hyperoxia). High oxygen concentration also increases material flammability hazards. To ensure oxygen levels are maintained within normal levels, oxygen is added to the habitable atmosphere from a stored gas (pure oxygen or compressed air), chemical, or liquid oxygen supply.

Carbon Dioxide Concentration

The carbon dioxide concentration in the standard sea-level atmosphere is 0.039 per cent. Once this concentration rises three percent or more crew members will typically begin to exhibit symptoms that may affect the ability of crewmembers to perform safety critical functions, such fatigue, impaired concentration, dizziness, faintness, flushing and sweating of the face, visual disturbances, and headache. Exposure to 10% or greater concentrations at 1 atm can cause nausea, vomiting, chills, visual and auditory hallucinations, burning of the eyes, extreme dyspnoea, and loss of consciousness. To ensure carbon dioxide levels are maintained, your spacecraft will probably be fitted with continuous monitoring equipment, such as non-dispersive infrared photometers that use light-emitting diodes as the infrared sources. Such instruments have acceptable accuracy for CO_2 concentrations of 100–50,000 ppm (0.01–5 percent by volume).

Concentration of Hazardous Vapors or Gasses

In the enclosed space of a suborbital vehicle, most materials have the potential to produce gas or vapor contaminants, which could create hazardous environmental conditions. Materials entering the cabin could be the result of leaks of fluids or vapors from internal vehicle systems. For example, carbon monoxide concentrations from 120 to 180 parts per million (ppm) can result in a headache and breathlessness, while loss of consciousness results from concentrations above 300 pm. Another example is the decomposition of fire suppressants during a cabin fire, which may produce significant quantities of hazardous contaminants. For example, Halon is one of the most effective fire suppression agents in use, and even though it is considered to have low toxicity, safety and health problems can occur from its release in confined spaces comparable to those expected on suborbital launch vehicles. Decomposition of halogenated agents occurs upon exposure to flame or surface temperatures above approximately 900 °F, and may include hydrogen fluoride, hydrogen bromide, hydrogen chloride, bromine, or chlorine. A suborbital spacecraft also contains all sorts of volatile organic compounds (VOCs), some of which may have short and long-term adverse health effects, including eye, nose, and throat irritation, headaches, loss of coordination, nausea, liver damage, and central nervous system damage. Typical signs or symptoms associated with exposure to VOCs include nose and throat discomfort, headache, allergic skin reaction, dyspnoea (labored breathing), nausea, emesis (vomiting), epistaxis (nosebleed), fatigue, dizziness. Chances are your operator will provide you with VOC countermeasures such as goggles or face masks which will be incorporated into emergency procedures. You should realize that if you're wearing goggles and you need to effect an egress, then your egress procedures may be affected by reduced sight.

Particulate Contaminants

Airborne particulates such as dust may contain minerals, metals, textile, paper and insulation fibers, non-volatile organics, and various materials of biological origin such as hair, skin flakes, dander, vomitus, and bacteria. Contaminants such as dense smoke can impair situational awareness by obscuring vision, or causing intense bouts of coughing, choking, and extreme eye irritation. In a microgravity environment, metal or plastic shavings from machining of the onboard materials can become ingested or cause significant eye injury after becoming dislodged during launch. Fine particles (less than 2.5 micrometers) are of health concern because they easily reach the deepest recesses of the lungs, and have been linked to a series of significant health problems, including aggravated asthma, acute respiratory symptoms, aggravated coughing and difficult or painful breathing, chronic bronchitis, and decreased lung function that can be experienced as shortness of breath.

One way your operator will monitor for particulates is by using a nephelometer (a continuous monitor of light scattered by suspended fine particles) to monitor cabin air for particulates during recirculation. Your operator will also employ various control techniques to minimize particulates floating around the cabin. Some of these control techniques will include vacuuming the cabin pre-flight, periodic ground checks, material selection, and flight suit cleanliness. Preventative measures may include a Foreign Object Damage (FOD) program designed to prevent the circumstances that place foreign objects within functioning systems or occupied areas before hazards can occur. Another preventative measure will likely be the use of HEPA filters for the cabin air return duct inlets - HEPA filters remove 0.3-micron particles with a minimal efficiency of 99.97%.

Ventilation and Circulation

In microgravity, convection is reduced or non-existent which means air stagnancy can be a risk. NASA has determined that the minimum linear air velocity for maintaining crew comfort is 10-15 feet per minute, and it is likely your operator has adopted similar flow rates, accomplished using flow-meters or through the direct monitoring of fan speed.

SPACESUIT OPERATIONS AND BIOMETRIC MONITORING

Module Objectives

➤ Understand safety precautions associated with spacesuit use
➤ Learn how to prepare for use and store a spacesuit
➤ Learn how to put on (don) and remove (doff) a spacesuit
➤ Learn how to pressurize a spacesuit
➤ Learn how to operate effectively using a spacesuit

Contributing Authors:

Ted Southern
Nikolay Moiseev
Ravi Komatireddy, Ph.D.

IIAS uses Final Frontier Intra Vehicular Activity (IVA) spacesuits for flight crew training. An Intra Vehicular Activity (IVA) space suit is the personal survival garment for astronauts and high-altitude pilots. IVA suits are worn inside spacecraft as a safety precaution to protect a human from the harsh environment of space and upper atmosphere in the event of cabin depressurization. An IVA suit must therefore protect against vacuum and extremes in temperature while providing sufficient oxygen, visibility, and mobility for escape and rescue. This section describes the safety precautions associated with spacesuit use, suit preparation, suit donning (putting on), pressurizing the spacesuit, operating in the spacesuit, doffing (removing) the spacesuit, and storing the spacesuit.

FINAL FRONTIER SPACE SUIT SPECIFICATIONS

- Approximate Weight (incl. Gloves): 14 lbs / 6.35 kg
- Rated for terrestrial training purposes only; pending certification for flight and high-oxygen
- Environments
- Intended Operating Pressure: <5 PSID / .34 ATM
- Estimated suit usage time is +100 hours
- Approximate adjustment range: 5'2" to 6'4"

Safety Precautions

- Before use of a space suit under pressurized conditions, it is recommended that the intended wearer should be cleared by a physician who is knowledgeable about hyperbaric medicine.

- This suit is rated for use up to <5 PSID / .34 ATM. Do not exceed this pressure as doing so poses a risk of injury to the wearer and damage to the suit.

- Do not don or operate this suit around any sharp objects or in a dirty environment where damage to the suit may occur.

List of Delivered Items

(1x) Pressure Suit
(2x) Pressure Gloves

(1x) Communications Cap

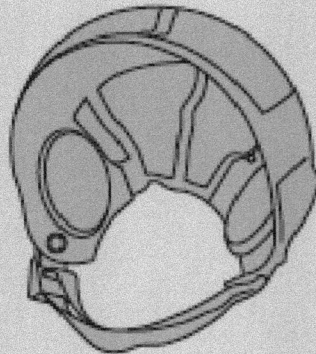

(1x) Protective Visor Cover

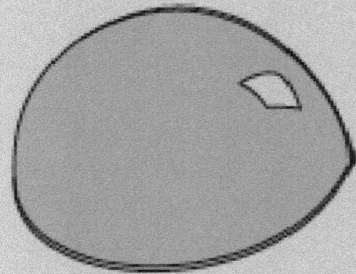

(4x) Comfort Gloves

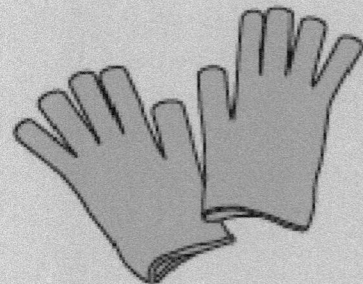

(1x) Comfort Underwear Top
(1x) Comfort Underwear Bottom

(1x) Space Suit Hanger

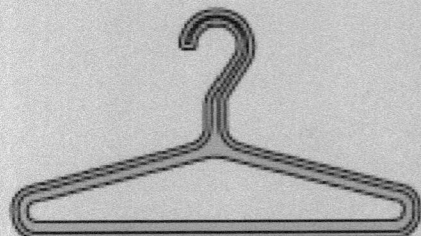

8

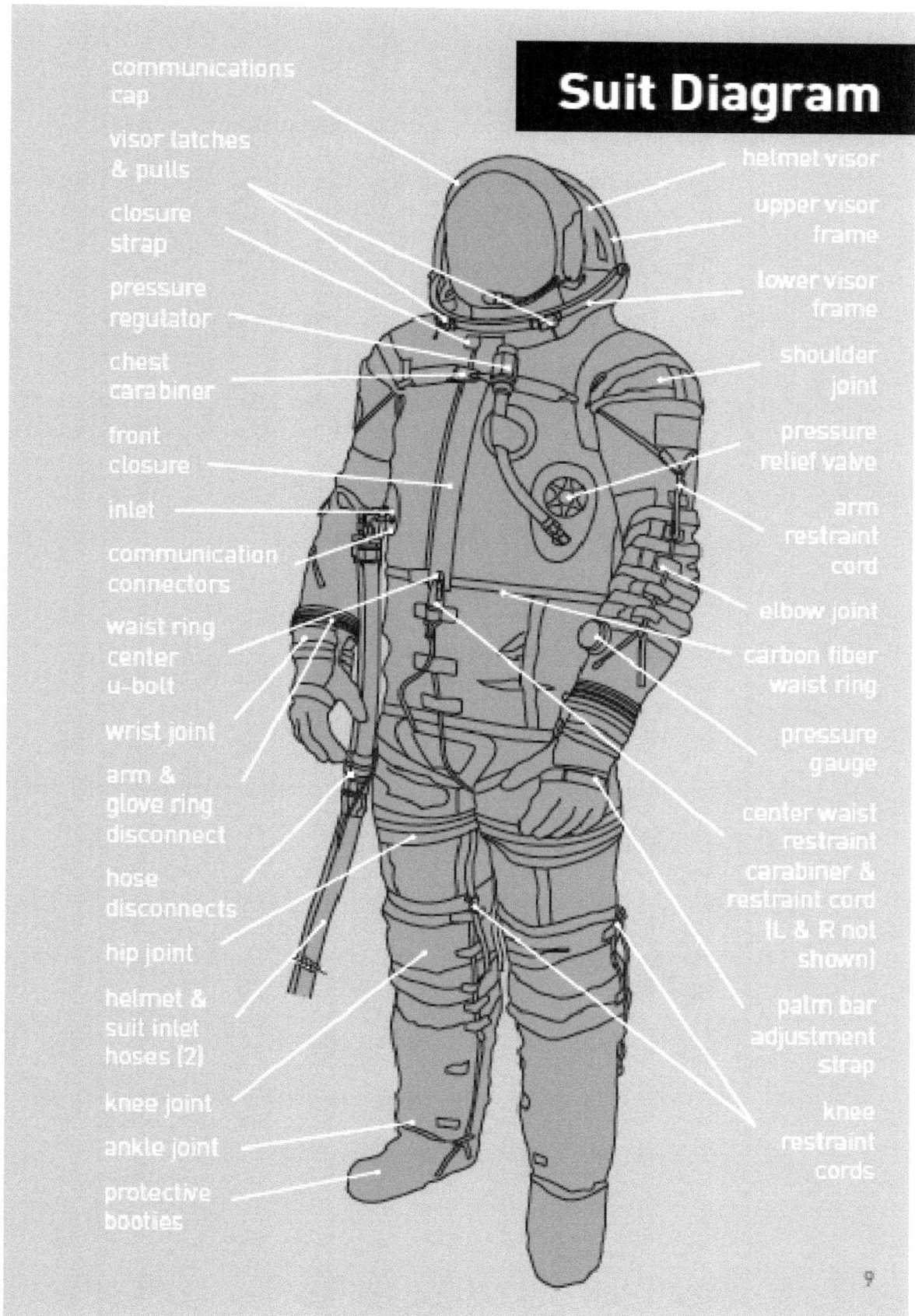

Suit Diagram

communications cap

visor latches & pulls

closure strap

pressure regulator

chest carabiner

front closure

inlet

communication connectors

waist ring center u-bolt

wrist joint

arm & glove ring disconnect

hose disconnects

hip joint

helmet & suit inlet hoses (2)

knee joint

ankle joint

protective booties

helmet visor

upper visor frame

lower visor frame

shoulder joint

pressure relief valve

arm restraint cord

elbow joint

carbon fiber waist ring

pressure gauge

center waist restraint carabiner & restraint cord (L & R not shown)

palm bar adjustment strap

knee restraint cords

Suit Preparation Checklist

1. Before laying out the suit, clean the preparation environment.

2. Check for any sharp objects or items that may pierce or otherwise damage the suit. Remove sharps from the body as well including rings, watches, jewelry, etc.

3. Remove **protective covers** from wrist disconnects.

4. Open **visor latches** and retract visor if not already done

5. Clean visor with **cloth & plastic cleaner***
   *FFD recommends Novus #1 Plastic Clean & Shine for general cleaning and #2 or #3 for scratches. Other supplies, incl. water and paper towels may scratch the visor.

6. Disconnect torso restraint **chest carabiner**

7. **Open front closure & spread open bladder**

8. Disconnect (3) **waist restraint carabiners**

9. Loosen all (11) **suit restraint cords**
   4 on sleeves, 3 attached to waist restraint carabiners, 4 on legs

10. Lay out **comfort gloves**

11. Lay out **pressure gloves** with palm bar adjustment straps open (1 per glove)

12. Test in-suit **communication connectors**

13. Close valves on **external air supply** (not provided) so no air can flow to suit

14. Prepare & pressurize **external air supply** (do not exceed 50 PSI on primary regulator)

15. Completely open **pressure regulator** on suit

16. Connect air supply to **helmet and suit inlet hoses** and initiate air flow into suit

17. Check ventilation flow from air supply into suit

DONNING THE FINAL FRONTIER SPACE SUIT

Initial Donning Procedures

1. Lay suit flat on floor in front of wide, comfortable chair in preparation environment (hanger removed)

2. Don comfort underwear top & bottom and socks

3. Insert legs into suit via chest bladder opening

4. Position feet snugly into suit feet, attach booties (if not already attached) and adjust for comfort

5. Stand up, grasp waist ring and pull waist ring up to align with wearer's waistline

6. Note: During donning, the bladder has a tendency to slide into the suit. If this occurs, gently pull bladder out and spread evenly around torso

7. Place one arm through sleeve and position that side of suit onto shoulder (straighten bladder as needed)

8. Repeat with other arm (straighten bladder as needed)

9. Grasping waist ring once more, pull suit as high as comfortable to facilitate rest of donning procedure

10. Remove protective cover from visor

11. With both hands, reach behind head, grasp lower visor frame and, while ducking, pull enclosure up, forward overhead, and down around neck (straighten bladder as needed)

12. Open air supply to suit approx 4 SCFM / 120 LPM. Adjust flow for cooling and comfort inside suit

13. Don communications cap & headset (not provided)

14. Connect headset to internal suit connectors, stow connection in internal pocket (on wearer's right side), and test comms to verify connection

15. Close bladder as detailed on following pages

16   Note: Proper bladder closure is critical for maintaining a low leakage rate

a.   Fully extend bladder outward from the body and allow the two elastic cords to dangle freely

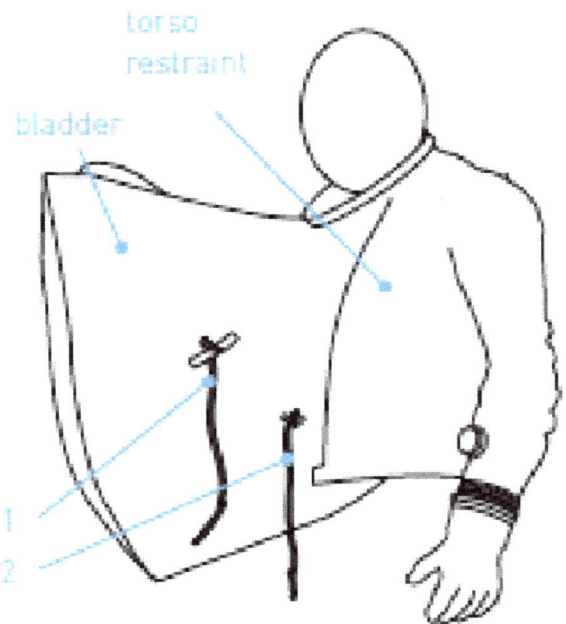

b.   Gather the bladder, matching edges and folding or pleating the material to about a 3cm circumferance when gripped

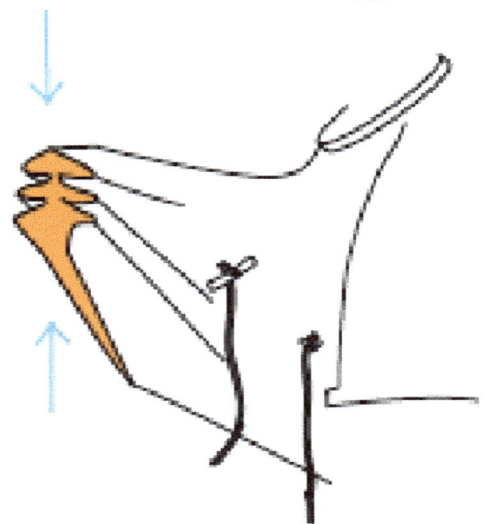

c.   Continue to gather material and finish with grip just beyond elastic cord #1

torso restraint

bladder

elastic cord #1

elastic cord #2

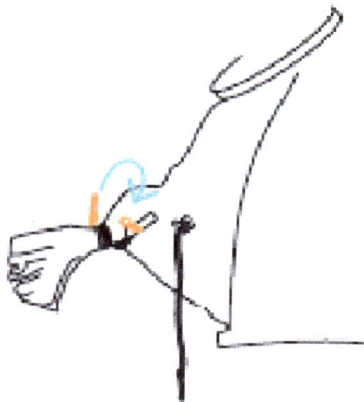

d. With free hand, wrap elastic cord #1 around gathered material a minimum of four (4) times, stretched tight. Avoid overlapping any portions of the cord and secure cord by inserting through the loop provided

e. Twist excess material and fold just beyond elastic cord #1 as shown in image

f. Wrap elastic cord #2 around gathered material a minimum of four (4) times, stretched tight. As with cord #1, avoid overlapping any portions and secure by inserting cord through the loop provided

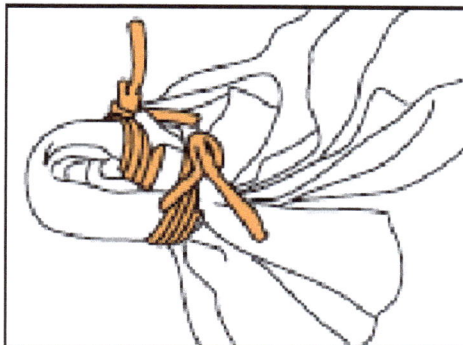

g. Securely closed bladder should appear similar to image at left. Tuck bladder where comfortable, away from internal air inlets and communication connectors

Torso Restraint Closure

17. Place right side of torso restraint through waist ring center u-bolt

18. Start torso restraint zipper, one-third of the way up

19. Place left side of torso restraint through waist ring center u-bolt

20. Place center waist restraint carabiner through waist ring center u-bolt to secure torso restraint

21. Close torso restraint zipper two-thirds of the way up

22. Secure right side of torso restraint to five corresponding neck bolts on lower visor frame

23. Close torso restraint zipper completely

24. Secure left side of torso restraint to corresponding neck bolts. Ensure latch pulls do not get caught in restraints.

25. Affix velcro to finish securing front closure

26. Connect chest carabiner

27. Adjust restraint cords at inside and outside of knees evenly (accessible through outer cover layer pockets if installed). Center of suit knee joint should align squarely with wearer's knee cap

28. Don comfort gloves

29. Connect suit gloves as detailed on the following page and check for even contact at all points along rings

finger stop   tooth   latch   lock ring   arm ring   glove ring

inside detail

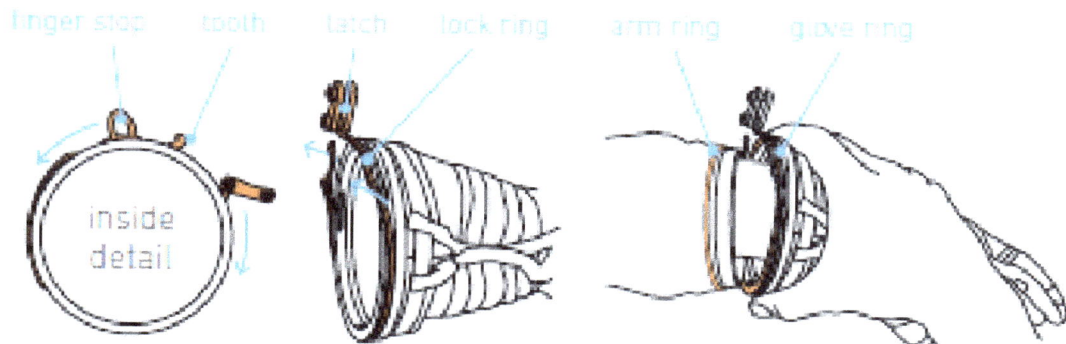

a. Open latch & unseat lock ring, out of its track, towards wearer

b. With palm bar strap open, insert arm ring completely & firmly into glove ring

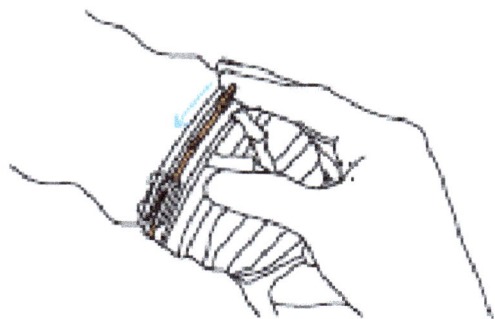

c. Reseat lock ring in its track and slide finger stop towards latch

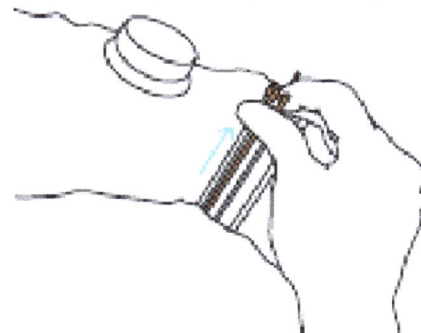

d. Finish reseating by pulling latch towards tooth

inside detail

e. If glove and arm rings are securely in contact, lock ring will clamp as shown

f. Snap latch closed

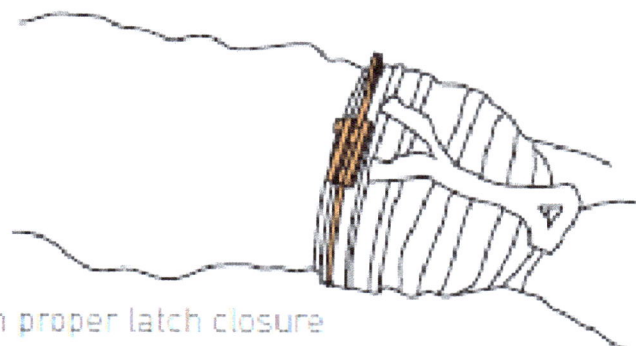

g. Confirm proper latch closure & secure, even connection

31.   With arms extended, adjust restraint cords at inside and outside of arms evenly (accessible through outer cover layer pockets if installed) so that wearer's fingertips bottom out firmly against inside the glove fingertips

32.   Once arm lengths are properly adjusted, tighten palm bar adjustment straps as well

33.   If not already standing, do so and do basic movements to test suit fit, comfort, and range of motion. If there are no concerns at this point, you have fully donned your space suit and are ready for operating preparation and initial pressurization

OPERATING THE FINAL FRONTIER SPACESUIT

Before use of a suit under pressurized conditions, it is recommended that the intended wearer should be cleared by a physician who is knowledgeable about hyperbaric medicine.

Before operating your 3G space suit, please review the basic safety precautions:

• This suit is rated for use up to <5 PSID / .34 ATM. Do not exceed this pressure as doing so poses a risk of injury to the wearer and damage to the suit.

• Do not don or operate this suit around any sharp objects or in a dirty environment where damage to the suit may occur.

While standing, visually locate the following key features of your space suit:

• Pressure Regulator (left side chest)

• Pressure Gauge (left forearm)

• Pressure Relief Valve (left side lower chest, just above regular hose connection to suit)

• Valsalva Device (mounted on inside of upper visor frame)

Operating Preparation

1.   With visor open, open pressure regulator completely to allow air to flow freely

2.   Connect external suit communications wires to the radio source. If using hand-held radio, clip to chest carabiner strap

3.   Test communications

4.   To close the visor, rotate the upper visor frame downwards towards the lower visor frame. Transfer grasp to back edge of visor and, with both hands, continue to rotate visor frame downwards so that upper and lower visor frames come together. Latches should snap audibly when locked

5.   Confirm that both latches are securely engaged

6.   Confirm that Valsalva device remains properly mounted on inside of visor

7.   Confirm air flow to helmet with test personnel and adjust flow until comfortable. If air supply is insufficient, unlatch helmet and assess

8.   If air flow is sufficient and comfortable and all closures are secure, you are now ready for initial pressurization

Initial Pressurization

NOTE: When fully enclosed, initiating air flow into your suit will result in a pressure environment greater than what you experience at sea level. This increased pressure is not unlike that when scuba diving or in the deep end of a pool. Some discomfort is expected; however, it is critical you notify your technicians or fellow trainees if you experience pain and/or suspect any off-nominal conditions.

10.   To initiate pressurization, slowly close suit regulator to begin increasing suit pressure

11.   While pressurizing, monitor the pressure gauge on the left arm. Recommended operating pressure is between 2-3 PSID /.13-.20 ATM)

12.   NOTE: DO NOT attempt to exceed recommended suit operating pressure rating (<5 PSID / .34 ATM)

13.   NOTE: For your safety, your suit features a Pressure Relief Valve which is designed to open at 5 PSD / .34 ATM. This is secondary fail safe and should NOT be relied on as a means to avoid excess pressurization

NOTE: You will likely experience some discomfort in your inner ear airways due to pressurization. To alleviate this discomfort, open your mouth as you would while flying or you can also perform the valsalva maneuver using the valsalva device mounted on the inside of your visor, which is designed to block the airways to your nose in lieu of your fingers

The valsalva maneuver is a "moderately forceful attempted exhalation against a closed airway, usually done by closing one's mouth, pinching one's nose shut while pressing out as if blowing up a balloon."

Secondary Fit Evaluation

13.   As you approach your desired operating pressure, be mindful of comfort and fit.

14.   At desired operating pressure, pressure in the suit should hold relatively constant. If so, the suit is properly pressurized. (Note: As with all pressure garments, some slow leakage is expected)

15.   Again, while standing, perform basic movements to test suit fit, comfort, and range of motion. Pay special attention to fit at the arms, legs, and waist and adjust as necessary by first, pressurizing down, making the necessary tweaks, and then pressurizing once more to evaluate.

If there are no concerns at this point, you have successfully performed initial pressurization of your space suit and are ready to continue operation and training.

DOFFING THE FINAL FRONTIER SPACE SUIT

Doffing Checklist

1.   Open pressure regulator completely to depressurize suit

2.   To release remaining pressure, disconnect gloves by opening latches as detailed below:

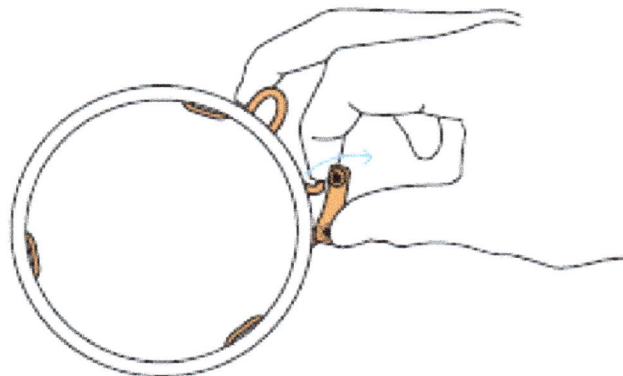

3.   Remove pressure gloves and comfort gloves

4.   Disconnect comms

5.   Open visor by simultaneously pulling both latch pulls out, away from the visor frame. Retract visor

6.   Disconnect chest carabiner and center waist restraint carabiner from the u-bolt

7.   Begin opening torso restraint be removing closure strap from its bolt on right side of lower visor frame

8.   Open front closure and free torso restraint from rest of neck bolts and waist ring u-bolt

9.   Unfold bladder, release elastic cords, and open bladder

10.   Disconnect communications links from internal connectors and remove communications cap and headset

11.   Grasp waist ring and pull suit upwards

12.   Begin doffing suit by grasping lower visor frame with both hands, then pulling up and back while ducking head down and forward through bladder opening

13.   Remove arms, one arm at a time

14.   Work suit down around legs

15.   Remove feet from booties and step out of suit

16.   Reinstall protective visor cover

17.   Log suit usage with date, wearer, and length of time suit was in use

Drying and Storage

After each use:

1.   The suit should be wiped down and dried before storing. Alcohol-based wipes are sufficient for brief, manned pressurizations. After prolonged use, Final Frontier Designs recommends a more thorough wipe down, especially the torso and helmet, using a lightly dampened cloth with minimal soap. To dry, Final Frontier Designs recommends laying suit flat or seated with the bladder and visor open, gloves off, air flowing (if possible), for at least one (1) hour after use.

2.   Spray visor with anti-fog cleaner.

3.   Lay pressure gloves and comm cap flat, and open to dry overnight.

4.   Wash comfort gloves and comfort underwear as recommended on clothing care labels.

5.   Store suit hanging or horizontal, chest, wrists, and helmet open.

When hanging for storage, be sure the hanger contacts with the suit's shoulders as opposed to supporting the suit from the bladder or other less-reinforced areas that may stretch or otherwise sustain damage. Once the hanger is in place and the suit is dry, close the restraint fully (as if donning), and hang.

DO NOT store the suit in a box.

BIOMETRIC MONITORING SYSTEMS

Biometric monitoring has long been an essential part of space missions. Even in the futuristic vision of Star Trek, the need for non-invasive monitoring was never questioned. Biometric monitoring has played an essential part of manned spaceflight from the first Mercury and Vostok flights; and even the non-human space voyagers that preceded Gagarin and Shepard were well instrumented with biometric monitoring systems.

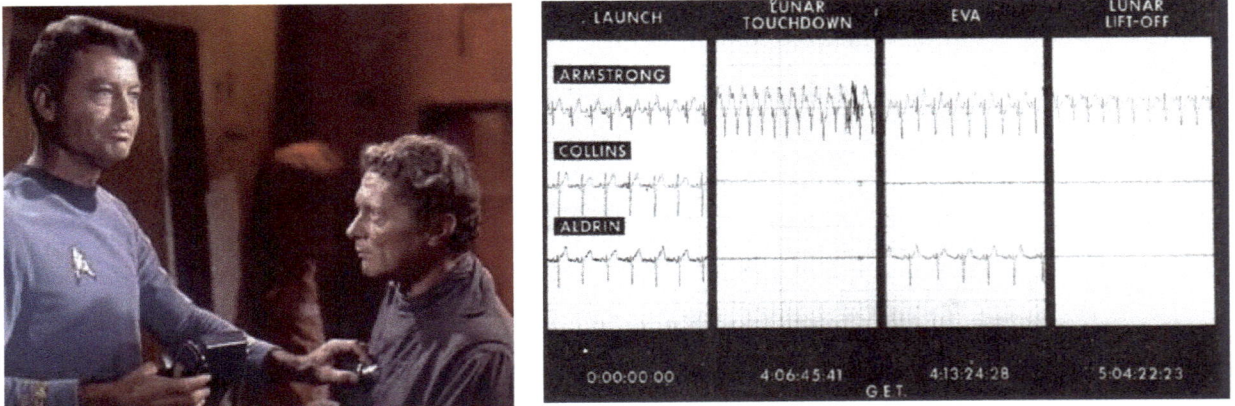

Figure 36. Doctor McCoy scans a patient's biometrics (L), The biometrics of the crew of Apollo 11 during key mission phases (R). Notice the difference in Armstrong's pulse compared to those of Collins and Aldrin (credit: NASA)

The Vital Space System

IIAS works in conjunction with Vital Space to provide biometric monitoring to its Scientist-Astronaut trainees. Illustrated in Figure 37, the Vital Space system integrates a 3-lead electrocardiogram (EKG), non-invasive blood pressure, and pulse oximetry into a simple interface that is wirelessly connected to a central monitoring station. Heart rate, respiratory rate, and temperature may also be inferred from these data.

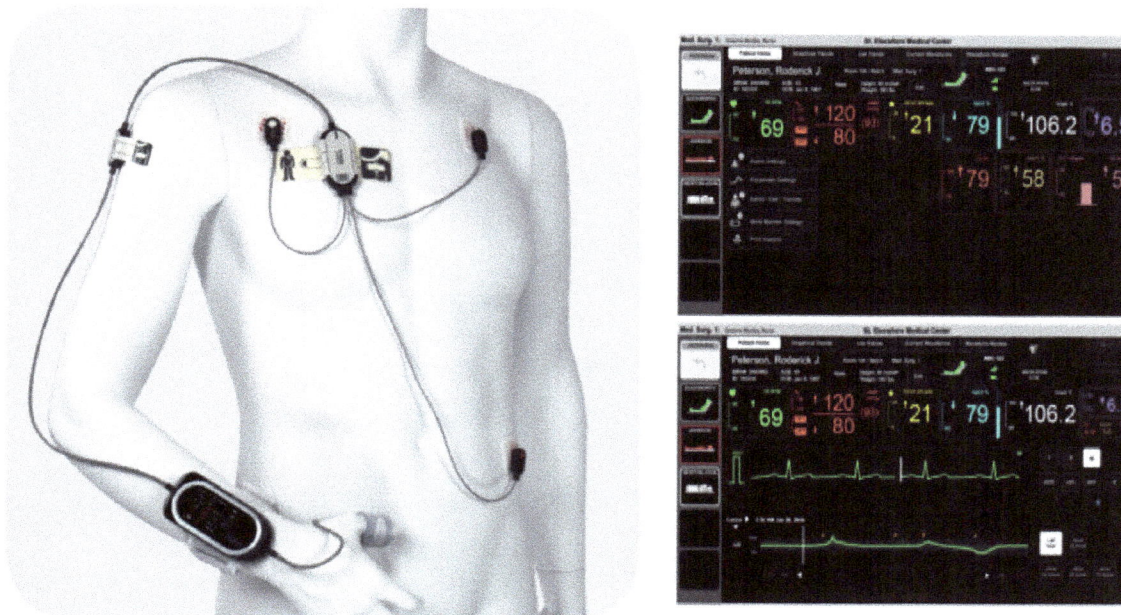

Figure 37. The Vital Space biometric monitoring system (credit: Vital Space)

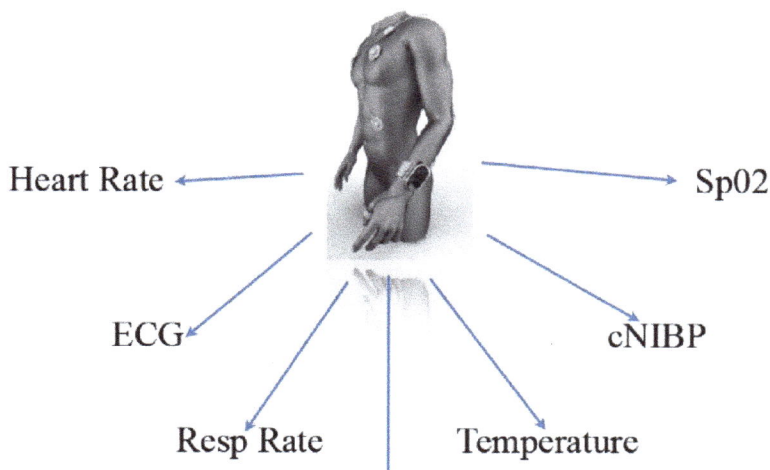

Figure 38. The data products produced by the Vital Space system (credit: Vital Space)

Validating the Vital Space system in Analog Spaceflight

In conjunction with Astronauts4Hire (A4H), Vital Space validated their biometric monitoring system (ViSi) in microgravity through a series of 160 microgravity parabolas. The objectives of these tests were to 1) assess successful basic operation of the ViSi with respect to continuous physiologic data capture under varying gravity loads, and 2) assess the ease of use and interface between the ViSi hardware, software, and subject under varying gravity loads.

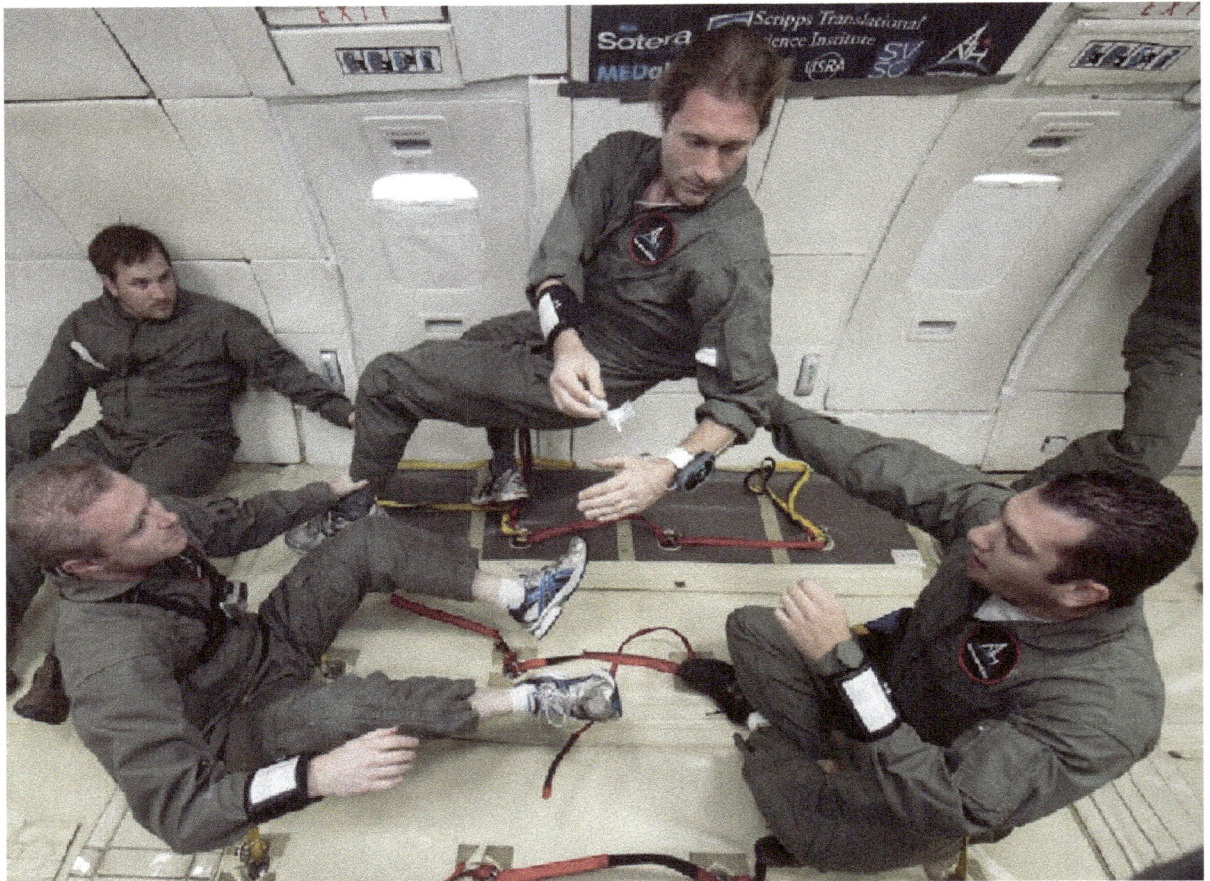

Figure 39. Vital Space and Astronauts4Hire validate the ViSi biometric monitoring system in microgravity (credit: NASA Flight Opportunities Program)

The test subjects demonstrated that the ViSi system could be donned in under four minutes in microgravity and stowed in under 100 seconds. Results from test data are presented in the following figures.

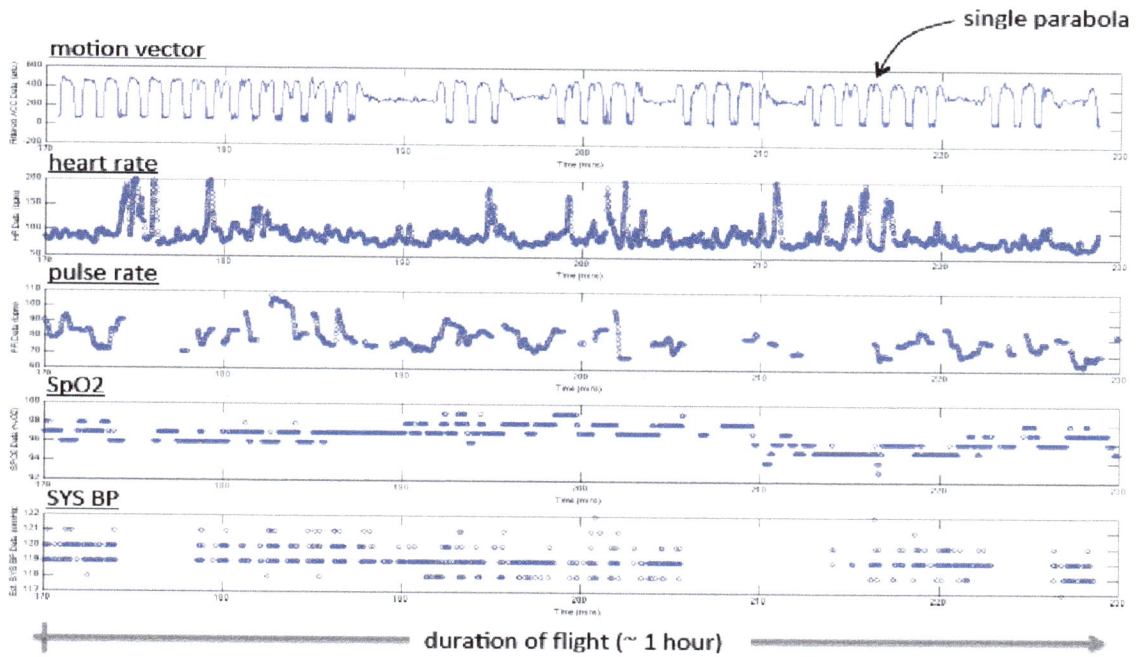

Figure 40. In-flight data products obtained in the microgravity testing of the ViSi system (credit: Vital Space)

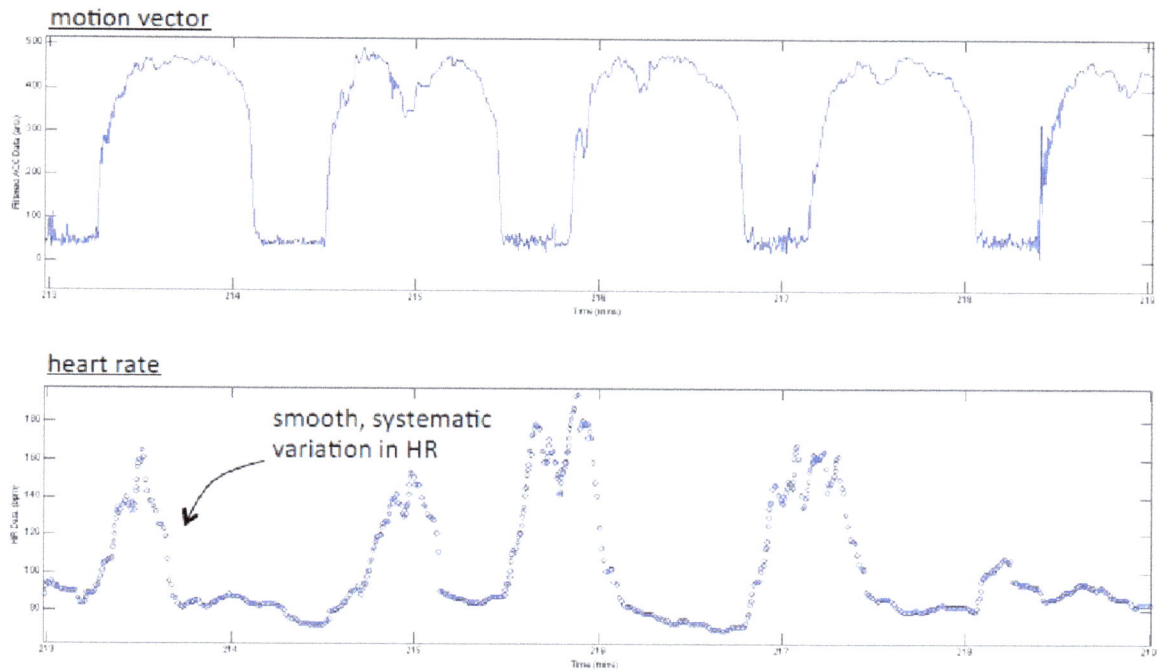

Figure 41. Correlation between motion and heart rate on an A4H test subjects (credit: Vital Space)

motion vector

ECG waveform

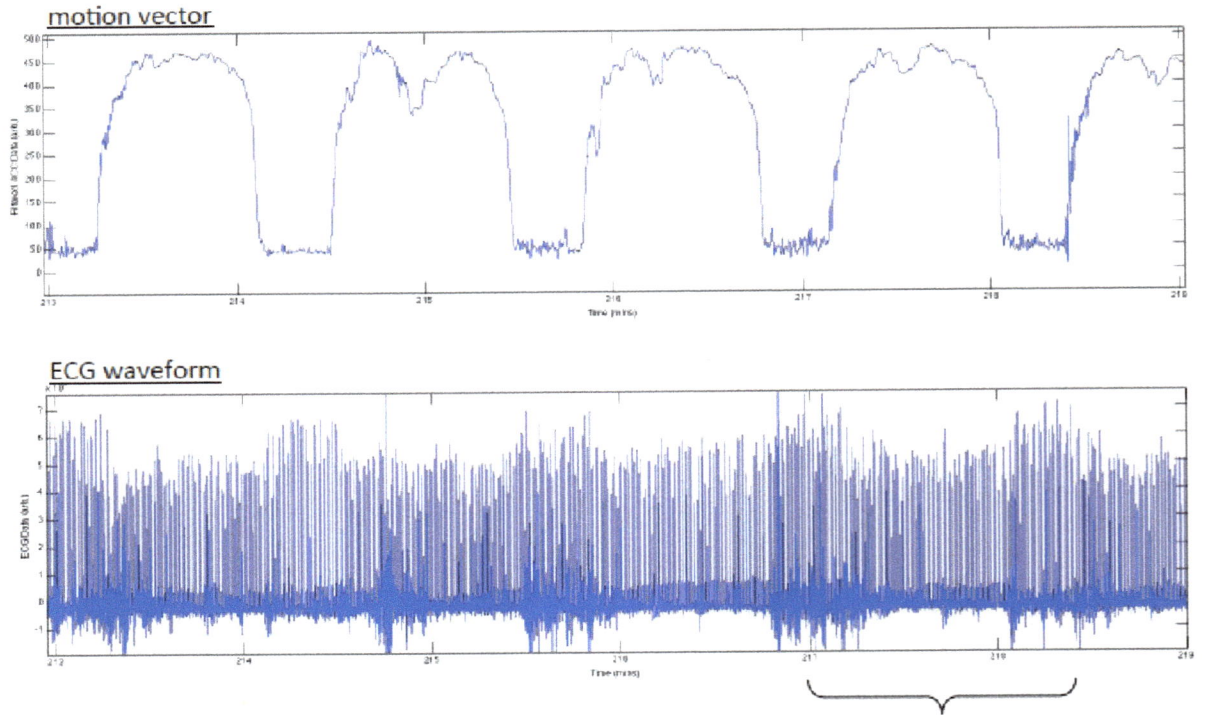

Figure 42. Correlation between Motion and ECG Waveform on an A4H test subject (credit: Vital Space)

HYPOXIA

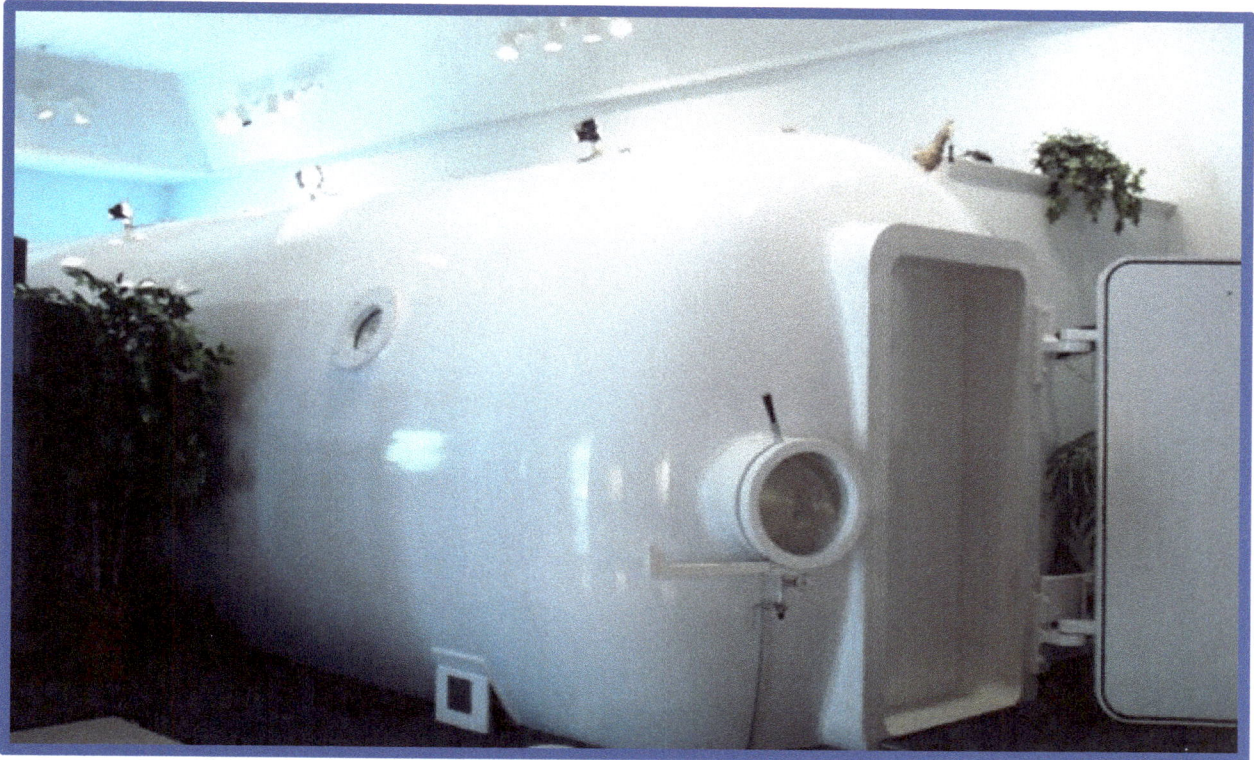

Module Objectives

- ➢ Understand the environment in which hypoxia can occur
- ➢ Understand the concept of Effective Performance Time
- ➢ Recognize the signs and symptoms of hypoxia
- ➢ Recognize the primary behavioral changes that result from hypoxia

Contributing Author:

Paul Buza, M.D.

THE ENVIRONMENT

Our atmosphere exerts a certain pressure on our bodies that we take for granted. Without this pressure we would quickly "die". The great majority of all life on the planet resides within a "narrow envelope of pressure" from sea level up to approximately 10,000ft of altitude. After that there is a significant drop off in pressure and oxygen within the "thin air" leading to low levels of oxygen. These low *pressures are completely incompatible with life due to the lack of pressure and oxygen within the gases we breathe.* The results are catastrophic if we do not quickly identify these changes within our cabin.

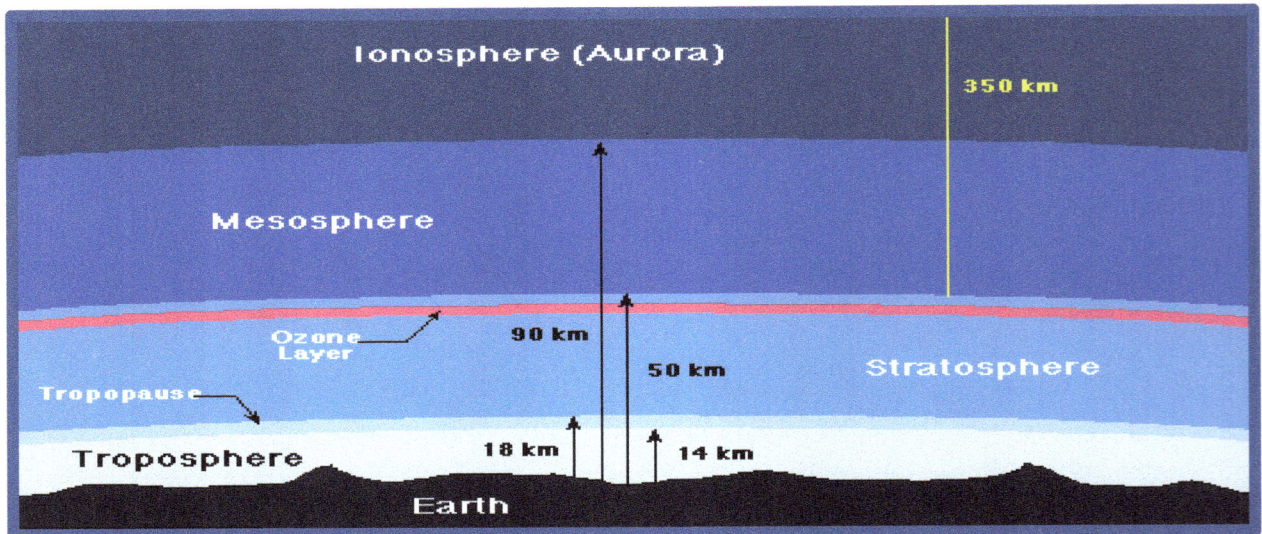

Figure 43. The weight of our atmosphere exerts 14.7 psi. of pressure on our bodies.

Once our planes clear through FL100 (10,000 ft) we are entering hostile territory from the perspective of the atmospheric characteristics. Our pressures, levels of oxygen, ambient temperatures and humidity decrease dramatically. As a result, we need to control our cabin environment so that we are comfortable and safe. Once we level off at FL400 (40,000 ft) the temperature outside is -70°F, humidity is zero, pressure is only 2.73 psi and the oxygen available to breathe is essentially non-existent. We might as well be on "Mars" when one considers the hostility of the atmosphere at these altitudes. So, while it may be quite comfortable inside the cabin we should always be aware of the potential dangers posed to us in the event of a cabin depressurization.

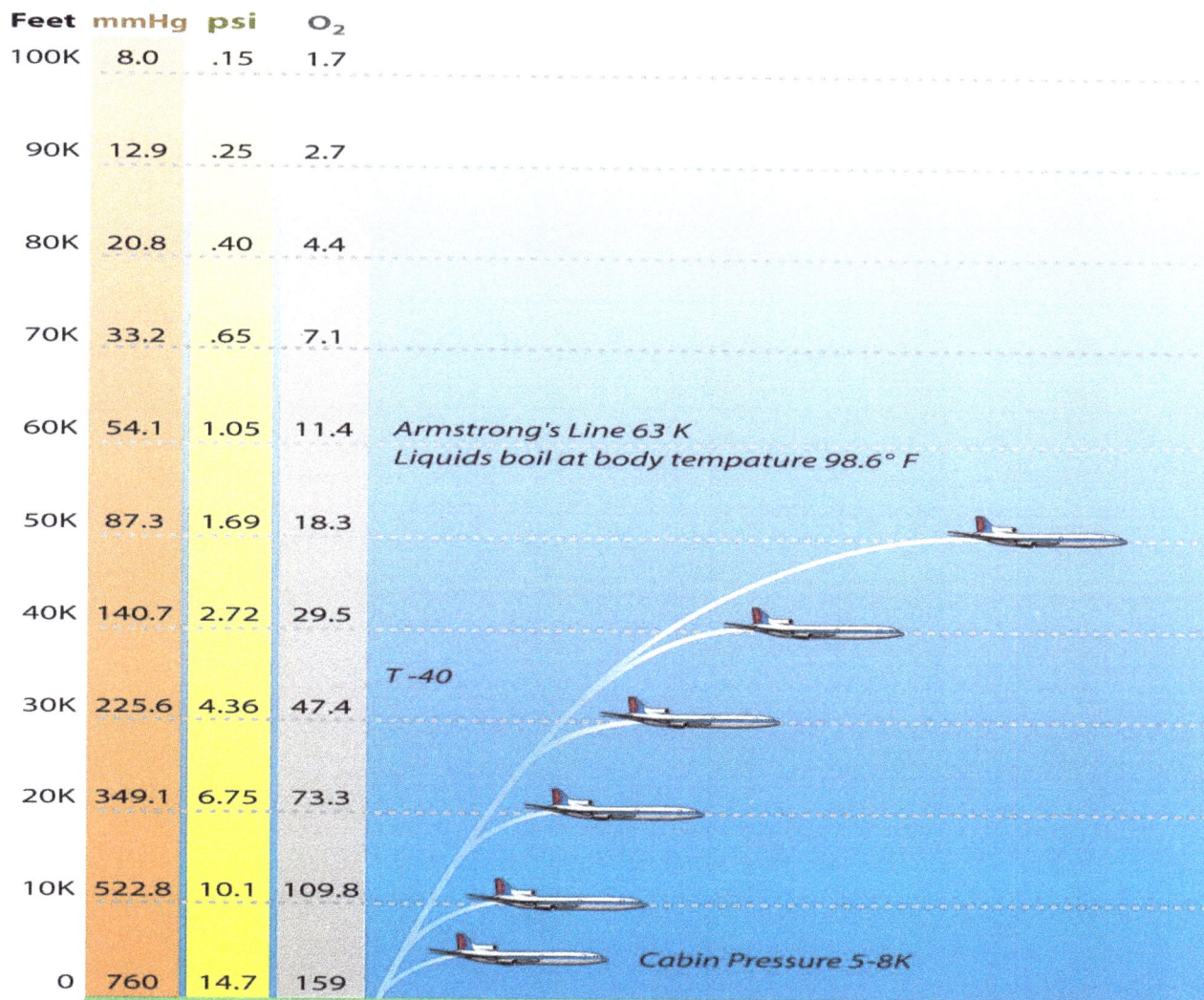

Feet	mmHg	psi	O₂
100K	8.0	.15	1.7
90K	12.9	.25	2.7
80K	20.8	.40	4.4
70K	33.2	.65	7.1
60K	54.1	1.05	11.4
50K	87.3	1.69	18.3
40K	140.7	2.72	29.5
30K	225.6	4.36	47.4
20K	349.1	6.75	73.3
10K	522.8	10.1	109.8
0	760	14.7	159

Armstrong's Line 63 K
Liquids boil at body tempature 98.6° F

T -40

Cabin Pressure 5-8K

Figure 44. The amount of oxygen we breathe decreases dramatically as we ascend

 For the purposes of this book, we will arbitrarily use psi for convenience. So, the entire atmosphere of the planet is pushing on our bodies with a significant force but we simply don't "feel it" because it is distributed evenly in all directions and it is what we are used to. Another words, it is our "physiological home" and we easily take it for granted. We walk freely through the air and pressure as if there was no resistance. Yet we know that the aerodynamics of lift can generate tremendous forces. Therefore, it is safe to say that we live within an "ocean of air" but most of it resides within the first 18,000ft of altitude. This is worth remembering as a rule of thumb since half of all the pressure exerted by the atmosphere is found within the first 18,000 feet of altitude. At 18,000 feet the pressure is 7.34 psi, which is half of 14.7 psi. If we remove this pressure from our bodies, we begin almost immediately to feel ill effects. Therefore, we need to spend some time to understand why we are so dependent on pressure to sustain life.

Feet	Inches of Mercury	Millimeter Mercury	PSI	Centigrade	Fahrenheit
0	29.92	760	14.69	15	59
2,000	27.82	706.7	13.66	11	51.8
4,000	25.84	656.3	12.69	7.1	44.7
6,000	23.98	609.1	11.91	3.1	37.6
8,000	22.22	564.6	11.77	-0.8	30.5
10,000	20.58	522.7	10.1	-4.8	23.4
12,000	19.03	483.4	9.34	-8.8	16.2
14,000	17.57	446.5	8.63	-12.7	9.1
16,000	16.21	412	7.96	-16.7	1.9
18,000	14.94	379.7	7.34	-20.7	-5.1
20,000	13.75	349.5	6.76	-24.6	-12.3
22,000	12.63	321.3	6.21	-28.6	-19.4
24,000	11.59	294.9	5.7	-32.5	-26.5
26,000	10.62	270.3	5.22	-36.5	-33.5
28,000	9.72	237.4	4.78	-40.5	-40.7
30,000	8.89	226.1	4.37	-44.4	-47.8
32,000	8.1	206.3	3.99	-48.4	-54.9
34,000	7.38	188	3.63	-52.4	-62
36,000	6.71	171	3.3	-55	-69.7
38,000	6.09	155.5	3	-55	-69.7
40,000	5.53	141.2	2.73	-55	-69.7
42,000	5.03	128.3	2.48	-55	-69.7
44,000	4.56	116.6	2.25	-55	-69.7
46,000	4.15	105.9	2.05	-55	-69.7
48,000	3.77	96.3	1.86	-55	-69.7
50,000	3.42	87.4	1.7	-55	-69.7

Table 6. Table of US Standard Atmosphere

When pressure is distributed equally onto our bodies, we simply do not feel it. When we dive underwater and increase the pressure, we do not feel this as well. As a matter of fact, we as humans are quite capable of withstanding very high pressures such as diving to 100 feet. At these depths we do not feel a "crushing weight" on our bodies because the pressure is evenly distributed. The same holds true if we ascend to high altitudes as again, we do not feel the low pressure on our bodies because the pressure is evenly distributed. However, if someone pushes their finger against your arm you can easily detect this pressure because it is a unidirectional force and not evenly distributed. So, when our cabin depressurizes we will not be able to rely upon our senses to know there has been a pressure change.

So, to protect ourselves as we fly through high altitudes we have to bring along pressurization systems to maintain" cabin pressure". Hence our aircrafts act like *environmental physiology units* (EPU) as shown in Figure 46 These units can control for pressure and achieve high pressure or low-pressure environments. Current pressurized aircraft are designed to maintain a cabin pressure equivalent of 5000 ft and is not allowed to rise above 8000 ft. This is the outer limit of what we consider to be normal without the use of supplemental oxygen. But as we will soon see, our oxygen levels are lower even at these altitudes but usually do not pose a problem unless we have certain medical conditions that we will discuss later.

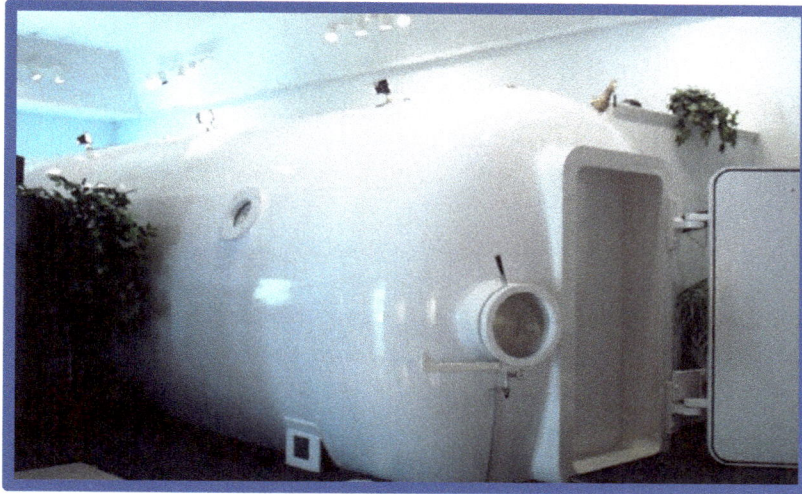

Figure 45. The great majority of all life resides in less than 10,000 ft. of altitude (credit: SAMI)

This is why it is imperative that oxygen and pressurization systems be checked before each flight. These functions include supply, pressure, regulator function, oxygen flow, mask fit, cockpit and air traffic control (ATC) communication using mask microphones. Numerous NTSB reports are available revealing weaknesses in these systems. Pilots who correctly identified cabin depressurization were further compromised because there was no oxygen available to them from faulty oxygen delivery systems. The majority of these cases was attributed to human error and could have been prevented.

RESPIRATION AND EFFECTIVE PERFORMANCE TIME (EPT)

Respiration refers to our breathing patterns in normal and abnormal environments while *effective performance time* (EPT) refers to our ability to think and perform normal duties within a certain time frame. The two are so closely dependent upon each other that we will study them together. We breathe to obtain oxygen, as it is the "fuel" for our bodies. If we were as mindful about our "oxygen levels" as we are about our remaining jet "fuel" we could easily identify the problem and respond accordingly. Let's take a moment to understand how it is possible to lapse into a state of unconsciousness while we think we are otherwise breathing in a normal fashion.

Decreasing pressure directly affects the amount of oxygen that we can obtain for each breath we take. For example (fig.2), at sea level the pressure is 760 mmHG of pressure and oxygen is 21% resulting in 159 mmHG. This is the normal amount of oxygen available to us and it is what we are accustomed to. At 18,000 feet in altitude the pressure is reduced in half so the pressure is now only

380 mmHg where again oxygen is 21% resulting in 80mmHg of oxygen. So for each breath you take, you are only receiving half of the normal amount of oxygen and it will only be a matter of minutes before you start to feel its effects. At this altitude the average time of useful consciousness (TUC) or effective performance time (EPT) is approximately 20 to 30 minutes. Now compare this to 25,000 feet where we begin to see a rapid drop off in EPT. At this altitude the pressure is only 280 mmHG of which oxygen is 21% resulting in a net pressure of 59mmHG of oxygen. This is extremely low and EPT is now only 3-5 minutes. Ascend only 10,000 more feet to 35,000 feet and EPT is now only 30 to 60 seconds. At 40,000 feet EPT is approximately 10 to 15 seconds. Pressurization systems at this altitude are critical for life support.

The average times of EPT at different altitudes are shown below. It is worthwhile to commit these particular altitudes to memory emphasizing how little time we have. Keep in mind that these times are just an average and are not exact by any means. For example, at 25,000 feet the EPT average is 3-5 minutes. Each pilot is different so one pilot may perform well for a full 5 minutes while the other loses EPT in 2 minutes. Also keep in mind that the EPT is progressively diminishing over this same time period but does not become obvious until a certain "threshold" is achieved resulting in the above-described averages. Do not make the assumption that "all is normal" for at least 3-5 minutes at 25,000 feet before you decide to don oxygen.

Important to remember is the fact that these average times of useful consciousness assume that you are breathing at a normal and comfortable rate. During an emergency you will be anxious and as a result your breathing and heart rates will dramatically increase using up oxygen. So your effective performance time will decrease from the average and you'll make it even worse if you or your crew is standing or walking as your body uses more oxygen when compared to a seated position.

Because of such low oxygen availability even at these low altitudes, it is required to don your oxygen at 12,500 ft. to 14,000 ft MSL for that portion of the flight that is at those altitudes for more than 30 minutes. All flight crew must don oxygen at altitudes above 14,000 ft. and all occupants must don oxygen at 15,000 ft. Pressurized aircraft require at least a 10 minute additional supply of oxygen for each occupant above FL250. At flight altitudes above FL 350, a single pilot must wear and use an oxygen mask that is secured and sealed. The oxygen mask must supply oxygen at all times or automatically supply oxygen when the cabin pressure exceeds 14,000 feet. An exception to this regulation applies for two-pilot crews that operate at or below FL410. One pilot does not need to wear and use an oxygen mask if both pilots are at the controls and each pilot has a quick donning type of oxygen mask that can be placed on the face with one hand from the ready position and be properly secured, sealed and operational within 5 seconds. If one pilot of a two-pilot crew is away from the controls, then the pilot that is at the controls must wear and use an oxygen mask that is secured and sealed.

FAA Air Regulations (§121.331) regarding supplemental oxygen

Altitude	Oxygen Requirement
10,000-12,000' MSL	Supplemental oxygen for **all crewmembers** for portions of the flight **over 30 minutes** at these altitudes
>12,000' MSL	Supplemental oxygen for **all crewmembers** during entire flight at these altitudes.
>15,000' MSL	Supplemental oxygen for **all crewmembers and passengers** during entire flight at these altitudes.

DECOMPRESSION ILLNESS

Hypoxia is not the same as *decompression illness* (DCI) although both occur as a result of sudden lowering of pressure. Hypoxia occurs due to the breathing of "thin air" where the oxygen levels are low and the brain is unable to function normally. DCI occurs because a sudden lowering of pressure allows the nitrogen stored in your body to come out too quickly creating bubbles that hurt your body tissues. Imagine a can of cola that you know is under pressure (like your body under pressure from the atmosphere), suddenly you open the can and bubbles come out of the fluid as a gas (like your body if there is a sudden explosive decompression).

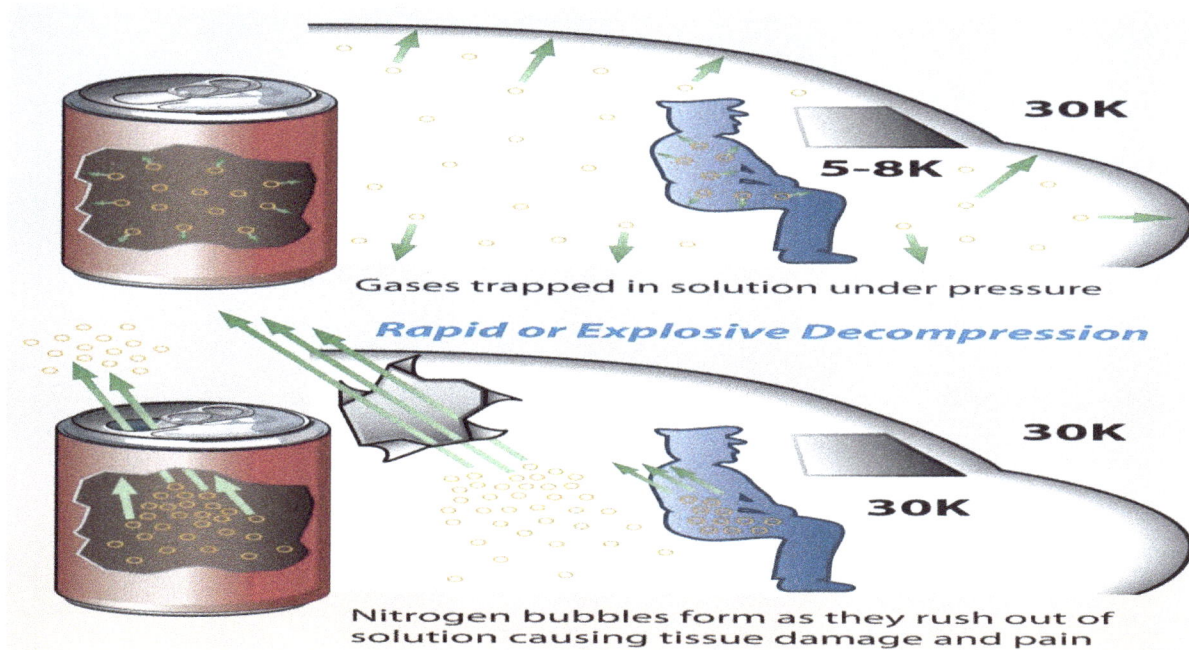

Figure 46. Sudden Decompression

Hypoxia and DCI are two different medical conditions that can occur together but should not be confused as being the same. Hypoxia is a lack of oxygen while DCI results from excessive nitrogen dissolving out of our tissue creating bubbles that can lead to pain and tissue damage, especially the nervous system.

At altitudes greater than 20,000 feet the risk for DCI increased dramatically. Donning oxygen at these altitudes will prevent hypoxia long enough to descend to safer altitudes, but should never be considered appropriate for flying to achieve your flight plan as DCI may occur. DCI is to be considered a very serious condition and not taken lightly. Serious long lasting injury can occur if symptoms are ignored and not properly treated right away. In some cases the bubbles can damage nervous tissue in your spinal cord leading to paralysis (stroke like condition).

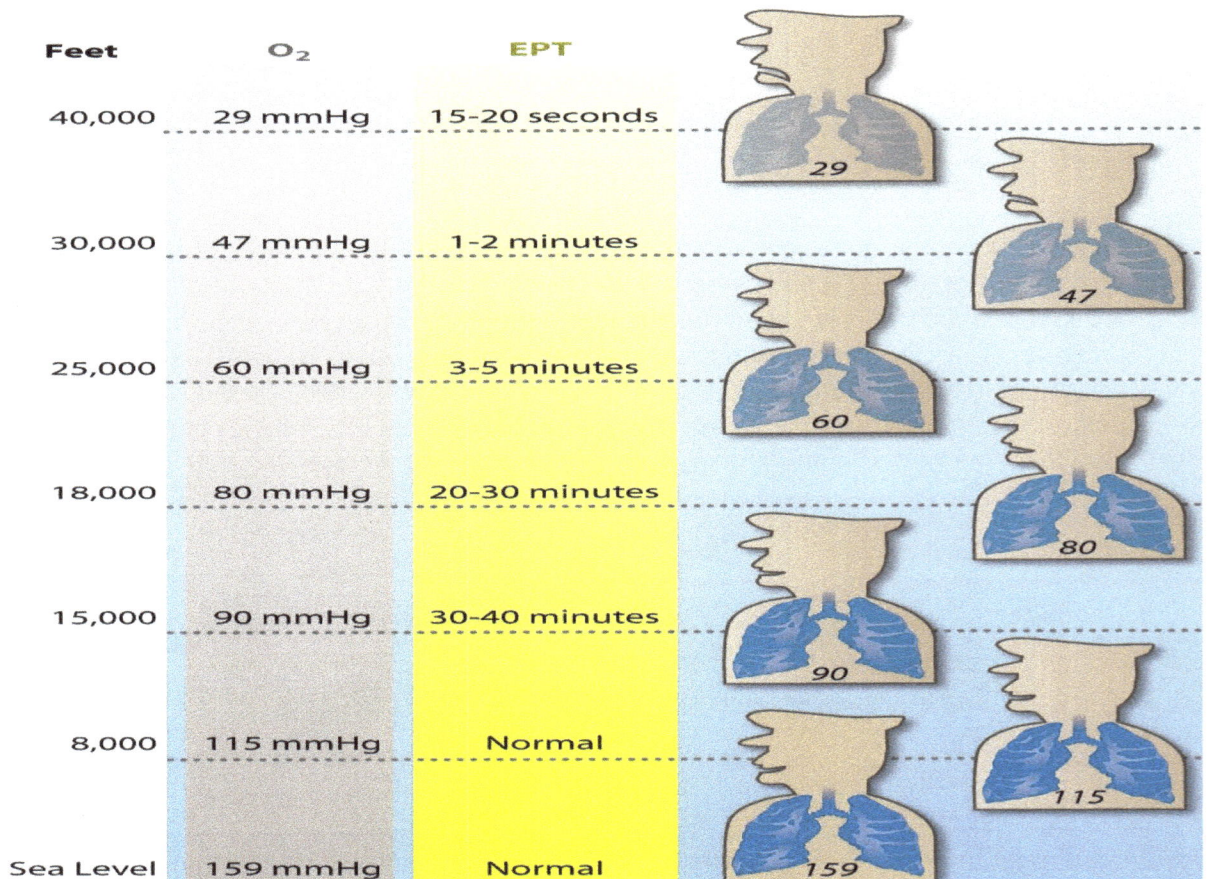

Feet	O$_2$	EPT
40,000	29 mmHg	15-20 seconds
30,000	47 mmHg	1-2 minutes
25,000	60 mmHg	3-5 minutes
18,000	80 mmHg	20-30 minutes
15,000	90 mmHg	30-40 minutes
8,000	115 mmHg	Normal
Sea Level	159 mmHg	Normal

Figure 47. The Primary Physical Signs and Symptoms of Hypoxia

Most pilots find this portion of the training to be rather easy and so we should be able to get through this section quickly. It involves sensations that we can feel or see in ourselves or amongst our crew.

Sensory Disturbances

Numbness and Tingling of the Hands, Feet, Face, and Scalp

This is one of the earliest and most common symptoms experienced during hypoxia. Suddenly you will feel tingling like sensations over your feet, hands, face and scalp. For some it will only be one part of their body while others feel it vigorously throughout.

Sense of Warmth or Heat over the Upper Chest and Neck

This phenomenon is rather unique to hypoxia and can be a valuable symptom almost exclusive to hypoxia. Not all will experience this symptom but those that do can feel rest assured that if they feel this sensation that there should be a very high index of suspicion. This unusual sensation is a result of dysfunction of our autonomic system (not on exam). This system controls the inner functions of heart rate, sweating and other functions for normal body functions. I describe these phenomena as "autonomic flush". To the pilot it feels as if someone put a warm towel over our upper chest and neck.

Sense of warmth or coolness over the hands and feet

In addition to numbness and tingling is the sense of temperature change over your feet and hands. Suddenly your hands and feet will feel hot or even cold as if it were winter. This is due to the fact that the hands and feet are the farthest from the heart and blood flow to these regions is affected first.

Visual Changes

Loss of Color Vision

Until now we have focused primarily on the effects of hypoxia on the brain. Another vulnerable area is the eye. In particular, the back portion of the eye is called the retina where our vision is mediated. From the retina the optic nerve carries our vision back to the brain and is considered part of the brain proper. Color vision requires a very high level of oxygen to process correctly so it is the first to go during hypoxia. The most common observation is the loss of distinction between blue and green, red and yellow. This area is so sensitive that it begins to occur at altitudes as low as 5000 ft and oxygen is required for prolonged night flights.

Pilots have also described other changes in color perception. Some will describe a "graying" like effect where everything seems to be shades of black and gray (color desaturation). Or they can sense a difference in color but can't recognize it. Less common but occasionally reported is where the entire visual field turns to a light shade of purple. These changes are important as they may impair our ability to interpret color-coded directions during landing and in the past have been associated with significant aviation events.

Tunnel Vision

As color begins to diminish some pilots will notice a "tunneling" like event. The peripheral portions of our vision will diminish and we tend to focus only on the central parts as if looking into a tunnel. This impairs our ability to "scan" the instrument panel and requires more head movement to obtain needed information.

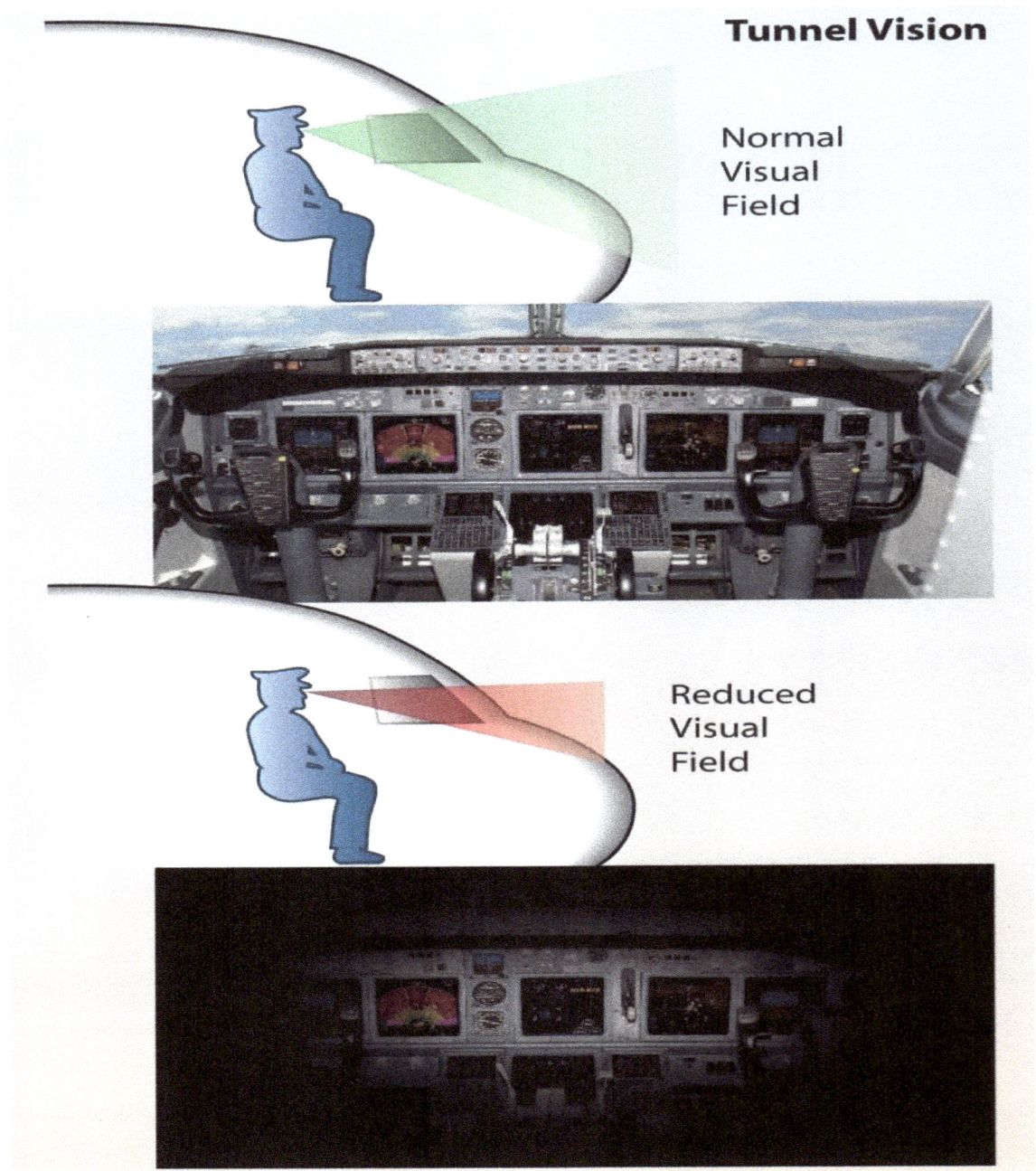

Figure 48. Tunnel Vision

Less Common and Unusual Visual Patterns

As hypoxia progressively worsens the visual field can change even further. Bizarre and unusual patterns can occur that are often difficult to describe. Some examples include a change in depth of field where objects appear farther than they really are. Some describe unusual distortions of objects or difficulty in recognizing common objects. Chances are when this occurs the effects are in part due to the brain as well as ultimately all vision is mediated within the brain itself.

Disturbance of Balance (disequilibrium)

This is probably one of the most commonly reported symptoms during hypoxia. The sensation is that of light-headedness without associated spinning or nausea. Although reported the spinning sensation (vertigo) is rare and when it does occur is almost always associated with nausea. True vertigo with nausea is more likely to be related to inner ear dysfunction as a result of barotraumas (discussed later in the next chapter). The sense of light-headedness during hypoxia is usually non-threatening and often is considered mild. However, the sensation is so common and it tends to occur so early in hypoxia that its recognition can be very useful for early detection of hypoxia. This sensation is likely to be amplified in the presence of significant turbulence.

Changes in Skin Color (Cyanosis)

Fingers and Lips: By the time you notice this particular phenomenon you are dangerously close to becoming incapacitated. Cyanosis is when your fingernails and lips become blue or purple. This is due to the lack of oxygen in your bloodstream and this can be seen most easily in the nail beds of your fingertips and lips. These areas are richly supplied with blood vessels and are easily seen due to the thinness of skin allowing us to visualize color changes within the blood. Hemoglobin is the main carrier of oxygen in our blood stream and when rich with oxygen possesses a bright red color. Once the body has used up this oxygen the hemoglobin returns to the lungs for additional supplies and has a blue or purple like color. This can be seen best in the fingernails, lips and face. When noted this is considered to be an emergency until proven otherwise as the oxygen levels are so low that incapacitation is about to occur.

Facial Pallor

Another sign is the sudden loss of color in our faces as oxygen levels continue to drop. This is a very reliable and rather early finding, even before the cyanotic changes of our fingertips. Most people have a rich sense of color in their faces due to generous amounts of blood vessels within our face. In the presence of hypoxia, we begin to lose this color and there tends to be "whitening or paleness" as if we had seen a ghost. This is very useful as we can easily recognize this phenomenon simply by looking at the faces of our crewmembers and or vice versa, our crewmembers looking at us.

107

Sudden Muscle Fatigue

Sudden arm weakness during instrumentation

We take the strength and weight of our arms for granted. If you were able to remove one of your arms and hold it with the other you would be amazed at its weight. Yet we move our arms through space almost as if it were weightless. During hypoxia there is a sudden sense of heaviness that is most unusual and should alert us to the possibility of hypoxia. This is due to the fact that arm movement requires a high proportion of energy expended and demands ample supplies of oxygen. During routine instrumentation there will be a sense of "heaviness and arm fatigue" that you would not usually expect.

General sense of heaviness or fatigue

In addition to our arms we can also suddenly just feel weak or tired out of character in our immediate environment. There is a general sense of "slowing down" that you would not expect especially during takeoff and climbing to the desired altitude. This in addition to sudden arm heaviness should alert us to the possibility of cabin depressurization.

Less Common Signs and Symptoms of Hypoxia

Anxiety

Anxiety does not tend to be a primary finding in hypoxia. Most of the time the effect is quite the opposite with the effect being mostly associated with a lack of concern. Chances are the crewmember was anxious for other reasons before the event and the hypoxic environment amplified the degree of anxiety.

Tinnitus (Ringing of the Ears)

This phenomenon can occur in one or both ears and consist of a high-pitched ringing that can be quite disruptive. Again, chances are this is not related to hypoxia but is more likely to be a consequence of barotraumas (discussed in the next chapter).

Double Vision

This is a rare occurrence and is not a primary finding of hypoxia. Chances are the crewmember has an underlying medical condition and the hypoxia event precipitated weakness of muscular control of the eyes. Medical attention and further evaluation is needed if double vision occurs.

Headache

Headache is rare during hypoxia. The great majority of crewmembers who experience hypoxia find it to be painless and non-stressful. When headaches occur in the hypoxic setting, chances are the crewmember has underlying medical conditions such as Migraines that may need further attention.

Hyperventilation

Noticeable hyperventilation during hypoxia is a late finding. Hopefully with the multitude of signs and symptoms discussed thus far we will not witness this and already be descending our aircraft while wearing our oxygen masks.

Convulsions and Loss of Consciousness

This is why we need the training. Failure of early identification of hypoxia will result in convulsions, loss of consciousness and death. Need I say more?

One of the remarkable phenomena noted in aviation physiology is the rapid recovery of normal function immediately upon donning of oxygen. In some cases within several normal breaths the "lights are back on" and function has almost completely recovered. Within a minute even the most severe cases will completely recover. You do not need to hyperventilate on your mask to recover; only resume normal breathing and almost immediately you will recover. Do not pass up this opportunity.

A misconception about hypoxia is that a pilot could take a deep breath from the mask and hold his breath to attend to someone unconscious in the back. ABSOLUTELY NOT. This is not the same as trying to swim underwater where most of us can hold our breaths for approximately one minute or a little longer. This cannot occur at altitude and the reasons will be discussed in the upcoming chapter as we review the characteristics of the atmosphere and gas laws.

THE PRIMARY BEHAVIORAL CHANGES OF HYPOXIA

We take our brain for granted when all seems normal and well. A tremendous amount of sophistication for processing is occurring while we feel normal in our duties even during complex multi-tasking. When we are confident in our abilities and feel well we can process a tremendous amount of information in a seemingly relaxed and normal function. However, this ability " goes out of the cabin" along with the pressure and oxygen. The loss of mental abilities is directly related to the loss of cabin pressure and oxygen. Our many facets within our multi-tasking environment become a " house of cards" and we lose our ability to make proper decisions in a timely fashion. The most vulnerable behavior to suffer during hypoxia is our ability to speak and listen with fluency while following complex commands issued from ATC.

Disruption of Communication

The most sensitive behavior that we can notice easily during hypoxia is the change in communication while we are actively engaged in speech. Our ability to listen and respond quickly and appropriately deteriorates rapidly while the cabin oxygen levels are dropping. This of course is not as noticeable when all is quiet in the cabin.

Changes in Vocal Patterns and Quality of Speech

The quality of speech begins to change at altitude. Normal speech is characteristic to the individual. Each person has a unique quality of tone and style of delivery. The individual style is learned amongst friends or work associates to the point of immediate recognition, such as a single word mentioned on the phone leads to immediate recognition of the identity of the source. When two friends call each other and the caller says " hey it's me" often it is sufficient to know who the caller is. This is remarkable learning between humans who can pick up very subtle characteristics due to repetitive learning over time. This represents an opportunity for crewmembers that know each other well to pick up on subtle changes that will occur at altitude. The better we know the verbal and behavioral characteristics of our crew members the more sensitive we can be to early changes associated with cabin depressurization even before the development of more obvious signs and symptoms. A common change in speech is the tendency to become monotone with a resultant loss in intonation (accents that give character to the natural auditory quality of speech.)

Prolonged Verbal Response Time (latency)

As oxygen levels in the cabin fall our ability to respond to communications is delayed. Typically we would answer questions in a rapid fashion that is smooth and continuous. Disruption of fluent dialogue between the pilots, crew and ATC is easily identified especially when it is "out of character" for the individual.

Increased latency occurs primarily because of the fact that the pilot is having difficulty processing the information in a timely and relevant fashion. Simply put, it is taking longer to think or react to a situation. As hypoxia ensues, many neurological functions are impaired making it more difficult at first to answer complex questions that require multiple thought processes. As hypoxia worsens it becomes very difficult to answer even simple questions. One of the first behavioral changes that occur is unusual delay in the answering of simple questions. This represents an opportunity to suspect the possibility of cabin depressurization even before other signs and symptoms that are more obvious occur. If more than one crewmember begins to "slow down" in normal speech patterns this should significantly raise suspicion amongst the crew.

Due to the delay in verbal reaction times, there is a tendency to "fill the gap" with the word "um". When we are not sure how to answer a question, it is normal for us to say "um". For some people this is a normal part of their speech pattern and we are accustomed to it. However, many pilots do not have this characteristic and if suddenly they start using "um" frequently this should raise awareness and be considered.

Speech Interruption by Frequent deep "Sighs or Yawning"

Another change in verbal characteristics that can alert the crew to the possibility of hypoxia is the sudden development of frequent "sighs and yawns" amongst the crew. In early hypoxia the initial reaction by the body to dropping oxygen levels is deeper breaths that eventually lead to deep sighs that often trigger yawning. These sighs will often precede the answer to a question or simple routine conversation. It is considered an early sign of hypoxia. It is the first compensation of the body to try and increase its oxygen levels. The sighs attempt to expand the lungs to its full capacity to compensate for lower oxygen availability. This occurs before obvious hyperventilation or rapid breathing which tends to be more noticeable. As hypoxia worsens the deeper breaths are insufficient resulting in an increase of the number of breaths per minute. By the time that a pilot is noticeably hyperventilating he or she is usually already neurologically impaired and is considered to be a late finding of hypoxia.

In summary, when a pilot and or multiple crew members show the sudden development of monotone and slowed speech, delayed verbal reactions times often filled with "ums" and frequent sighs the suspicion of hypoxia should strongly be considered.

Decreased Spontaneity of Movement

Sudden Decrease in Facial Expressions (Staring)

An early finding of hypoxia is the sudden development of "facial staring". Pilots have different personalities and some are energetic with expressive qualities while others are quiet and reserved. Pilots who are energetic will suddenly develop a quiet like characteristic and begin to "stare" as a result of hypoxia. There will be less smiling, rising of the eyebrows, frowning etc. In medical terms this is known as bradykinesia (not on the exam). In Latin this term means, "slow movement". This is not restricted to the face but refers to all body movement in general. Changes in facial expression with sudden decrease in spontaneous movement are the easiest to identify, as pilots are rather restricted in their seated positions within the cockpit.

Sudden Decrease in Head and Arm Movements

In addition to sudden changes in facial expression are the development of slowed head movements and the natural use of our hands while we are communicating. Again, all pilots have different personalities where some are rather energetic and colorful in their style of expression frequently using their hands and moving their heads in combination with frequent facial expressions. During hypoxia there will be a sudden slowing of this spontaneity

Decreased Postural-Seated adjustments and Leg movement. (Slumping)

Along with decreased facial expression and arm movement are decreased postural-seated adjustments. Given the restrictions of the cockpit this is less obvious but can be identified in some individuals. Some pilots tend to be "fidgety" in their seats and often will make positional changes and rotate their bodies as part of their style of communication. During hypoxia this will diminish and can be noted along with the other changes described above.

Motor – In Coordination (Clumsiness)

Lack of Hand-finger Coordination during Instrumentation

Certain portions of the brain are more sensitive to sudden hypoxia. One of the regions is the cerebellum that is in the lower and back part of the brain (not on the exam). The cerebellum controls for smooth movements such as fine hand and finger control. During sudden hypoxia in coordination is noted especially with arm, hand and finger movement. This often results in overshooting, such as past pointing or undershooting. When reaching to perform routine instrumentation the finger may miss the target or not quite reach it in a fluent manner. This can result in mistakes for altitude and or velocity adjustments. The general observation in the crew during hypoxia will have the appearance of clumsiness of hand movement. Items may be dropped and or there will be a sense of shakiness (tremor).

Lack of Foot Precision during Pedal Instrumentation

Less noticeable is the clumsiness of foot controls during hypoxia. There will be fewer tremors noted but the possibility of pressing too hard or not hard enough is increased. As hypoxia worsens during decompression this in coordination worsens to a state of involuntary jerking (myoclonus– not on exam). This is very serious signaling that the pilot is about to go unconscious. Myoclonus can involve both the arms and legs and or may occur independently. Regardless, any involuntary jerking noted among the crew is an immediate sign of severe hypoxia and must be dealt with immediately.

Loss of Prior Learned Routine Tasks

Failure to Recognize the Geometric Configuration of the Instrument Panel

In almost everything we do on a daily and routine basis is pre-programmed into the brain from a prior learning process. This is due to the fact that we can't afford to "reinvent the wheel" every time we want to perform a simple task such as brushing our teeth, picking up a cup of coffee, or entering data into our computers or writing into a log book. We take these simple tasks for granted so that we may perform multiple simultaneous functions (multi-tasking). It is beyond the scope of this training session to appreciate the complexities of the neurological requirements to perform even the simplest of tasks. An analogy that helps us understand is that of the computer. A routine function such as saving a document only requires a single entry yet the computer performs effortlessly a series of complex pre-programmed commands to execute the function. The brain is similar in that you only need to think of the task and effortlessly you perform it without much additional thought. This phenomenon is impaired during hypoxia. Let's look at some different forms in how this may impair our abilities during flying.

Our brains like to memorize geometric configurations within our visual field to ease the recognition of different objects (constructional apraxia- not on the exam). This includes the cockpit. The instrument panel is essentially a fixed geometric configuration so that over time you recognize the location and function of each component. While multi-tasking and scanning you easily pick off the information you need quickly and you do it easily without giving it much thought. In other words, you are not searching around the panel looking for the information you need; it is right there at a moment's notice. The cockpit can be a series of circular gauges while the central panel is rectangular with different levels where each row has different functions. You are very comfortable with your instruments and can easily scan and identify the proper location without having to spend time searching for it. This is impaired during hypoxia for a number of reasons. Because your brain does not have enough oxygen you begin to decrease the number of multi-tasking duties and you tend to focus on only one or two functions because it is all that you can remember - "Fixation". You may want to only focus on one of the 6 gauges and when pressed by your crew you will resist thinking of other functions. Obviously this is unacceptable in the cockpit but represents one of the most important early findings in hypoxia. If you notice a crewmember that resists thinking about different related functions while scanning your flight profile a high index of suspicion should be considered that "something isn't right".

Failure to Perform Routine Sequence of Motor Functions

We often perform a multitude of simple tasks in a repeatable sequence. For example, after leveling off at the assigned altitude we may have a series of basic check offs checking a series of different functions. This is performed as routine on every flight. Assume there is failure of pressurization and the cabin altitude continues to climb with resulting hypoxia. Those series of steps that you normally perform become difficult to execute or you just forgot the normal sequence. This is another form of apraxia known as "sequential apraxia". You may have enough oxygen to remember the functions of each task individually but you can't remember which one should go first.

To summarize, a well-trained crew that knows each other well will easily identify when a fellow crew member begins to execute established tasks out of sequence and this should raise awareness "that something isn't right".

Loss of Short-Term Memory

Inability to repeat immediate sequence of numbers and letters during routine transmission

Verbal communication in the cockpit possesses a unique set of characteristics. Flight numbers and letters for identification with changing altitudes and velocities are expressed succinctly with minimal "verbal waste". Minimizing the number of words and condensing the data in a concise fashion is essential for proper and efficient transmission. New instructions are easily picked up and repeated for verification. This is severely impaired during hypoxia. Repeating up to seven words and numbers in a rapid fashion becomes difficult and demonstrates the loss of short-term memory that is essential for communication. Most of us can easily remember a phone number without having to write it down if we plan to immediately dial that number. During hypoxia we are unable to recall that sequence of information. This can be a valuable sign of early detection when one of the crewmembers begins to "repeat" basic verbal communications and indicates again "something isn't right".

Inability to Approximate Passage of Time

An interesting phenomenon that occurs during hypoxia is the inability to approximate how much time has transpired from a recent event. For example, from the time of takeoff till leveling at the desired altitude may be 15 minutes. We don't need to measure time with our watches to closely approximate the amount of transpired time between basic events. During hypoxia this is impaired. A 15-minute ascent may seem like five minutes or in some cases even 30 minutes. This is relevant as it impairs our ability to approximate real time events. During hypoxia time seems "compressed or shortened". This may lead to inappropriate approximations for decision-making. During hypoxia time is vital. With each passing minute the effects on the crew are accelerating. Waiting even one additional minute may lead to complete incapacitation.

Sudden Giddiness

One of the most important findings of hypoxia is the absence of anxiety or pain. As the oxygen levels drop, we do not feel short of breath, discomfort or even concern that something is gravely wrong. As a matter of fact, we may suddenly start to feel giddy and just start kidding around with the crew almost as if we had a drink at the local pub. We begin to behave recklessly and disregard important issues.

Sudden Lackadaisical Attitude

A no caring attitude may develop where we just don't seem to care much about what is going on in the cockpit. For example, we might just want to stare out the window and look at all the pretty clouds while the other crewmembers are busy with their flight plan. When pressed, the pilot may answer "whatever", "not right now", "why don't you do it", etc. This should represent a serious change in the pilot's behavior and raise the suspicion that "something isn't right".

Sudden Reclusive Characteristics

Probably the most common change in personality associated with hypoxia is sudden reclusive like characteristics "quiet like state". Some pilots are quite verbal and energetic. During hypoxia their mood will change and suddenly they become quiet as if lost in thought. They don't want to respond and engage in conversation and appear as if they want to be left alone. In rare cases, they become angry and defiant when pressured to perform their normal duties. Any sudden change in mood out of characteristic of the individual should raise the awareness that "something isn't right".

PART III: REMOTE SENSING AND INSTRUMENTATION

Concepts of Remote Sensing

The PMC-Turbo Instrument Suite

PoSSUM Suborbital Tomography Instrument Suite

CONCEPTS OF REMOTE SENSING

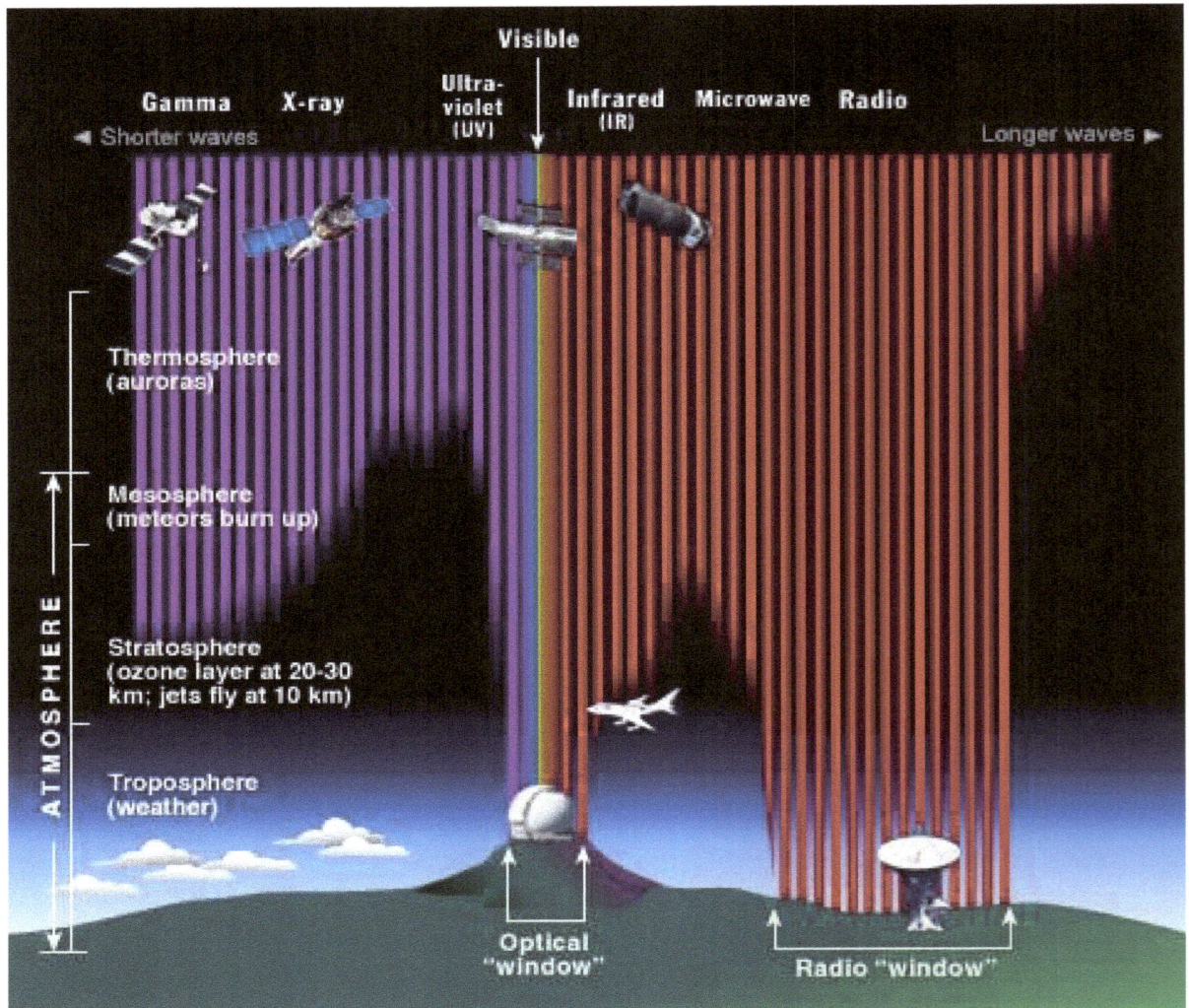

Module Objectives

> Become familiar with the fundamental techniques of remote sensing
> Understand the underlying physics behind radiative transfer
> Understand the different ways that light is scattered in the atmosphere

Contributing Author:

Jason Reimuller, Ph.D.

Fundamentals of Remote Sensing

All scientific knowledge derives from two sources of observables: *in-situ* data and *remote-sensed* data. In-situ data is data derived from direct interrogation of the subject being studied: a social scientist may go to particular neighborhood to interview subjects; a botanist might measure distributions of a certain species of plant in the field. Some in-situ data is obtained by use of instruments that are placed in direct physical contact with the subject matter, such as a thermometer might be used to study air temperature or an anemometer might be used to measure wind speed. These in-situ measuring instruments are collectively termed *transducers*.

Remote sensed data, on the other hand, is obtained from some remote distance. Science would indeed be very limited if we could only derive conclusions about subject matter that could be directly measured, and remote sensed data derives information of a subject matter based largely upon the radiative qualities it exhibits when exposed to a remote sensing instrument.

The first data collection of information that would later be regarded as 'remote-sensed' came from aerial photography, and the science that grew from the interpretations of these images was termed photogrammetry. Yet as more and more technologies became available and subject objects could be viewed in different and specific wavelengths, a broader definition was needed. Finally, in 1983, the American Society for Photogrammetry and Remote Sensing (ASPRS) published a formal definition of remote sensing as follows:

> *Remote Sensing is the measurement or acquisition of information of some property of an object or phenomenon, by a recording device that is not in physical or intimate contact with the object or phenomena under study.*

Some have proposed (e.g. Colwell 1984) a more simplistic definition such as "the acquiring of data about an object without touching it", yet this definition excluded little. Regardless, remote-sensing is a tool that follows the scientific method: 1) stating the problem, 2) forming the research hypothesis, 3) observing the subject, 4) interpreting the data, and 5) drawing conclusions. In the context of remote sensing, a problem is stated, data is collected, the data is processed and analyzed, and data products are produced from which conclusions may be drawn.

Remote-Sensed Data Collection

Remote sensing techniques are produced using a *sensor* generally sensitive to electromagnetic radiation (EMR) and can be categorized as either *passive* or active systems. Passive systems are unobtrusive and do not disturb the object of interest, relying instead on the natural patterns of scattering and absorption of light from an external light source (e.g. the Sun). On the other hand, active systems emit their own electromagnetic radiation at wavelengths best calibrated to study a specific characteristic of the object being studied. These systems include lidar systems, radar systems, and sonar systems and they can give us valuable range data as well as other data that can only be produced by having a very precise knowledge of the incident radiation that the subject object is exposed to.

Remote sensing systems can collect analog data, such as through a frame camera or video camera using film, or through digital methods, such as with a digital camera, multispectral or hyperspectral scanners, or through multispectral or hyperspectral linear or area arrays. The amount of electromagnetic radiance, L, recorded within the field-of-view of a remote sensing system is a function of several variables:

$$L = f(\lambda, s_{x,y,z}, t, \Theta, P, \Omega)$$

Where:

λ = wavelength of radiation being studied
$s_{x,y,z}$ = location of the pixel and size of the pixel
t = time of exposure of the pixel to the radiation
Θ = the angle between the radiation source (e.g. the Sun), the target of interest (the NLC), and the sensor
P = polarization of back-scattered energy recorded by the sensor
Ω = precision at which the data are recorded by the sensor system

Spectral Resolution

Various characteristics may be determined through these data, including spectral, spatial, and temporal resolution, as well as radiometric, angular, and polarization information. The *spectral resolution* of an instrument is defined as the number and bandwidth in the electromagnetic spectrum that the sensor is sensitive to. Multispectral instruments record information in multiple bands of the electromagnetic spectrum. A typical multispectral system might record several bands throughout the visible and infrared spectrum at bandwidths that might range from coarse (~300 nm) to fine (~100 nm). Bandwidth is typically defined in terms of the Full Width at Half Maximum (FWHM) of an idealized Gaussian shape.

Hyperspectral imagers are capable of recording data in hundreds of spectral bands. The Airborne Visible and Infrared Imaging Spectrometer (AVIRIS) has 224 bands spaced at intervals of 10nm spanning a spectral range of 400nm (blue region of visible) to 2500 nm. Improvements in technology in recent years has led to the introduction of *ultraspectral imagers*, which collect data in many hundreds of bands.

In all cases, bands are generally chosen to obtain information by maximizing the contrast between the object of interest and the background, and careful selection of spectral bandwidths increase the probability that desired information may be extracted from the images.

Spatial Resolution

The *spatial resolution* of an imager is determined by the line array sensor (commonly termed a 'pushbroom' sensor) or the area sensor (commonly, a Charged Coupled Device, or CCD). Spatial resolution is the smallest angular resolution between two objects that can be resolved by the remote sensing system. In other words, it is the size of the smallest possible feature that can be detected. Spatial resolution is defined by the size of a pixel in a CCD projected onto the object being studied, and is thus a function of the distance of the sensor from the object and the *Instantaneous Field of*

View (IFOV) of the lens system coupled to the CCD; the wider the IFOV or farther the distance from sensor to object, the larger the spatial resolution will be.

Digital Globe's QuickBird satellite, for example, has a spatial resolution of 61cm for its panchromatic band (panchromatic images encompass all visible wavelengths, such as images obtained from a typical commercial digital camera). This means that ground imagery obtained from typical orbital altitudes may be resolved into 61cm x 61cm pixels. When observing terrestrial features, aircraft imagers intrinsically have greater spatial resolution because they take images much closer to the ground than satellites do. PoSSUM's suborbital flights will be able to obtain imagery of noctilucent cloud structures with greatly enhanced spatial resolution that could be obtained from aircraft, balloon, or satellite due to the proximity of the imaging system as the suborbital spacecraft fly through and penetrate the cloud layer.

It is important to note that pixel size and spatial resolution are not interchangeable. It is possible to display an image with a pixel size different from the resolution. Images may have their pixels binned to represent larger areas and to reduce the file size of images being stored; the image may have a reduced spatial resolution even though the original spatial resolution of the sensor that collected the imagery remains the same.

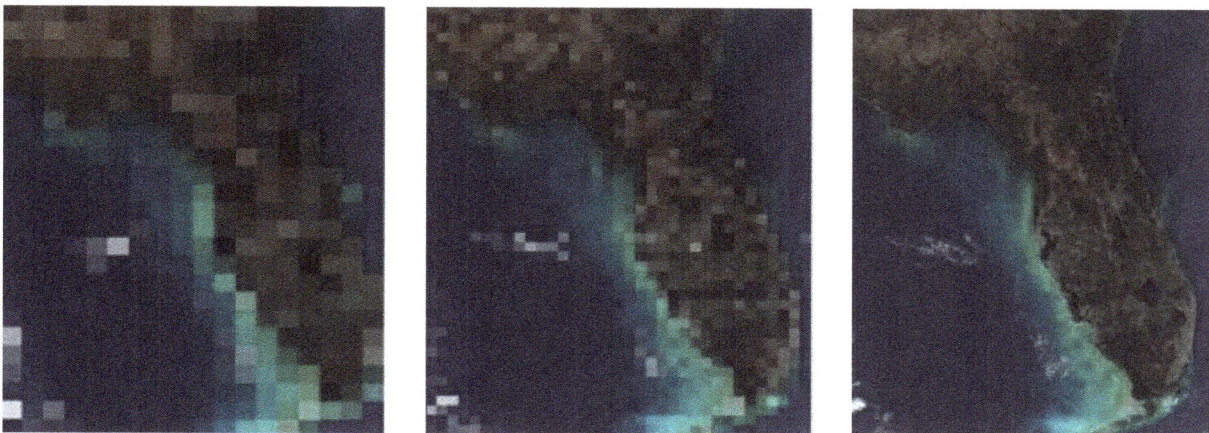

Figure 49. Images of Florida obtained at varying Spatial Resolution

Temporal Resolution

The *temporal resolution* of a sensor generally refers to how often that sensor records imagery of the area being studied. For satellites, the temporal resolution is determined by the groundtrack. Typically, satellites in Low Earth Orbit (LEO) overlap terrestrial regions every two to three weeks, so the temporal resolution is defined by this parameter. *Off-nadir pointing* (pointing at some angle that is not directly below the spacecraft) allows a satellite to improve temporal resolution of a region of interest; however, off-nadir oblique viewing also introduces bidirectional reflectance distribution function (BDRF) errors, which can introduce errors in the spatial and radiometric resolution of the object being studied. Temporal resolution of aircraft remote sensing campaigns can be defined by the aircraft flight trajectory.

Earth-observing satellites in geostationary orbit (GEO) maintain a fixed point over the surface of the Earth. Their temporal resolution is very high and defined only by the rate at which the sensor may image the region of interest and transmit the data. The NASA GOES satellite is very effective at monitoring time-varying weather phenomena such as hurricanes, but due to the greatly increased distance of a satellite in GEO (35,786km) versus a satellite in LEO (as low as 250 km) the spatial resolution is significantly reduced. For example, GOES produces images at 4 km to 8 km spatial resolution whereas satellites in LEO can now produce images at sub-meter resolution.

A great benefit of using balloons and aircraft to produce remote sensed image products are their ability to produce images at both high spatial and temporal resolution. Aircraft, especially helicopters, may observe a region of interest from a relatively stationary position for a period of time limited only by the aircraft's endurance (fuel capacity) or operational constraints (e.g. safe weather conditions). Balloons may stay aloft for many weeks, though high-altitude winds cause drift of the sensor platform. Balloon payloads currently being developed by Project PoSSUM produce images at intervals between 5 and 30 seconds. Since the objects being studied (noctilucent cloud structures) are wide-spread and far, they have a temporal resolution at these intervals. The PoSSUM suborbital tomography experiment will take images at a rate in excess of 60 frames per second and will produce image sequences at very high spatial and temporal resolution, but for a period of only five minutes.

Radiometric Resolution

Digital images, or analog images that have been subsequently subdivided into pixels with each pixel assigned a digital number representing its relative brightness. The computer displays each digital value as different brightness levels. Digital imagers record electromagnetic energy as an array of numbers in digital format, as shown in Figure 50.

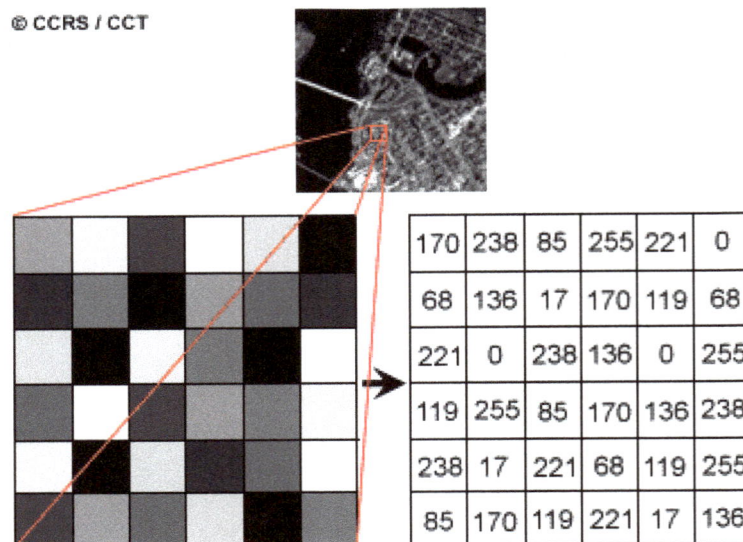

Figure 50. 8-bit radiometric resolution of airborne imagery (credit: NRC)

Radiometric resolution is defined as the sensitivity of a remote sensing detector to differences in signal strength as it records the radiant flux reflected, emitted, or back-scattered. Often referred to as *quantization*, high radiometric resolution increases the probability that the object being studied will be remotely sensed more accurately. Earth-observing sensors used in the 1980's typically had an 8-bit radiometric resolution, meaning that each pixel could differentiate intensities received into one of 256 bins. Modern sensor systems have radiometric resolution in excess of 12-bit, differentiating received intensities into one of 4096 bins. A comparison of coarse versus fine radiometric resolution of airborne imagery over Ottawa is shown in Figure 51.

Figure 51. Comparison of coarse versus fine radiometric resolution (credit: NRC)

Polarization

Polarization characteristic of electromagnetic energy may also be recorded by sensor systems. Incident sunlight is weakly polarized, but when this light strikes a non-metal object, it becomes largely or entirely depolarized as the incident energy is scattered differentially. Generally, the smoother the surface of the object being observed, the greater the degree of polarization the reflected light exhibits.

Noctilucent clouds maintain polarization, and PoSSUM flights use linear polarizers to record light at various angles. The degree of polarization may be observed by comparing the intensities received by two different but otherwise identical camera systems, each coupled with a linear polarizer cross-polarized with respect to the other. The difference of received intensity of the two camera systems is the magnitude of polarization and can reveal important information about the particle sizes and shapes being studied.

Angular Resolution

Angular data is recorded by remote sensing instruments, and *angular resolution* is an instrument's ability to record the angle at which the data was taken relative to an illumination source. Of each exposed pixel, the relative angle of the sensor and the illuminating source is recorded. In the case of a passive system, such as the imagers controlled by the PoSSUMCam system, the illumination source is the sun. The location of the orbital or suborbital remote sensing system is recorded relative to its azimuthal and zenith angles.

Angular data recorded from observations is essential to determining the particle characteristics off of which the light is scattered. Particle shapes can be inferred through the careful recording of the relative angles of the scattering. The CIPS imager on the AIM satellite used a range of scattering angles available from an orbiting spacecraft to estimate the sizes of the NLC particles it was observing.

Angular information is essential to the use of remote sensed data used in photogrammetric applications. Stereoscopic image analysis assumes that coordinated imagery has been obtained from two different locations, observing the same region of interest from two different angles. This process introduces stereoscopic parallax in the same manner that the human eye can determine depth through use of both eyes. If we close one eye, we lose that ability; however, we still infer these distances because our brain is able to assume properties based on years of previous observations.

Image Processing

Remote sensed imagery data are analyzed using analog (or visual) image processing techniques, or digital image processing techniques. Digital imagery enables greater improvements in discriminating colors or intensities than would be possible with the eye alone, and data obtained in this way is easier to quantify and geo-locate. The video camera system, wide-field visible imager, and infrared imagers that will be used on PoSSUM campaigns all record data in RAW digital format, meaning that the files have not been compressed in any way to reduce file size. RAW formats consume much memory per image, but the data recorded is available in a 'pure' format and post-processing of those data may be performed after the mission.

The major steps of the modern post-processing of digital imagery include: 1) radiometric corrections, 2) geometric corrections, 3) image enhancement, 4) photogrammetry, 5) parametric and non-parametric information extraction, 6) hyperspectral information extraction, and 7) change detection.

Step 1: Radiometric Correction

The first step of post-processing of digital data is the extraction of noise or error introduced by the sensor system, such as the unwanted introduction of scattered light off of the atmosphere by applying and subtracting atmospheric scattering models. Electronic noise created by the imager system can also be removed using data sets obtained from images taken devoid of external light. Non-zero results on the imaging CCD result from erroneous charges introduced in the CCD by the imagery system itself, and can be subtracted.

Step 2: Geometric Correction

Through data obtained by a Global Positioning System (GPS) and an accurate knowledge of the pointing parameters of the imagery system such as is possible through use of an Inertial Reference Unit (IRU), we can geo-locate each pixel obtained through the imagery system. These data are important for analyzing each pixel in the data set in the context of the exact object or region that it represents.

Step 3: Image Enhancement

Images may be digitally enhanced to identify information that might otherwise be overlooked. Remote sensed data may be linearly or non-linearly transformed so that the information is mire highly correlated with real-world phenomena.

Step 4: Photogrammetric Data Extraction

Angular data recorded through synchronous observations may be used to extract data on the vertical structures of NLC cloud layers in the same way that accurate Digital Elevation Models (DEMs) are constructed through triangulated aerial photogrammetry.

Step 5: Hyperspectral Data Extraction

Imagers that take data in spectral bands rely on special software to reduce the dimensionality of the multiple data sets while retaining the integrity of the original data. Through these methods, analysts are able to identify the type and proportion of the different elements within each pixel element.

Terrestrial remote sensing has the advantage that accurate collateral information exists, such as soil maps, topographic maps, and hydrological maps. The integration of all these data sets is often referred to as a *Geographical Information System* (GIS). Since atmospheric phenomena are highly variable, it is harder to accurately geolocate pixels. Observing a known star or set of stars, however, is a good means to geolocate imagery of NLCs. *Image maps* refer to imagery data obtained that has been geolocated and placed on a map projection, though it is essential to report all sources of error that have been introduced through the data collection process.

Principles of Radiative Transfer

Electromagnetic Radiation

Electromagnetic radiation consists of an electrical field, which varies in magnitude in a direction perpendicular to the direction in which the radiation is traveling, and a magnetic field oriented at right angles to the electrical field. Both these fields travel at the speed of light (c) in the direction of propagation, as shown in Figure 52.

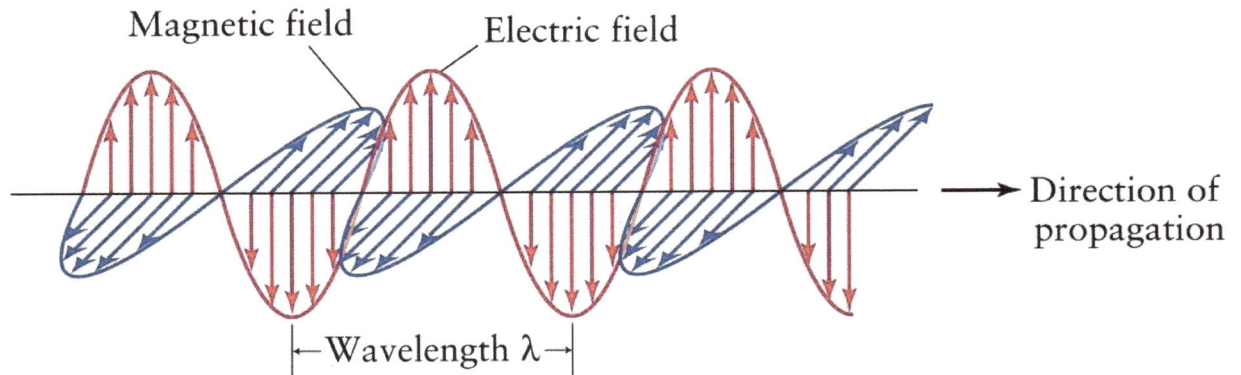

Figure 52. Electromagnetic Field Propagation

The *wavelength* is the length of one wave cycle, which can be measured as the distance between successive wave crests and denoted as lambda (λ) and measured in meters (m) or more commonly nanometers (nm, 10^{-9} meters) when describing radiation in the visible or infrared region.

Frequency refers to the number of cycles of a wave passing a fixed point per unit of time. Frequency is normally measured in *hertz* (Hz), equivalent to one cycle per second, and various multiples of hertz. Wavelength and frequency are inversely related to each other and expressed by the following formula:

$$c = \lambda v$$

Where λ = wavelength (in meters), v = frequency (in cycles per second, or Hz), and c= speed of light (3×10^8 m/s)

The electromagnetic spectrum ranges from the shorter and higher-energy wavelengths, such as gamma rays and x-rays, to longer wavelengths such as microwaves and broadcast radio waves. Remote sensing techniques are generally applicable for frequencies spanning from the *ultraviolet or UV* portion of the spectrum (of wavelengths shorter than what our eyes can perceive as blue, and thus invisible to the naked eye) to the infrared (of wavelengths longer than what our eyes can perceive as red, and thus invisible to the naked eye). An illustration of the regions of the electromagnetic spectrum of most interest for remote sensing applications is shown in Figure 53.

The visible spectrum spans a range of approximately 0.4 to 0.7nm. Specifically,

Violet: 0.400 - 0.446 μm
Blue: 0.446 - 0.500 μm
Green: 0.500 - 0.578 μm
Yellow: 0.578 - 0.592 μm
Orange: 0.592 - 0.620 μm
Red: 0.620 - 0.700 μm

The *infrared region* spans a range of 0.7 μm to 100 μm - more than 100 times as wide as what our eyes may perceive as visible light - and can be divided into two categories based on their radiation properties - the *reflected IR*, and the emitted or *thermal IR*. Radiation in the reflected IR region is used for remote sensing purposes in ways very similar to radiation in the visible portion. The reflected IR covers wavelengths from approximately 0.7 µm to 3.0 µm. The thermal IR region is quite different than the visible and reflected IR portions, as this energy is essentially the radiation that is emitted from the Earth's surface in the form of heat. The thermal IR covers wavelengths from approximately 3.0 µm to 100 µm.

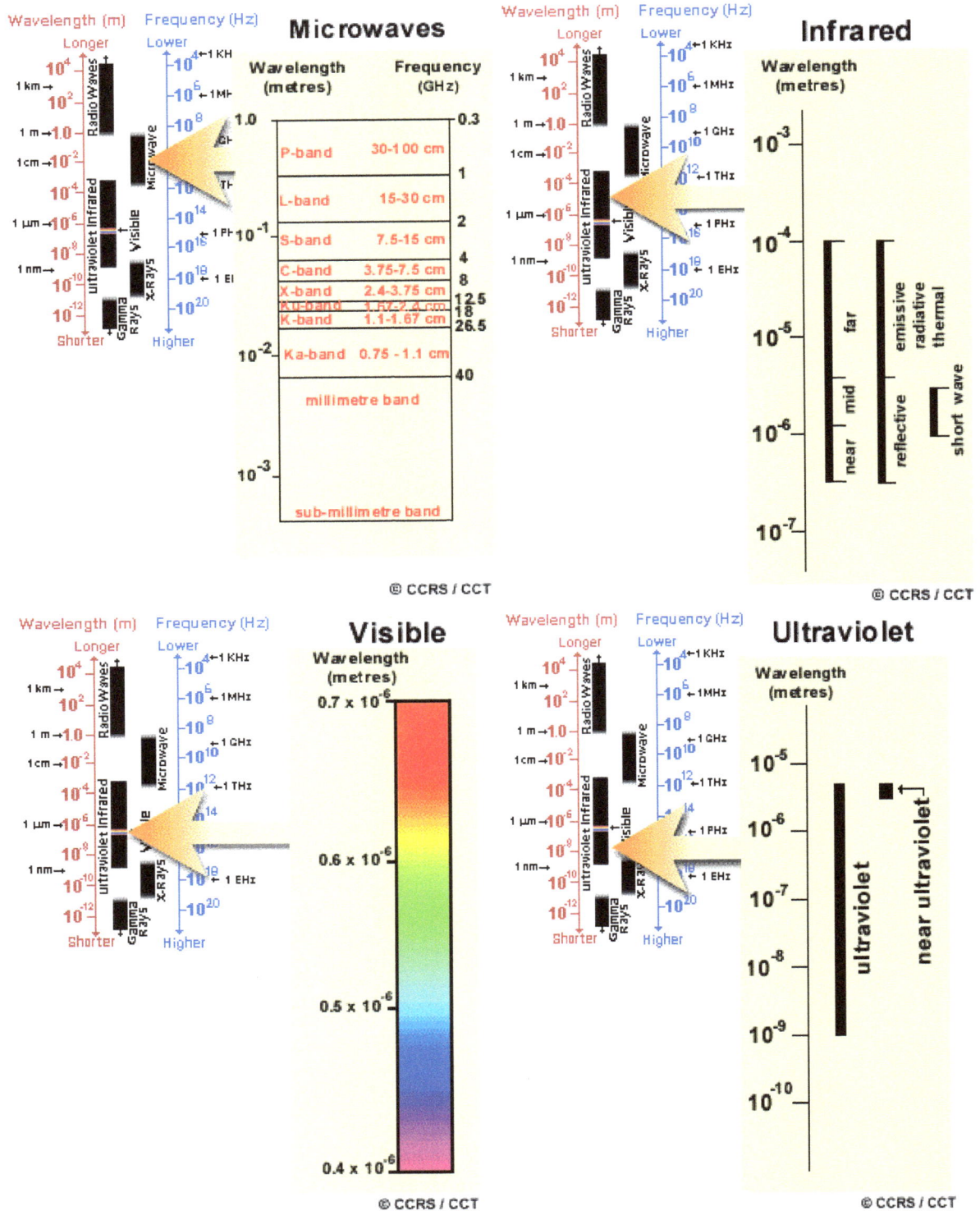

Figure 53. The Electromagnetic Spectrum (credit: NRC)

Blackbody Radiation

All objects emit electromagnetic energy. Though the sun emits the most radiation, all matter gives off radiation in proportion to its temperature. An ideal object that absorbs and radiates energy at its optimal rate per unit area and per wavelength is called a *blackbody*, and this idealized structure emits radiation (M_λ) in proportion to the fourth power of its absolute temperature (T), measured in Kelvins. This is called the *Stefan-Boltzmann law* and is expressed as:

$$M_\lambda = \sigma T^4$$

Where σ is the *Stefan-Boltzmann constant* with a value of 5.6697 x 10^{-8} Wm^{-2}K^{-4}. The greater the temperature of the sun or the object on the Earth, the greater the radiant energy will be produced, and at 6000K, the Sun produces far more radiation than the Earth.

The dominant wavelength, or wavelength exhibiting the maximum irradiance, is defined by *Wein's displacement law*, expressed as:

$$\lambda_{max} = \frac{k}{T}$$

Where k=2898 $\mu m\ K$. Although the Sun has a dominant wavelength ner 480 nm (in the green part of the visible spectrum), it produces radiation panning from the longest radio waves to the shortest, high-frequency gamma rays; however, as shown in Figure 55, as the temperature of an object increases, the dominant wavelength shifts towards the shorter and higher energy wavelengths of the spectrum.

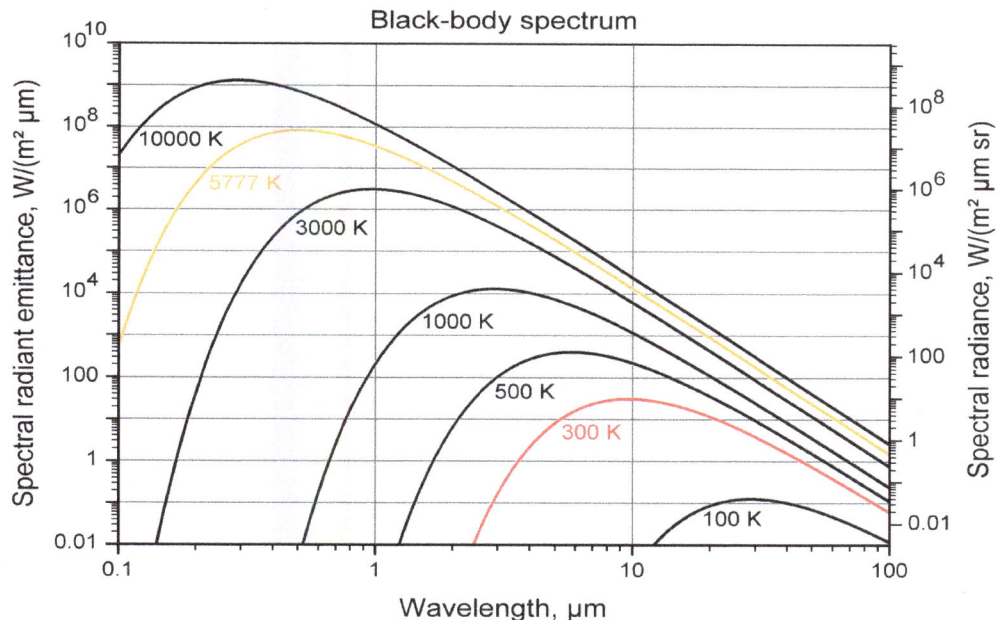

Figure 54. Blackbody radiation spectra (credit: Creative Commons)

One can observe this as coils of a stove heat; as the coils get hotter, they will eventually exhibit visible light. If the heating continues, yellow hues are introduced and eventually the color will turn 'white hot' as the whole visible spectrum is displayed.

Atmospheric Scattering

Scattering occurs when particles or large gas molecules present in the atmosphere interact with and cause the electromagnetic radiation to be redirected from its original path. How much scattering takes place depends on several factors including the wavelength of the radiation, the abundance of particles or gases, and the distance the radiation travels through the atmosphere.

Light scatters off of particles or molecules in the atmosphere in one of three methods: Nonselective (geometric) scattering for the largest particles, Rayleigh scattering for smallest molecules, and Mie scattering for particle sizes in between. These relations all depend upon the relationship between the incident wavelength and the diameter of the particle.

Nonselective Scattering

For larger aerosols suspended in the air, the laws of reflection and refraction can be used to approximate a solution using simple geometric methods. This type of scattering is termed *non-selective scattering* and derives its name from the fact that all wavelengths are scattered about equally. Water droplets and large dust particles can cause this type of scattering, and this scattering causes fog and clouds to appear white to our eyes because all wavelengths of light are scattered in approximately equal quantities, which will average to white light.

Rayleigh Scattering

Rayleigh scattering causes shorter wavelengths of energy to be scattered much more than longer wavelengths. Rayleigh scattering is the dominant scattering mechanism in the upper atmosphere. The fact that the sky appears "blue" during the day is because of this phenomenon. As sunlight passes through the atmosphere, the shorter wavelengths (i.e. blue) of the visible spectrum are scattered more than the other (longer) visible wavelengths. At sunrise and sunset, the light has to travel farther through the atmosphere than at midday and the scattering of the shorter wavelengths is more complete; this leaves a greater proportion of the longer wavelengths to penetrate the atmosphere.

For atmospheric particles whose diameter is much smaller than the wavelength of the incident light, such as atmospheric molecules, a dipole approximation may be made. Assuming the particles to be spherical, the intensity, *I*, of the scattered radiation is given by

$$I = I_0 \left(\frac{1 + \cos^2 \theta}{2R^2} \right) \left(\frac{2\pi}{\lambda} \right)^4 \left(\frac{n^2 - 1}{n^2 + 2} \right)^2 \left(\frac{d}{2} \right)^6,$$

where I_0 is the light intensity before the interaction with the particle, R is the distance between the particle and the observer, θ is the scattering angle, n is the refractive index of the particle, and d is the diameter of the particle.

It can be seen from the above equation that Rayleigh scattering is strongly dependent upon the size of the particle and the wavelengths. The intensity of the Rayleigh scattered radiation increases rapidly as the ratio of particle size to wavelength increases. Furthermore, the intensity of Rayleigh scattered radiation is identical in the forward and reverse directions.

<u>Mie Scattering</u>

The Rayleigh scattering model breaks down when the particle size becomes larger than around 10% of the wavelength of the incident radiation. For objects whose size is similar to the wavelength, a more exact solution is necessary. This solution was first calculated by the German physicist, Gustav Mie.

Mie scattering is roughly independent of wavelength and it is larger in the forward direction than in the reverse direction. The greater the particle size, the more of the light is scattered in the forward direction. Cloud particles, such as those in NLC, are modeled using Mie scattering models. In contrast to the Rayleigh scattering demonstrated by atmospheric molecules, the water droplets which make up clouds are of a comparable size to the wavelengths in visible light, and the scattering is described by Mie's model rather than that of Rayleigh. For this reason, clouds that contain larger droplets of water exhibit isotropic scattering, or scatter light equally in all directions, appear white or grey.

Figure 55 shows the Mie scattering profiles for three spherical particles of varying size. Note for all sizes there is a strong preference for forward scattering. As the particle grows in size, more lobes are introduced. If we understand the scattering geometry around a particle, we can infer its size. This concept is used by the Cloud Imaging and Particle Size Experiment (CIPS) on the AIM satellite. As the satellite passes over the NLC layers, it images the same area of cloud over a range of seven scattering angles. From these data, average particle size may be calculated.

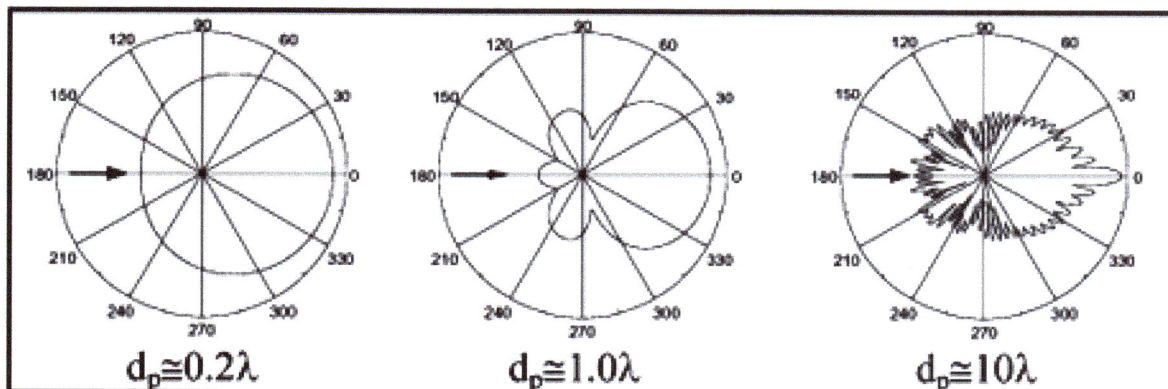

Figure 55. Mie scatter diagrams for different size seeding particles (dp is particle size)

A key assumption is that the Mie model assumes that particles are spherical. However, it is commonly believed that NLC particles are non-spherical. The LiDAR experiment being developed by IIAS will test the assumption that NLC particles are spherical by using a LiDAR depolarization method.

An online Mie calculator can be found here: http://omlc.org/calc/mie_calc.html

Atmospheric Refraction and Reflectance

Refraction occurs when electromagnetic radiation such as light encounters a change of density in the medium through which it propagates. Though the speed of light in a vacuum is a constant 3×10^8 m/s, the speed of light through other mediums will vary. The *index of refraction (n)* is a measure of the optical density of a substance and can be computed as the ratio of the speed of light in a vacuum, c, to the speed of light in the media in question:

$$n = \frac{c}{c_n}$$

The index of refraction for the atmosphere (in the evenly-mixed homosphere) is 1.0002926 and the index of refraction in water is 1.33. Light will always travel slower through a medium than it will through a vacuum.

Snell's Law states that for a given frequency of light, the product of the index of refraction and the sine of the angle between the incident ray and a line normal to the interface is constant:

$$n_1 sin\theta_1 = n_2 sin\theta_2$$

The amount of refraction is thus calculated if one knows the indices of refraction of the two materials as well as the angle of the incident ray of light.

Unlike refraction, *reflection* is the process where the incident radiation 'bounces off' of an object that exhibits threshold 'smoothness' of the surface. In this case, the incident radiation and the reflected radiation are coplanar and the angle of incidence and angle of reflection are approximately equal. Specular reflection occurs when the surface from which the radiation reflects is essentially smooth, such as light bouncing off of a smooth lake surface or off a shiny metallic surface. However, if the surface has a large surface height relative to the incident wavelength (if it is 'rougher'), then the reflected rays may be dispersed in many different directions. This sort of reflection, termed diffuse reflection, does not produce a mirrored image but rather a diffuse radiation. This is often seen in white materials that reflect light in a diffuse manner.

Much of the incident radiation from the sun is reflected from the top of clouds in the atmosphere. This diffuse reflection produces a white color to clouds, and a significant percentage of this incoming energy is radiated back to space. Noctilucent clouds are very thin and will exhibit scattering and absorption properties described later in this section. The fact that they do not appear to be diffuse white in color, like denser tropospheric clouds, show that diffuse reflections are not the driving interaction of incoming sunlight.

Atmospheric Absorption

Absorption is the other main mechanism at work when electromagnetic radiation interacts with the atmosphere. In contrast to scattering, this phenomenon causes molecules in the atmosphere to absorb energy at various wavelengths. Ozone, carbon dioxide, and water vapour are the three main atmospheric constituents which absorb radiation. Ozone, which exists predominantly in the stratosphere, absorbs harmful ultraviolet radiation from the sun. Carbon dioxide absorbs radiation strongly in the far infrared portion of the spectrum and thus traps heat within the atmosphere. For this reason, it is commonly termed a *greenhouse gas.* Lastly, water vapour in the atmosphere absorbs much of the incoming longwave infrared and shortwave microwave radiation (between 22μm and 1m). Knowledge of where and how these gases absorb electromagnetic energy helps us determine where (in the spectrum) we can "look", since areas of the spectrum which are not severely influenced by atmospheric absorption are called *atmospheric windows*.

The PMC-Turbo Instrument Suite

Module Objectives

➢ Understand the concepts of the PMC-Turbo Mission
➢ Understand the camera systems integral to the PMC-Turbo mission

Contributing Authors:

Jason Reimuller, Ph.D.
Dave Fritts, Ph.D

Origins of the PMC-Turbo Experiment

PMC-Turbo is a $1.4M NASA-funded experiment led by PoSSUM Chief Scientist Dr. Dave Fritts. The PMC-Turbo mission flew a set of seven camera systems on a high-altitude, uncrewed balloon the Arctic polar vortex in July 2017. Project PoSSUM works in partnership with GATS, Columbia University, and Integrated Spaceflight Services by contributing instrument development, testing, and educational outreach expertise through contributions of PoSSUM graduates. This novel experiment studied atmospheric dynamics that can only be viewed in exquisite detail through very high-resolution imagery techniques.

The images obtained during the campaign are being used to analyze how waves generated at lower altitudes dissipate via instability and turbulence processes. These processes account for the deposition of significant energy and momentum transported by the waves from lower altitudes. They also play key roles in weather and climate throughout the atmosphere, but are poorly understood at present. Imaging of noctilucent clouds provides a unique window on these processes that is not available at any other altitude. Thus, this largely inaccessible region has the potential to educate us about important processes occurring throughout the atmosphere.

The PMC-Turbo Mission

PMC-Turbo launched from Kiruna, Sweden in Summer 2017 at a time after the summertime polar vortex became stable. This vortex guided the balloon from Kiruna westward to Nunavut, Canada at an altitude of approximately 38 km for a six-day flight that captured 125TB of high-resolution imagery data. Coupled with temperature measurements obtained by a lidar payload provided by the German Space Agency (DLR), PMC-Turbo produced unprecedented insight into the complicated dynamics of our mesosphere.

The PMC-Turbo payload was fully tested at NASA's Columbia Scientific Balloon Facility (CSBF) in Palestine, TX approximately four months prior to its launch window. Observations were made continuously in the anti-sun direction, observing noctilucent cloud structures that drift through the field-of-view. The Summer generally guarantees good PMC brightness and high sensitivity by the imagers to gravity waves, instability, and turbulence structures at spatial scales extending from about 100 km down to the smallest turbulence inner scales near 10-20 m. Processed images were provided to the NASA Space Physics Data Facility (SPDF) for archiving and are now available through public archives (e.g., the CEDAR database at NCAR). Analyses of PMC-Turbo data continues to be performed on the Yeti computer cluster at Columbia University and the Penguin computer cluster at GATS, using the existing images from the 2012 EBEX flight as a baseline of automated algorithms for ranking images based on the likelihood that they contain noctilucent cloud structures.

PMC-Turbo Payload Design

The PMC-Turbo payload fits within a customized gondola that supports a camera array, a LiDAR system, power, communications, control systems, and associated interfacing software. The novel camera array to be used in the campaign will employ seven scientific camera systems to be configured for sustained operations in the Antarctic stratosphere. The camera enclosures were constructed and tested by Integrated Spaceflight Services and Columbia University and then shipped to NASA's Columbia Scientific Balloon Facility in Palestine, TX for testing and integration. The gondola architecture is shown in Figure 56.

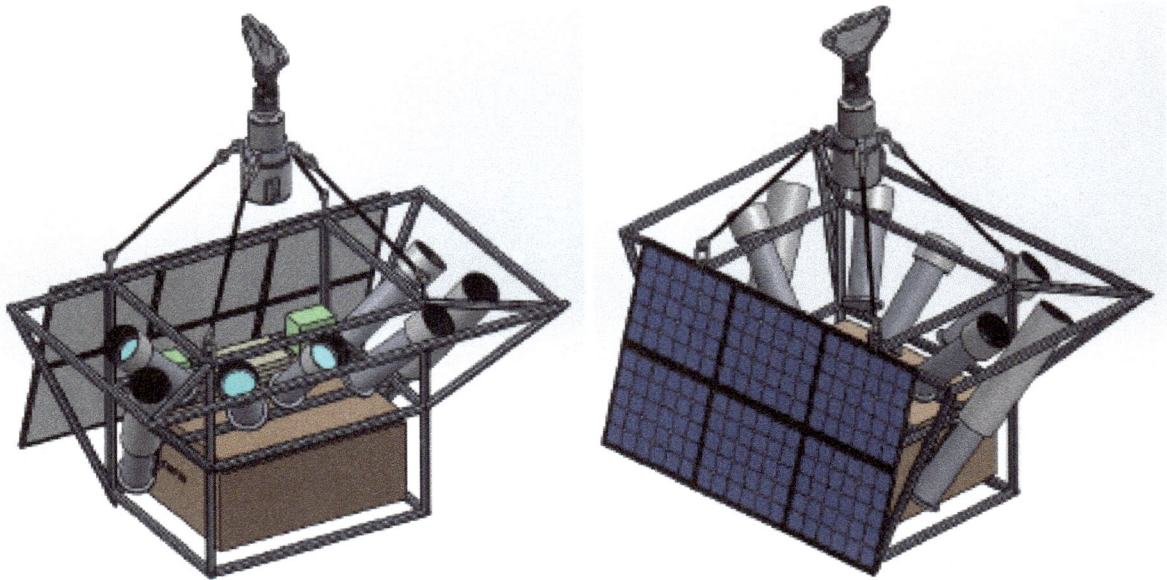

Figure 56. PMC-Turbo gondola, camera payload, and solar cells. The pointing system would maintain the FOVs in the anti-sun direction within 2° (credit Dave Fritts)

The camera systems leverage heritage obtained by Columbia University's 'EBEX' camera systems, launched in 2012 to study the Cosmic Microwave Background in a similar balloon mission. It was through serendipity that Noctilucent cloud structures were observed in this mission, revealing for the first time the fine signatures of noctilucent cloud small scale turbulence, driving the rationale for the PMC-Turbo mission.

The seven PMC-Turbo cameras are deployed in two configurations as narrow and wide FOV independent camera units, 'NFICU' and 'WFICU' respectively. The cameras are 16-megapixel Allied Vision cameras with 4864 x 3232 resolution and capable of a 3.5 frame per second sustainable framerate or 'burst capture' imagery techniques that may be used to obtain a rapid sequence of images. Four WFICUs are oriented with the long axes oriented vertically and FOVs of 39.6 x 26.9 degrees, yielding horizontal resolution at the PMC layer (~82 km) varying from about 4 to 15m per pixel from the smallest to the largest angle from zenith. Three NFICUs aligned horizontally would have 10 x 15.2 degrees, FOVs and yield resolution of less than 3m per pixel (Figure 57). The power system, flight control computers, and Support Instrumentation Package (SIP) leverage NASA data to be relayed to the ground during flight to evaluate performance and focus. This design thus provides significant redundancy and ensures against failure of most individual system components. A system-level schematic is provided in Figure 58.

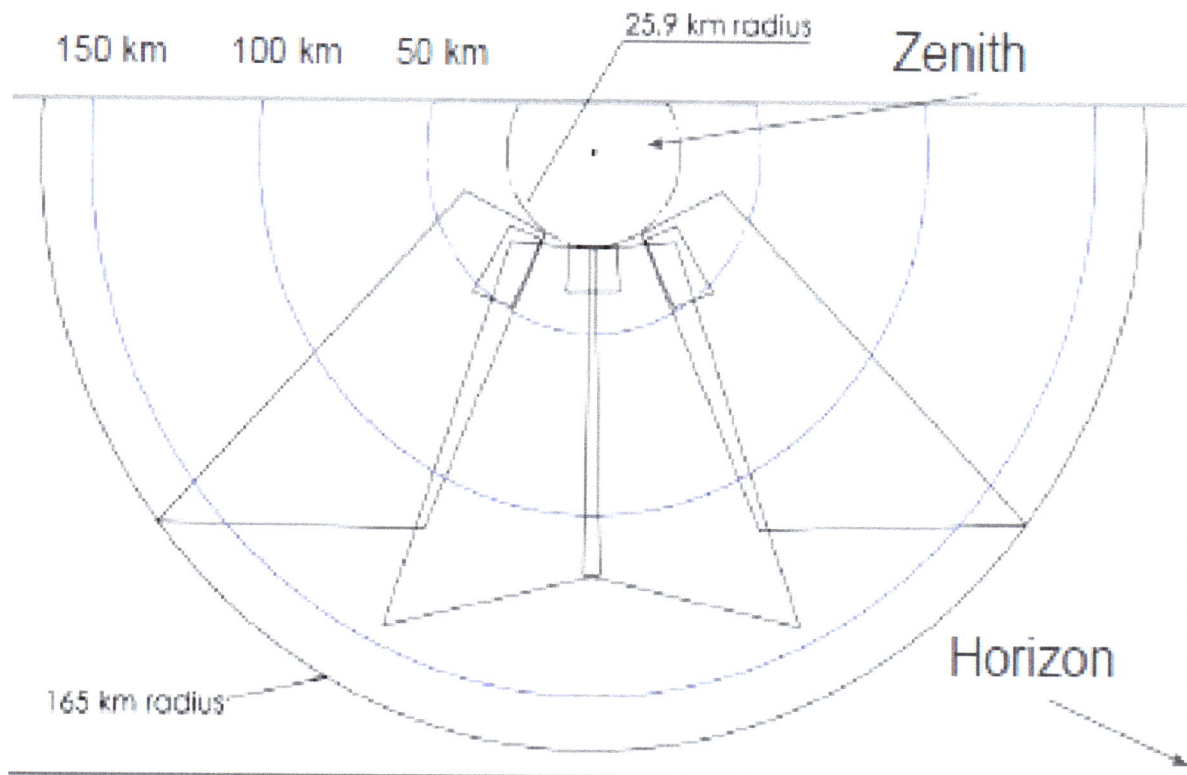

Figure 57. Field of views of both NFICUs and WFICUs superimposed (credit: Christopher Geach)

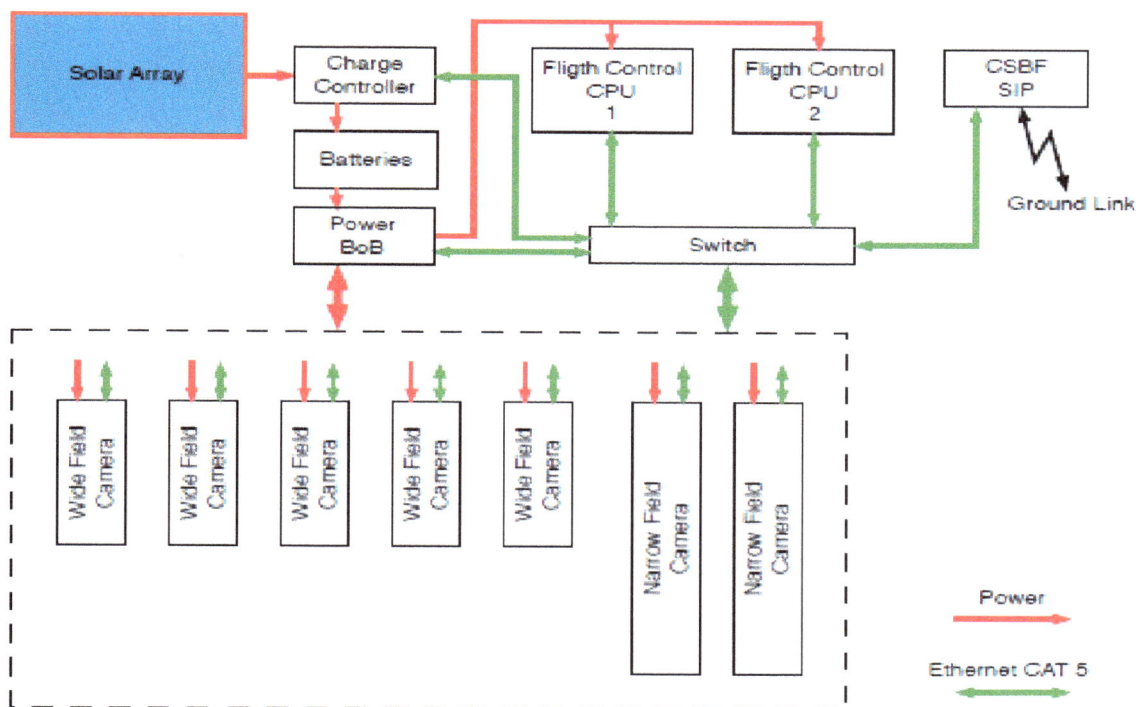

Figure 58. Schematic of PMC-Turbo camera suite with power, command, and control links.

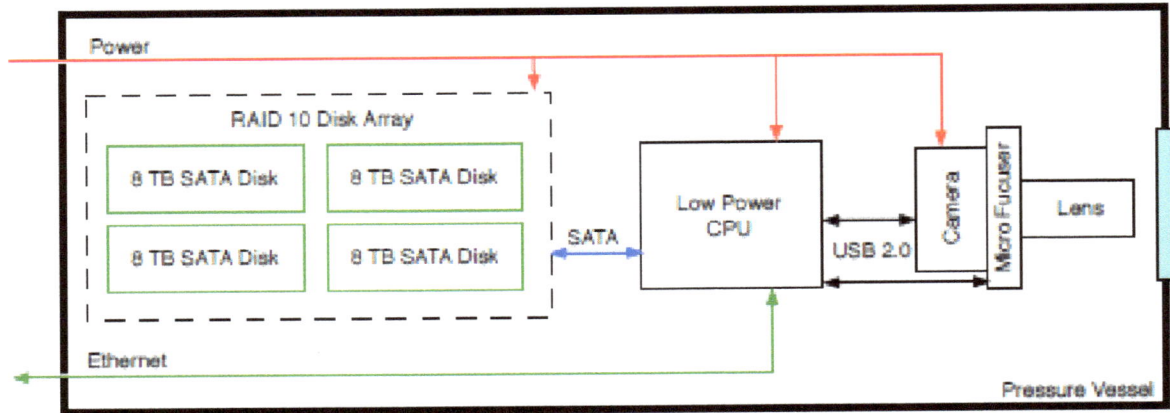

Figure 59. Schematic of individual PMC-Turbo camera system

The elevation of the camera systems has been fixed at bore sight elevation angles for WFICUs and NFICUs of 45° and 56°, respectively. Azimuthal pointing is achieved using the CSBF rotator pointing system that provides an anti-sun observing direction to an accuracy of ±2 degrees. Each WFICU and NFICU (see schematic in Figure 61) consists of a camera, lens, embedded control computer, and disk storage operating as an independent unit. Each would be connected to redundant flight control computers by Ethernet, providing a channel for commands from the ground and flight computers, and allowing small amounts of imaging data to be relayed to the ground during flight to evaluate performance and focus. This design thus provides significant redundancy and ensures against failure of most individual system components. Each pressure vessel is pressurized prior to flight. The flight computer systems integral to each camera system have been designed for industrial server use and may be managed via a single IPMI port.

For power, PMC-Turbo employs a 3x2 array of 115W per panel solar panels, modular battery packs, and a Morningstar charge controller. Telemetry, enabling monitoring of the health of the camera system, instrument temperatures, etc., would be achieved through a combination of line-of-sight communications during the first 24 hours followed by a combination of the TDRSS and Iridium satellite links. The total data set produced by this mission is expected to be approximately 110 TB. The completed PMC-Turbo camera systems are shown in Figure, along with the gondola onto which five of the cameras have been mounted so that a total FOV of 150 x 40 degrees is obtained.

Figure 60. Individual PMC-Turbo Camera system showing camera, computer, lens, and data storage units.

Figure 61. Set of camera systems configured on the gondola

References

- Baumgarten, G., and D. C. Fritts (2014), Quantifying Kelvin-Helmholtz instability dynamics observed in noctilucent clouds: 1. Methods and observations, J. Geophys. Res. Atmos., 119, 9324–9337, doi:10.1002/2014JD021832.

- http://lasp.colorado.edu/aim/

The PoSSUM Suborbital Tomography Instrument Suite

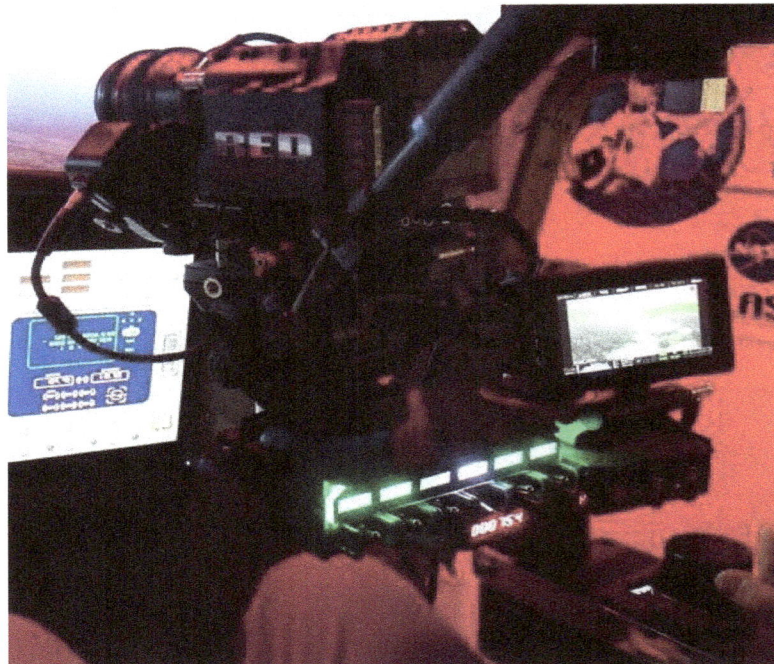

Module Objectives

- ➢ Understand the components of the PoSSUMCam System
- ➢ Familiarize with the controls of the PoSSUMCam System
- ➢ Learn how to configure the RED EPIC Camera for flight
- ➢ Learn how to configure the PoSSUM Wide Field Imager for Flight
- ➢ Learn how to program the Wide Field Imager Sequencer

Contributing Authors:

Jason Reimuller, Ph.D.
Van Wampler
Zoltan Sternovsky, Ph.D.
Gerald Lehmacher, Ph.D.
Kathy Mandt, Ph.D.

The PoSSUM Tomography Experiment is designed to obtain atmospheric data as well as visible and infrared imagery through a suite of instruments that will obtain data to be integrated post-flight and used to refine current state-of-the-art mesospheric models. Instruments developed for the PoSSUM Tomography Experiment include PoSSUMCam, the Mesosphere Clear Air Turbulence (MCAT) experiment, the Mesosphere Aerosol Sampling System (MASS) system, and the In-Situ Atmosphere Sampler.

- **PoSSUMCam:** PoSSUMCam is an integrated camera system control interface that houses 1) a RED EPIC Dragon camera coupled with a 5" monitor and an integrated zoom and iris controller, 2) the PoSSUM Wide-Field Imager (WFI), with a 70mm f/1.4 lens, linear polarizer, and control sequencer, and 3) an automated infrared video camera. The PoSSUMCam system also houses the Main Control Panel which regulates 1) power to MCAT, MASS, WFI, and the Atmospheric Sampler, 2) MCAT probe deployment and retraction, and 3) sample taking on ascent through the NLC layer and again upon descent through the NLC layer. There also is a user-programmable mission clock to assist in situational awareness.

- **Mesosphere Clear Air Turbulence (MCAT):** MCAT is an atmospheric pressure sensor coupled with a 3-axis accelerometer that is mounted external to the spacecraft. The MCAT instrument has a probe that is manually extended after Main Engine Cutoff and retracted just prior to re-entry. The MCAT and its associated probe are controlled through the Main Control Panel on the PoSSUMCam control interface and connected to the external payload via a wi-fi connection.

- **Mesosphere Aerosol Sampling System (MASS):** MASS is a neutral and charged-particle detector that calculates the flux of particles and bins them according to particle size. MASS is controlled through the Main Control Panel on the PoSSUMCam control interface and connected to the external payload via a wifi connection.

- **In-Situ Atmospheric Sampler (IAS):** The PoSUM Atmospheric Sampler is a sampling system with two reservoirs that can take a sample of the atmosphere at NLC altitudes for post-flight determination of $CO2$ and nitrogen oxide concentrations. This system connects directly to the PoSSUMCam control interface and is controlled via an operator actuated switch on the PoSSUMCam control interface.

The PoSSUM Tomography Experiment was originally designed around the XCOR Lynx Mark II vehicle. With a projected apogee of 103 km, the Lynx would have provided adequate performance to penetrate NLC layers on ascent and upon descent. Figure 64 shows where the PoSSUM suite of instruments was designed to integrate onto the Lynx vehicle. The In-Situ Atmospheric Sampler was designed to fit behind the pilot's seat in 'Payload Bay A' and serviced by an external air probe. Flight data is recorded internally to the PoSSUMCam system and recorded on a CF card so that the exact times when switch actuation is performed may be recorded. This assists in training as well as provides the means to extrapolate the exact altitudes when samples are taken and the probe is deployed.

In 2016, XCOR announced that it was ceasing development of the Lynx vehicle. The configuration of instruments was configured later that year for flight on the Virgin Galactic Spaceship Two. The current planning for utilizing Virgin Galactic's Spaceship Two for PoSSUM missions is:

- Vehicle One: Destroyed in 2015
- Vehicle Two: No external pods planned. Instruments to be flown include PoSSUMCam, WFI, and the PMC-Turbo LiDAR system for temperatures
- Vehicle Three and Subsequent Vehicles: No external pods planned. Instruments to be flown include PoSSUMCam, WFI, MASS, MCAT, and a spectrometer for measuring water vapor concentrations.

The PoSSUMCam system that was designed and constructed for flight on the XCOR Lynx is now used for training PoSSUM Scientist-Astronaut Candidates; the functionality is virtually identical to the PoSSUMCam system being designed for flight on Spaceship Two.

THE PoSSUMCam SYSTEM

PoSSUMCam is the heart of the instrument system on PoSSUM missions. This system houses a variety of camera controls that support the RED EPIC video camera, the PoSSUM Wide-Field Imager, and an automated infrared camera. The PoSSUMCam system also supports controls that regulate 1) power to the MCAT and MASS systems, the Atmospheric Sampler, and the Wide-Field Imager, 2) MCAT probe deployment and retraction, 3) Wide-Field Imager Sequence Start, and 4) atmospheric sample taking. There also is a user-programmable mission clock to assist in situational awareness.

Design Requirements

PoSSUMCam was designed with several top-down requirements. Specifically, the instrument had to be functional throughout all phases of nominal flight and conform to EMI and crash-loads requirements imposed by the FAA and the vehicle operator. Among the driving design requirements are:

- Operating Loads shall allow for effective operations while exposed to 4 G_x.
- Crash Loads shall comply with FAA requirements of +9G.
- The system cannot impede emergency egress operations.
- The system operator shall be able to manipulate all controls wearing space suit gloves.
- The mount shall produce negligible vibrations on the camera systems during the microgravity phase of flight.

PoSSUMCam Components

The PoSSUMCam system, as shown below in Figure 62, shows the major components of the system. The mount supports the RED Video Camera system and the associated monitor as well as the Wide-Field Imager and its sequencer. The armrest of the mount houses a zoom and iris controller that drives the lens of the RED video camera. A series of switches on the Main Control

Panel regulates the power to the MCAT, MASS, Sampler, and Wide-Field Imager instruments as well as control the MCAT probe deployment and retraction as well as the sampling system. The Power Panel houses the main power switch in addition to a chronometer to be used for situational awareness. A switch adjacent to the main power switch resets and starts the chronometer.

Interfacing PoSSUM Payloads with PoSSUMCam

The video camera system, along with the associated video monitor and zoom/iris controller, mount directly onto the PoSSUMCam control interface. The Wide-Field Imager connects to the control interface via two cables: a USB data connector and a camera control cable. The In-Situ Atmospheric Sampler connects directly to the interface using a 9-pin data cable. MCAT and MASS are external payloads controlled via a wifi link.

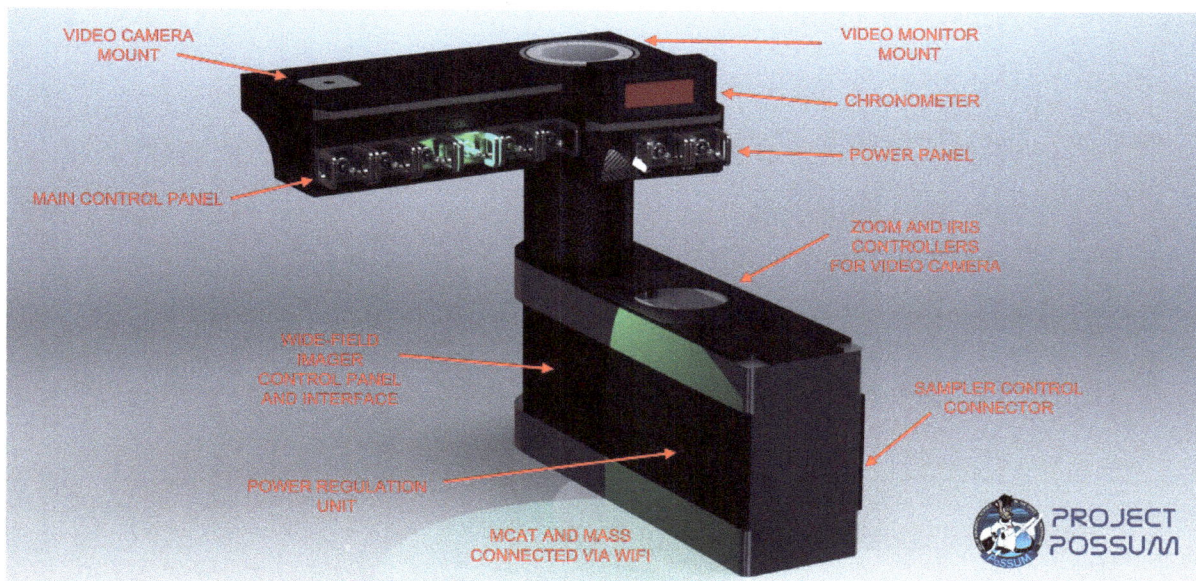

Figure 62. PoSSUMCam Components

PoSSUMCam Controls

All of the controls are integrated together on the PoSSUMCam system, which is mounted about the Scientist-Astronaut seated in the right seat of the spacecraft. In addition to the camera controls, the Main Control Panel of the PoSSUMCam system supports controls the1) power to MCAT, MASS, the Wide-Field Imager, and the Atmospheric Sampler, 2) MCAT probe deployment and retraction, and 3) sample taking. There also is a programmable chronometer to assist situational awareness.

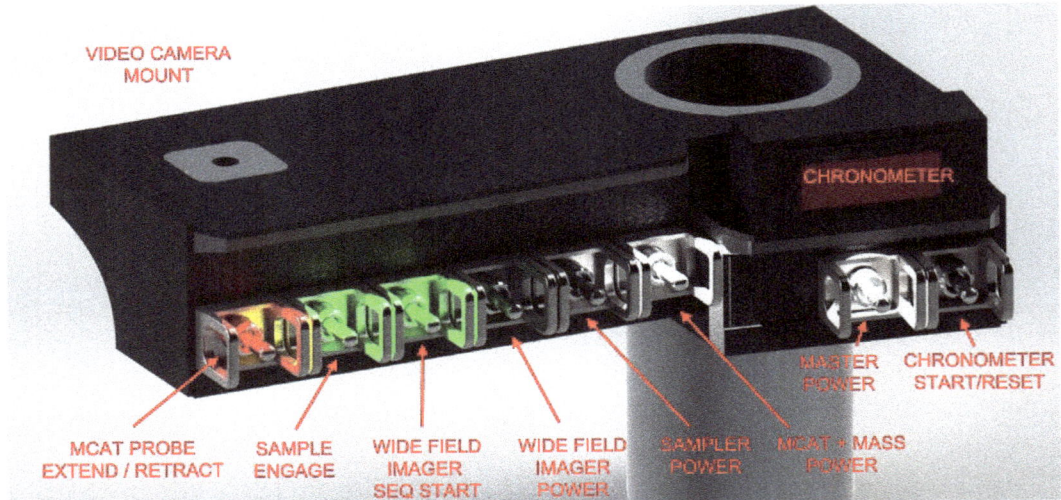

Figure 63. PoSSUMCam Controls

Video Camera Controls

- **RED EPIC POWER/REC** (on/rec/off): Touch once to power on, a second time to start recording. Hold for two seconds to power off.

- **RED EPIC IRIS CONTROL:** Rotate to adjust the amount of light entering the aperture

- **RED EPIC ZOOM CONTROL:** Rotate to zoom in/out

Control Panel (Right to Left)

- **MCAT and MASS POWER** (On/Off): Provide power to the MCAT and MASS System

- **SAMPLER POWER** (On/Off): Provide power to the atmospheric in-situ sampler

- **WIDE FIELD IMAGER POWER** (on/off): Push to power the Wide-Field Imager.

- **WIDE FIELD IMAGER SEQUENCE START** (Start): Push to start pre-programmed sequence on the Wide-Field Imager before launch.

- **SAMPLE** (Take): Momentary switch to take an atmospheric sample at NLC altitudes (during cloud penetration during both ascent and descent). The sampler switch will flash green three times when a sample is taken. Only two samples may be taken per flight, and the system will not accept any more sample votes after two samples have been taken unless the entire system is reset.

- **MCAT PROBE** (Up/Down): Momentary two-position switch to extend and retract MCAT Probe. While the probe is extended, the switch will be illuminated in red, indicating that the probe is extended and needs to be retracted before re-entry.

Power Panel

- **MAIN POWER** (on/off): Provide power to the PoSSUMCam system

- **MISSION ELAPSED TIME** (MET) (Count/Pause/Reset): Start, stop, and reset the mission clock

- **CIRCUIT BREAKER**: (5A) Protects system from power surges

THE PoSSUM HD VIDEO SYSTEM: RED EPIC-M DRAGON

The RED EPIC Dragon is used to capture video imagery in the visual spectrum on PoSSUM missions. The Dragon has a 19-megapixel sensor that captures video and stills at up to 6K resolution, nine times more resolution than standard high-definition video.

Figure 64. RED Epic Dragon HD Video Camera (credit: RED Camera)

RED EPIC Specifications

- Sensor: 19-Megapixel Red Dragon
- Pixel Array: 6144 (h) x 3160 (v)
- Dynamic Range: 16.5+ stops
- Max Image Area: 6144 px (h) x 3160 px (v)
- Lens Coverage: 30.7 mm (h) x 15.8 mm (v) x 34.5 mm (d)
- Acquisition Formats: 6K raw (2:1, 2.4:1), 5K raw (full frame, 2:1, 2.4:1, and anamorphic 2:1), 4.5K raw (2.4:1), 4K raw (16:9, HD, 2:1, and anamorphic 2:1), 3K raw (16:9, 2:1, and anamorphic 2:1), 2K raw (16:9, 2:1, and anamorphic 2:1), 1080p RGB (16:9), 720P RGB (16:9)

PREPARING THE RED EPIC FOR FLIGHT

The following steps are used to prepare the camera for flight. It is assumed that both the battery to which the EPIC camera is connected is fully charged and the iKAN controller within the PoSSUMCam armrest is fully charged.

Turning on the RED Epic camera

Turn the camera on by pressing the primary power/recording button on the right side. This button has been modified so that it can easily be pressed through the glove of a spacesuit. The camera will take 5-10 seconds to start up. Check that the monitor to the right of the camera is displaying an image and settings to confirm proper startup. If the display hasn't turned on, wait 10 seconds and press the power button again. If after another 5-10 seconds the camera hasn't booted up, further troubleshooting will be required.

If the display is on but not displaying image/information properly (most commonly, only half of the image or a checkerboard type pattern will appear), check the cable connection to the monitor. If the problem is still not resolved, turn the camera off by holding the power button down for ten seconds. Wait 10 seconds after releasing the power button, and press the button again to re-start the camera. If after 5-10 seconds the display still isn't working, further troubleshooting will be required.

Checking the Settings

Now that the camera has started up, check that all of the following settings across the top of the display are as follows:

29.98fps / iso800 / -- / 10deg / 5600k / 5K / RC 5:1

If any of these settings are wrong, they cannot be easily changed when wearing a spacesuit. Report any settings problems and continue with the following steps:

Confirming the Focus is set to Infinity

The focus should be pre-set to infinity, but check to make sure it is. This information is clearly marked on the lens on the front of the camera. On the lens, next to where it reads "Focus", there are two arrows. Check to see that the arrows are perfectly aligned, facing each other. If they are not, use the focus knob in the armrest. Rotate the knob until the two arrows align on the lens. Once the arrows are aligned, the focus is set. Do not alter it further.

Setting the Exposure

To set the exposure, first make sure the camera is pointed at the sky using the small handle on the camera base plate. Once the camera is pointed at the sky, use the iris control knob in the armrest to adjust exposure. To attain proper exposure, look at the histogram at the bottom left of the display. The histogram looks like this:

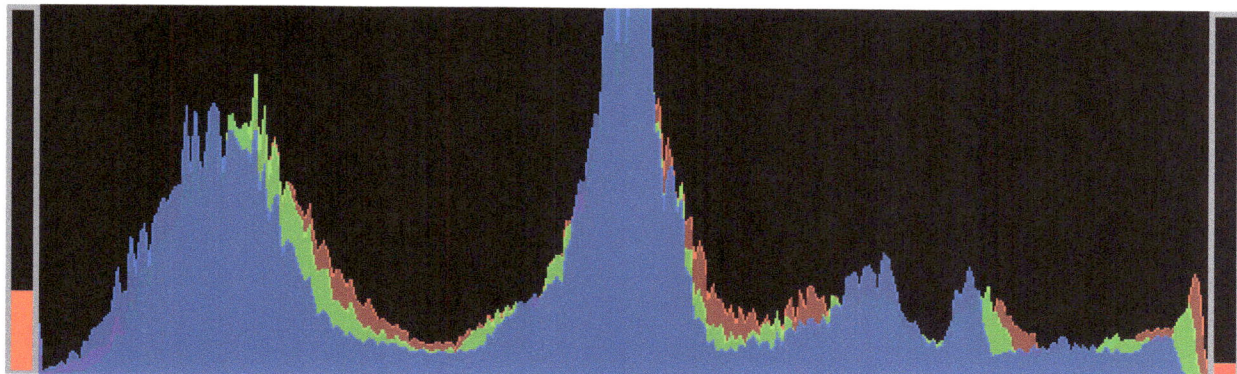

Figure 65. RED EPIC Histogram

For proper exposure, the mass histogram should be centered. Move the iris control knob left or right until the histogram is in the center of the chart. There are "goal posts" on either side of the histogram - neither of these bars should have any red in them. To the right of the histogram, there are three lights that look like this:

The lights should be gray (left). If any of the three lights are on (right), the exposure is too high. It is very important that all of the lights remain off. If any of these lights are on, close down the iris by turning the iris control knob until the histogram is centered and all of lights are off.

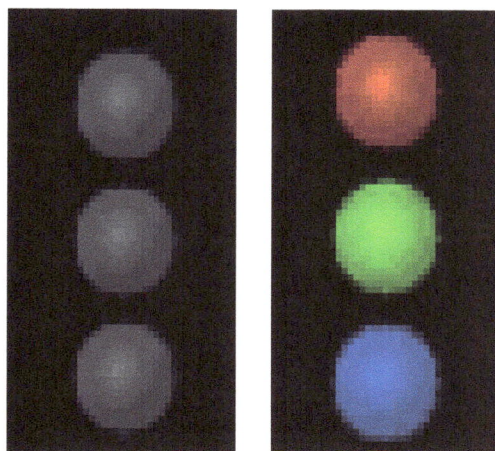

This in an example of a properly exposed histogram:

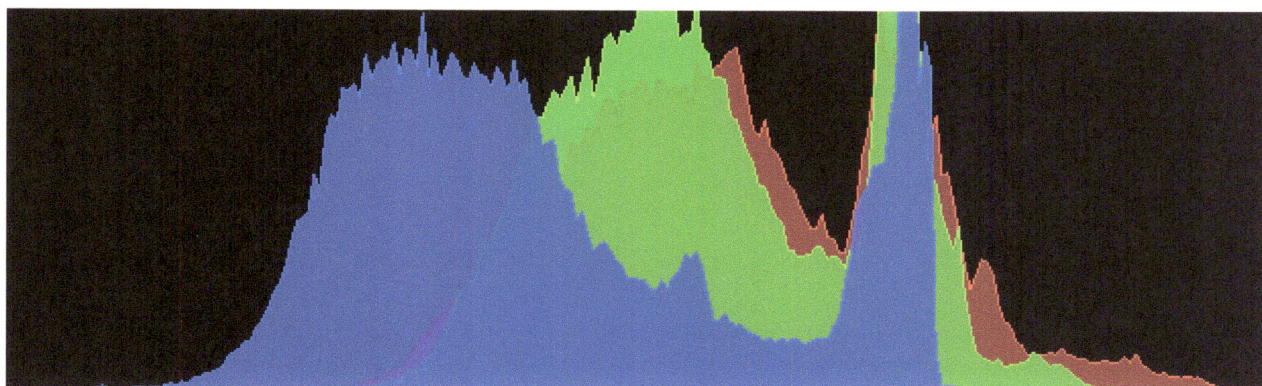

It the histogram looks like this; it is risking overexposure:

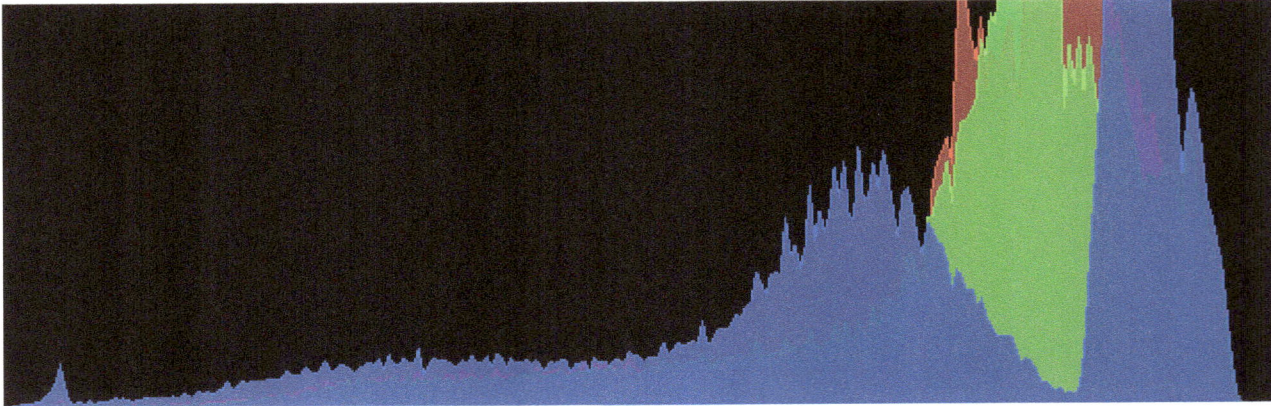

If the histogram looks like this; it is risking underexposure:

Figure 66. Overexposed and Underexposed histograms

If the sun is in frame, there may be a small part of the histogram that extends to the right and off the chart. If this is the case, the lights on the right will likely turn on. If the sun is in frame, just make sure that the mass of the histogram is centered in the chart. This is the only acceptable time for the lights on the right to turn on. If the sun isn't in frame, they should all be off. Once the exposure has been set, point the camera back at the horizon.

Preparing the iKAN Follow Remote Air System for Flight

The PoSSUMCam System used the iKAN Remote Air to control iris and zoom on the camera lens. This controller is mounted in the 'armrest' of the PoSSUMCam system and the operator controls these two settings using two concentric knobs. The outside knob (the knob of larger circumference and closer to the base) controls the iris setting; use this to regulate the amount of light that the camera is exposed to. The inner knob (the knob of smaller circumference) is used to control the zoom of the lens, enabling the operator to 'zoom in' to NLC micro-features.

To set up the system, make sure it is fully charged for flight and that the FREQUENCY = 2 and the Current Channel = 4GHZ. The signal connection should register (if not, restart the system). The normal range for the battery is between 3.4 and 4.2V.

When first turning on the motor, it will need to do an automatic scan. To do this, hold the 'NERVE' button for three seconds until the motor starts turning. The NERVE button for the Zoom control motor is on the left side of the controller. The NERVE button for the iris control is on the right side of the controller. After the scan, the motor will align in the middle of its range. Confirm operation by turning the knob in each direction. Repeat this step for the other motor.

FLIGHT OPERATIONS WITH THE RED EPIC

Not that the RED camera system is configured for flight, make sure the battery is fully charged. The system is now ready for flight. You will refine general skills throughout your training and simulation practice, but in actual flight, you will only have several minutes to identify and track the features of most interest.

Starting Recording of the Camera

Prior to launch, start the camera. The battery will power the camera for approximately 30 minutes, which will last the duration of a flight. Engage the camera once the spacecraft is in position for takeoff but before the launch commit decision has been made.

Once the camera has been properly set, press the power/record button to start the camera recording. DO NOT HOLD THE BUTTON DOWN. Holding down will turn the camera off. To verify that recording has started, check to see that there is a red circle in the top right corner of the display. If you do not see the red circle, hit the button again. If you still do not see the red circle, repeatedly hit the button until you locate the circle toggling on and off, then leave it on.

Once the camera is recording (again, a red circle will be visible in the top right corner of the display), you're good to go!

Checking the Exposure

Be sure to check the exposure from time to time during flight. There is a possibility that exposures might change during the flight as the aircraft passes through clouds. Keep an eye on the exposure and make necessary adjustments with the iris knob (the outer knob) if the histogram goes off the chart in either direction.

Pointing and Zoom

Use the handle at the base of the camera to move it up/down or side to side to face it in the direction of cloud structures. To zoom in on microfeatures of interest, use the zoom (the innermost knob) control.

Powering Down the System

After landing, hold down the power/record button to turn the camera off. Data may then be transferred from the camera for storage and processing. Be sure to back-up the data at the earliest possible opportunity.

Further Reading

There is a helpful article about exposing with a RED available here:

http://www.red.com/learn/red-101/red-camera-exposure-tools

THE PoSSUM WIDE-FIELD IMAGER SYSTEM

The PoSSUM Wide Field Imager (WFI) will be a flight-heritage system built for the PMC-Turbo mission. For training purposes, this camera system is simulated through use of a commercial Canon DLSR camera with a wide-angle lens and a microprocessor-controlled portable control system. Like the PMC-Turbo camera system, this camera is used to take sequential wide-field visual images of the noctilucent cloud layer so that the micro features observed by the video camera system can be properly geo-located in the context of the larger cloud structure. The following section details use of the commercial Canon EOS30D camera system, currently configured to the PoSSUMCam system for training purposes.

Configuring the Camera

The Canon camera will require a minor set up process prior to using the Promote Control:

> 1. Set your camera to Manual Exposure mode.
>
> 2. Set your camera to Manual Focus mode.
>
> 3. Set 1/3EV shutter speed steps in your camera setup menu. This is normally a default setting for your camera.
>
> 4. It is recommended to also set 1/3EV ISO steps in your camera setup menu, if available. This is normally a default setting for your camera.
>
> 5. Set your camera to Single drive mode. Operating your camera in Continuous drive mode with Promote Control may lead to unexpected results, such as extra pictures being taken. Alternatively, you can use Quiet or Silent drive mode, if available on your camera.

Determining the Exposure Value (EV)

The Exposure Value (EV) is a number that represents a combination of a camera's integration time and f-number such that all combinations that yield the same exposure have the same EV value. The EV is a base-2 logarithmic scale, defined as:

$$EV = \log_2 \frac{N^2}{t},$$

where

- N is the relative aperture (f-number) – 1.4 for the PoSSUM Wide Field Imager
- t is the exposure time, or integration time, in seconds.

Table 7 shows characteristic EV values for different lighting conditions. Airborne observations of NLC activity at 23,000 feet generally were shot using an EV of between 4 and 5. Since there will be negligible lower atmospheric scattering at NLC altitudes, the EV will be lower.

EV	TYPE OF LIGHTING SITUATION
-6	Night, away from city lights, subject under starlight only.
-5	Night, away from city lights, subject under crescent moon.
-4	Night, away from city lights, subject under half moon. Meteors (during showers, with time exposure).
-3	Night, away from city lights, subject under full moon.
-2	Night, away from city lights, snowscape under full moon.
-1	Subjects lit by dim ambient artificial light.
0	Subjects lit by dim ambient artificial light.
1	Distant view of lighted skyline.
2	Lightning (with time exposure). Total eclipse of moon.
3	Fireworks (with time exposure).
4	Candle lit close-ups. Christmas lights, floodlit buildings, fountains, and monuments. Subjects under bright street lamps.
5	Night home interiors, average light. School or church auditoriums. Subjects lit by campfires or bonfires.
6	Brightly lit home interiors at night. Fairs, amusement parks.
7	Bottom of rainforest canopy. Brightly lighted nighttime streets. Indoor sports. Stage shows, circuses.
8	Las Vegas or Times Square at night. Store windows. Campfires, bonfires, burning buildings. Ice shows, football, baseball etc. at night. Interiors with bright florescent lights.
9	Landscapes, city skylines 10 minutes after sunset. Neon lights, spotlighted subjects.
10	Landscapes and skylines immediately after sunset. Crescent moon (long lens).
11	Sunsets. Subjects in deep shade.
12	Half moon (long lens). Subject in open shade or heavy overcast.
13	Gibbous moon (long lens). Subjects in cloudy-bright light (no shadows).
14	Full moon (long lens). Subjects in weak, hazy sun.
15	Subjects in bright or hazy sun (Sunny f/16 rule).
16	Subjects in bright daylight on sand or snow.

Table 7. Typical Exposure Value (EV) settings

Setting the Camera ISO

In digital photography, ISO measures the sensitivity of the image sensor. Higher ISO settings are generally used in darker situations to get faster shutter speeds, such as in an indoor sports event when you want to freeze the action in lower light. However, higher ISO settings will introduce more noise. However, it is important to first mitigate blur that might be caused by motion of the vehicle relative to the cloud layer, and then accept any noise introduced by the higher ISO setting. Remember that *noisy sharp shots are always better than blurry clean shots*, so don't hesitate to crank up your ISO to get whatever shutter speed you need. The bottom line is that we want to determine the lowest ISO setting that does not cause any reduction in sharpness that results from vehicle motion or vibration, so we recommend starting conservatively; fly first with high ISO settings and back off as needed.

Figure 67. Sharpness is affected by higher ISO settings

Recent advances in camera CCD technology have enabled shots at quite high ISO settings without excessive noise; the Canon EOS5D Mark III camera can be set to ISO settings up to 25600! This is greatly enhanced over the EOS30D used for PoSSUM mission training, which has a maximum ISO setting of 1600.

Setting the Camera Focus

The camera focus is fixed at infinity throughout PoSSUM missions. This can be done by simply selecting an object at a far distance from the ground (a star field or mountain range) and focusing the camera on it. Once set, it is important to make sure the camera lens is not accidentally adjusted before flight.

Setting the Camera Aperture

Within any exposure factor (Exposure Value, ISO, aperture, shutter speed) each step is double (or half of) the preceding step. We need a relation between ISO and shutter speed for the PoSSUM set aperture of 1.4. Recall that longer shutter speed will introduce more blur and higher ISO will introduce more noise, but noise is better than blur. Exposure values in Table 7 are reasonable general guidelines, but they should be used with caution. For simplicity, they are rounded to the nearest integer, and they omit numerous considerations. The exposure values in

Table 7 is for ISO 100 speed ("EV_{100}"). For a different ISO speed, S, increase the exposure values (decrease the exposures) by the number of exposure steps by which that speed is greater than ISO 100, formally:

$$EV_S = EV_{100} + \log_2 \frac{S}{100} \, .$$

For example, ISO 400 speed is two steps greater than ISO 100:

$$EV_{400} = EV_{100} + \log_2 \frac{400}{100} = EV_{100} + 2 \, .$$

To photograph at an ISO 400 setting, add 2 to the setting estimated on Table 8 to get an EV400 setting. Figure 68 shows an exposure chart showing the relationships between EV, f-number, and shutter speed.

ISO	Shutter Speed (EV = 2)	Shutter Speed (EV = 0)
800	1 / 2 s	2s
1600	1/4 s	1s
3200	1/8 s	1/2s
6400	1/15s	1/4s
12800	1/30s	1/8s
25600	1/60s	1/15s

Table 8. Shutter Speed and ISO selection at EV=2 and EV=0

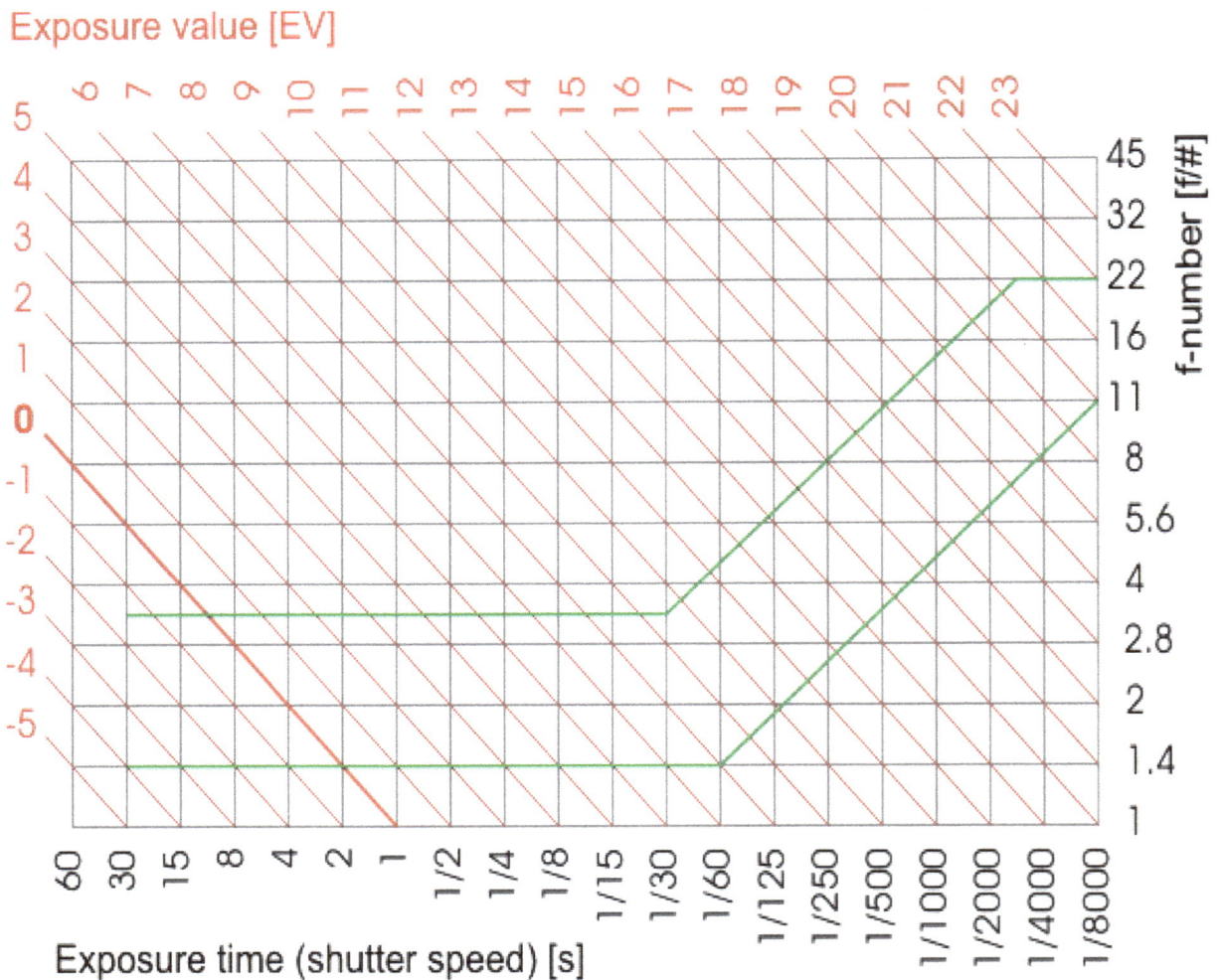

Figure 68. Popular exposure chart type, showing exposure values EV (red lines) as combinations of aperture and shutter speed values.

SUBORBITAL IMAGERY USING THE POSSUM WIDE-FIELD IMAGER

Suborbital Approximations for NLC Photography

ISO 800, 1/10s, f=1.4 settings were used on airborne observations of NLCs. This corresponds to an EV of 4 at an ISO of 800. To scale this into an approximation of settings for suborbital flight:

1. Consider the reduced background light as there will be negligible atmospheric background. For now, estimate this to reduce the EV by 4, so EV at NLC altitudes is estimated to be 0 at ISO 800.

2. Compensate for higher ISO capabilities of the camera. At ISO of 25600, this corresponds to 5 additional EV steps, or an EV of 5.

3. With an f-stop of 1.4, this corresponds to 1/15s

Flight Manifest of the POSSUM Wide Field Imager

These are just an initial estimation of what we expect to see at suborbital altitudes; these assumptions need to be refined through actual flight data. We anticipate two Wide Field Imagers to fly on each PoSSUM flight. Flights 1-3 will be used to refine the settings of the cameras. Flight 4 will be used to take the optimized settings and apply to a polarization experiment that will use cross-polarized filters to infer particle size distributions in NLC micro-features. The specifics of this experiment are beyond the scope of this text.

PoSSUM Flight	Camera 1	Camera 2
PL1	HDR Calibration	ISO 25600, Shutter Speed 1/15s
PL2	Refined from Flight PL1	ISO 12800, Shutter Speed 1/15s
PL3	Refined from Flight PL2	Refined from Flight PL2
PL4	Linearly Polarized from PL3 setting	Cross Polarized from PL3 settings

Table 9. PoSSUM Flight Manifest for the Wide Field Imager System

PL1 will use the High-Dynamic Range setting on the Promote Control controller that will be used to drive the PoSSUM Wide Field Imager system. The next section describes the programming and operations of this system.

CONFIGURING THE PoSSUM WIDE FIELD IMAGER FOR FLIGHT

For a basic understanding of the Canon EOS 5D Mark III camera and its functions, please refer to Appendix A3. This section supplies the minimal information required to configure the Canon EOS5D Mark III camera for flight. Please refer to the camera's user guide for a more detailed description of the camera's capabilities.

Installing the Battery

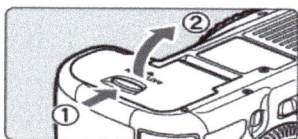

To install the charged battery prior to flight, follow these steps:

Installing the Battery

1 **Open the cover.**
- Slide the lever as shown by the arrows and open the cover.

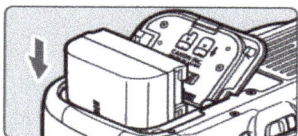

2 **Insert the battery.**
- Insert the end with the battery contacts.
- Insert the battery until it locks in place.

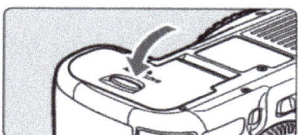

3 **Close the cover.**
- Press the cover until it snaps shut.

Installing CF Cards

The Canon Camera will record simultaneously to both a CF and an SD memory card, though PoSSUM sorties will use a 256GB CF card. A full format RAW image will use 27.1MB. If one image is taken each second for a flight of 30 minutes, the images will require approximately 49GB! It is important that a CF card of at least 64GB is used. To install the card, follow these steps:

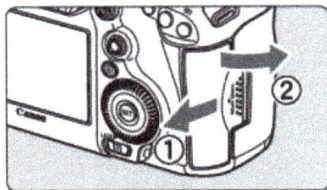

CF card

SD card

Write-protect switch

1 Open the cover.
- Slide the cover as shown by the arrow to open it.

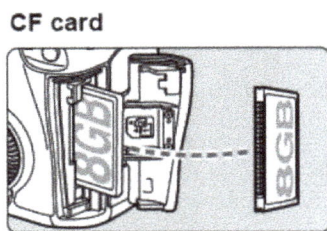

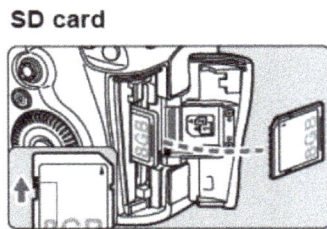

2 Insert the card.
- The camera-front side slot is for a CF card, and the camera-back side slot is for an SD card.
- **Face the CF card's label side toward you and insert the end with the small holes into the camera. If the card is inserted in the wrong way, it may damage the camera.**
- ▶ The CF card eject button will stick out.
- **With the SD card's label facing you, push in the card until it clicks in place.**

Credit: Canon

Affix the Lens and Turn on Camera

If a polarizer or lens hood will be used, affix these to the lens as well.

Set the Lens to Manual Focus

It is important to configure the lens to manual focus so that it can be locked at infinity. To focus the lens controller to manual operation, follow these steps:

Set the lens focus mode switch to <MF>.

- Turn the lens focusing ring to focus roughly.

Magnifying frame

Move the magnifying frame.

- Use <✥> to move the magnifying frame to the position where you want to focus.
- Pressing <✥> straight down will return the magnifying frame to the image center.

Credit: Canon

Magnify the image.

- Press the <Q> button.
- ▶ The area within the magnifying frame will be magnified.
- Each time you press the <Q> button, the view will change as follows:

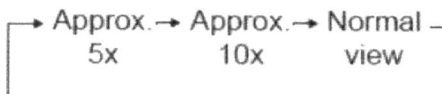

$$\rightarrow \text{Approx.} \rightarrow \text{Approx.} \rightarrow \text{Normal} \atop 5x \qquad 10x \qquad \text{view}$$

AE lock
Magnified area position
Magnification

Focus manually.

- While looking at the magnified image, turn the lens focusing ring to focus.
- After achieving focus, press the <Q> button to return to the normal view.

Take the picture.

- Check the focus and exposure, then press the shutter button completely to take the picture

Credit: Canon

Formatting the CF Card

If the CF card is new and has not been formatted, it will need to be formatted. Follow these steps:

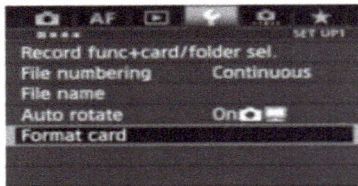

1 Select [Format card].
- Under the [✈1] tab, select [**Format card**], then press <☉>.

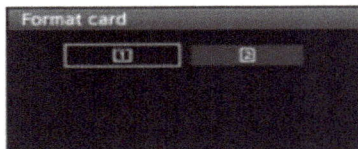

2 Select the card.
- [⚏] is the CF card, and [☒] is the SD card.
- Turn the <◯> dial to select the card, then press <☉>.

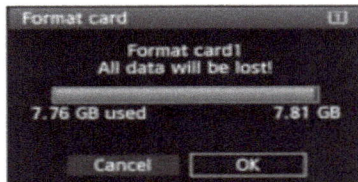

3 Select [OK].
- Select [**OK**], then press <☉>.
- ▶ The card will be formatted.
- ▶ When the formatting is completed, the menu will reappear.

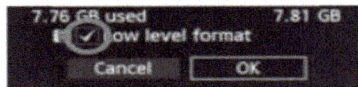

- When [☒] is selected, low-level formatting is possible (p.54). For low-level formatting, press the <🗑> button to append [**Low level format**] with a <✓> checkmark, then select [**OK**].

Credit: Canon

Selecting Full Format RAW imagery

PoSSUM missions require that the camera be set to full format RAW imagery. RAW images do not alter or compress the image data, and a pixel-by-pixel image is necessary to perform the most effective post-flight analysis of the images. To configure the camera for full-format RAW images, follow these steps:

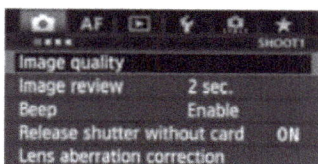

1 Select [Image quality].

- Under the [◻1] tab, select [Image quality], then press <(SET)>.

With [Standard / Auto switch card / Rec. to multiple] set:

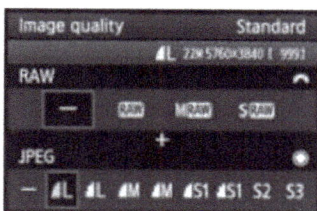

2 Select the image-recording quality.

- To select a RAW setting, turn the <🔆> dial. To select a JPEG setting, turn the <◯> dial.
- On the upper right, the "**M (megapixels) **** x ****" number indicates the recorded pixel count, and [***] is the number of possible shots (displayed up to 9999).
- Press <(SET)> to set it.

With [Rec. separately] set:

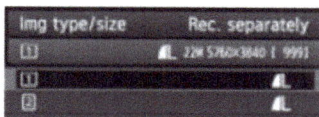

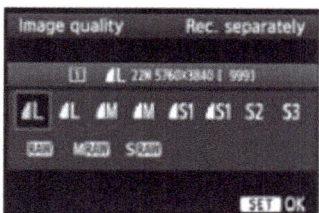

- Under [♈1: Record func+card/ folder sel.], if [Record func.] is set to [Rec. separately], turn the <◯> dial to select <①> or <②>, then press <(SET)>.
  On the screen that appears, turn the <◯> dial to select the image-recording quality, then press <(SET)>.

Configuring ISO for Manual Mode

For PoSSUM Sorties, the camera is configured for manual operations. This allows the camera not to 'fight' the instructions being given it by the sequencer. Make sure that the ISO setting is set to a manual selection as well. To configure for manual mode, select the shutter speed and aperture as calculated for the flight and configure as follows:

1 **Press the <ISO·🔲> button. (⏱6)**

2 **Set the ISO speed.**
- While looking at the LCD panel or viewfinder, turn the <✦> dial.
- ISO speed can be set within ISO 100 - 25600 in 1/3-stop increments.
- "**A**" indicates ISO Auto. The ISO speed will be set automatically

Engage High ISO Noise Reduction

This function reduces the noise generated in images taken at high ISO settings, like would be expected in PoSSUM flights. To configure the Canon EOS 5D Camera, follow these steps:

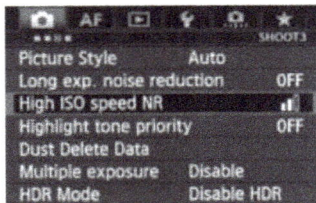

1 **Select [High ISO speed NR].**
- Under the [📷3] tab, select [**High ISO speed NR**], then press <SET>.

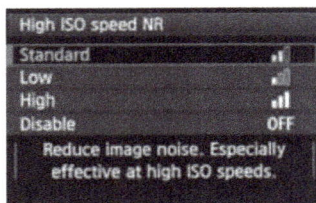

2 **Set the desired setting.**
- Turn the <◯> dial to select the desired noise reduction setting, then press <SET>.
- ▶ The setting screen closes and the menu will reappear.

3 **Take the picture.**
- The image will be recorded with noise reduction applied.

CONFIGURING THE PoSSUM WIDE FIELD IMAGER SEQUENCER FOR FLIGHT

The PoSSUM Wide Field Imager is controlled by a sequencer based upon a commercial camera controller called a 'Promote Control', modified and adapted for use in the POSSUMCam control interface. The Promote Control is an advanced, microprocessor-controlled portable unit, capable of automatically calculating complex exposure sequences, and providing extended camera control.

Wide Field Imager Control Unit Controls
The PoSSUMCam interface houses a modified controller so that it may be fully programmed before flight. The scientist-astronaut needs only power-on the system and initiate the sequence before take-off, which is done through two large switches easily engaged by an operator in a spacesuit.

1. Liquid Crystal Display with color backlight.

2. Power button. Press to turn on the Promote Control, then press briefly to toggle display backlight on and off. Press and hold to pause currently active image sequence, or if no sequence is active, turn off the Promote Control.

3. Mode button. Press to cycle through available modes. Hold "Left" navigation button to cycle mode list in reverse direction.

4. Focus button (Requires an optional shutter cable to function). Press and hold to pre-focus the camera lens on the subject.

5. Start button. Press to activate the selected photography sequence.

6. Four-way navigation keypad with center button. Navigates through available settings:
      Left / Right: switch the current setting.
          Up / Down: change the current setting value.
          Center button: confirm the change.

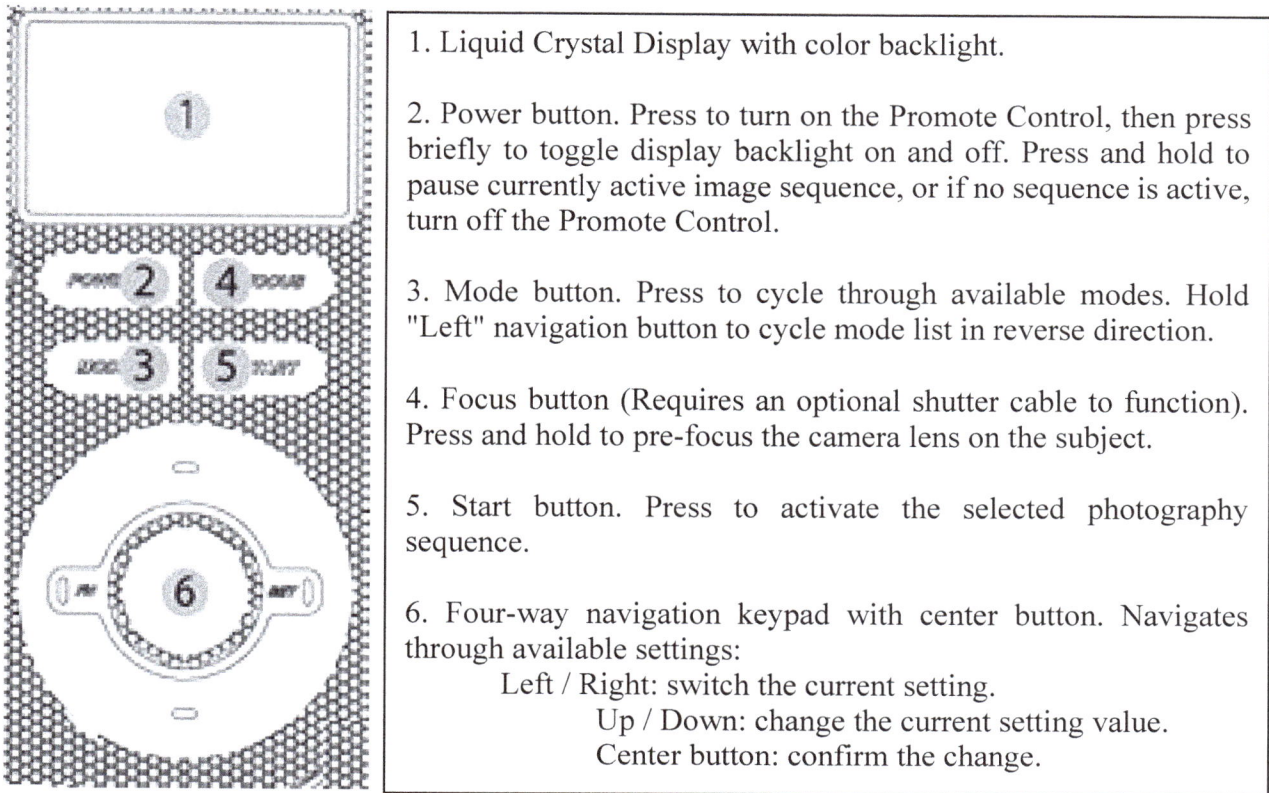

Figure 69. PoSSUM Wide-Field Imager Sequencer Controls

Wide-Field Imager Interfacing with PoSSUMCam

Two cables need to be connected to the Wide-Field Imager before flight: 1) USB connector (for camera control / device firmware update), and 2) the shutter cable port. These cables connect to their corresponding ports on the PoSSUMCam unit.

Preparing the Wide Field Imager Controller System for Flight

Make sure that fresh batteries are installed into the Promote Control system prior to flight. Then follow these steps:

1. Turn your camera on.

2. If the sequencer is on, turn it off the sequencer by pressing and holding the 'WFI PWR' switch on the PoSSUMCam main control array.

3. If your camera setup menu has an option for choosing different USB connection protocols (mostly called "USB"), set it to "PTP", "Print/PTP" or "PTP/MTP".

159

4. Use the supplied black USB cable to connect the Promote Control to your camera. Connect the rounded USB cable plug marked "P" to the Promote Control. Connect the other rectangular plug marked "C" to your Camera.

5. As the Promote Control is set up to use shutter cable, it needs to be connected every time you use your Promote Control. Connect the 3.5mm headphone stereo plug to the Promote Control shutter cable port located next to its USB port (see Interfaces). Connect the camera plug to the camera shutter release port.

IMPORTANT: PoSSUM missions use the HDR setting, so make sure to disconnect the Bulb Ramp Assist Cable. Leaving this cable connected in other modes may lead to improper sequence timing, taking extra images, or result in skipped frames.

7. Press the Promote Control "Power" button briefly to turn the unit on.

8. Press the "Power" button again briefly to toggle the display backlight on and off, as required.

All settings will be stored into Promote Control's internal memory every time you turn it off (or when it powers down automatically).

With your Promote Control on, press "Left" and "Right" buttons simultaneously to enter Promote Control Setup menu. "Left" and "Right" buttons scroll through available settings, while "Up" and "Down" buttons change setting values. For PoSSUM sorties, scroll to "Enable 1/3 EV ISO steps" setting and set that to Yes. When done, press "Center" button to return to your previously activated mode screen.

Programming the Promote Control for Flight

Navigate through available camera control modes using the "Mode" button. Hold the "Left" navigation key while pressing the "Mode" button to switch through the available modes in the reverse direction. First, use the "Mode" button to select the 'Time Lapse Mode' and set as per the following section. Use the four way navigation keypad to change settings. First, select a setting to be changed using the "Left" and "Right" keys, then use the "Up" and "Down" keys to change the value of the currently selected settings and then press the "Center" button to confirm your choices. Next, use "Mode" button to scroll to the "High Dynamic Range" mode and program as per the instructions of that section. Finally, make sure your camera is properly connected and that the camera is in Manual exposure and Manual focus.

Setting up a Time-Lapse Mode Program

Use "Time Lapse" mode to take a number of pictures separated by a preset time interval with an optional delay before starting the sequence. The resulting pictures can be used to create a fast motion video. The Time-Lapse mode will use your camera exposure settings. For exposure times less than two-seconds, use a shutter cable and set your camera to Bulb mode. Then instruct Promote Control to time an arbitrary exposure. Use of a shutter cable is highly recommended when using Time-Lapse mode with interval shorter than 2 seconds. Set the exposure settings as follows:

1. Set your camera to Bulb exposure mode.

2. Make sure the "Time-Lapse exposure setting mode" Promote Control Setup setting is set to "Arbitrary (Bulb mode)".

3. Connect the shutter cable to the camera.

4. Set your desired exposure in HH:MM:SS format.

Programming a Sequence in the Time Lapse Mode

Two subroutines need to be programmed before each flight: the Time Lapse Mode and the High-Dynamic Resolution Mode. The Time Lapse Mode established rapid automatic images to be taken. Be aware that the camera may not be able to take images as fast as the interval setting may suggest. The maximum frame rate is determined by the image type/size selected in the camera, memory card speed, presence of shutter cable and other factors. If your camera cannot keep up with the Time-Lapse interval you set, some images may be skipped. Using a shutter cable is highly recommended for any intervals shorter than 2 seconds. To program the Time Lapse Mode subroutine, follow these steps:

1. Use the "Interval" setting on Promote Control to specify how often a picture should be taken. The interval is specified in hours, minutes, seconds and tens of seconds. Set this to the shortest reasonable time, based on the settings established in the previous section (one-second is good for a training analog)

2. Specify how many "Frames", or images should be taken in total. For PoSSUM Sorties, set this to 'Infinity'. We certainly don't want to stop taking pictures until the end of the mission!

3. Use 'zero' for the "Start In" setting; we want the sequence to start immediately upon activation by the PoSSUM Scientist-Astronaut.

HIGH DYNAMIC RANGE TIME-LAPSE

Promote Control can repeatedly take a series of bracketed image sequences known as Time-Lapse High Dynamic Range imaging, or "Time-Lapse HDR". Every bracketed sequence is merged into an HDR image and then the series of resulting HDR images is used to create a High Dynamic Range Time-Lapse movie. The Time-Lapse portion of the "Time-Lapse HDR" functionality is controlled via "Time-Lapse" mode and the HDR portion is controlled via the "High Dynamic Range" mode screen. Creating a High Dynamic Range image from series of bracketed images requires an optional third-party software program. Many software applications are available to create High Dynamic Range images in manual or batch modes.

Two subroutines need to be programmed before each flight: the Time Lapse Mode and the High-Dynamic Resolution Mode. The Time Lapse Mode established rapid automatic images to be taken. Be aware that the camera may not be able to take images as fast as the interval setting may suggest.

The maximum frame rate is determined by the image type/size selected in the camera, memory card speed, presence of shutter cable and other factors. If your camera cannot keep up with the Time-Lapse interval you set, some images may be skipped. Using a shutter cable is highly recommended for any intervals shorter than 2 seconds.

The Promote Control automatically calculates High Dynamic Range (or HDR) image sequences for you depending on a number of easy-to-use settings. Promote Control varies the camera shutter speed to achieve different exposures on each image taken. The aperture always remains the same because this is important in obtaining the correct HDR image sequences. This procedure is also known as "exposure bracketing". The Promote Control allows you to completely and remotely control exposure bracketing parameters without manually operating the camera controls. By default, the Promote Control takes a "middle" shutter speed as a start, and then takes the requested number of pictures with the shutter speed stepped under and over the "middle" exposure, thereby creating an increasingly brighter sequence of images. To program the Time Lapse Mode subroutine, follow these steps:

1. Make sure the camera is connected via the USB cable.

2. Select "Time-Lapse" mode by holding the "Left" button down while pressing "Mode".

3. Use "Interval" setting on Promote Control to specify how often a picture should be taken. The interval is specified in hours, minutes, seconds and tens of seconds. Set this to the shortest reasonable time, based on the settings established in the previous section (one-second is good for a training analog).

4. Specify how many "Frames", or images should be taken in total. For PoSSUM Sorties, set this to 'Infinite'. We certainly don't want to stop taking pictures until the end of the mission!

5. Use 'zero' for the "Start In" setting; we want the sequence to start immediately upon activation by the operator.

6. Switch your Promote Control to the "High Dynamic Range" mode.

7. Enable the "T-Lapse" setting.

8. Confirm that the automatic focus is disabled and select the shutter speed + aperture + ISO speed combination selected previously.

9. Set the to "Manual" exposure mode, and manually adjust exposure settings until the exposure compensation indicator shows zero.

10. Enter the exposure using the "Mid" setting.

11.   Set the "Step" setting on the Promote Control to 1.0EV for general HDR photography. You may want to use a smaller "Step" value for a fine-grained tonal control.

12.   Begin with a "Total Exposures" setting set to "07". For example, setting this value to "07" with a step of "1.0EV" will result in 7 images taken as follows:

        -3 EV steps from the "Mid Exposure" shutter speed
        -2 EV steps from the "Mid Exposure" shutter speed
        -1 EV step from the "Mid Exposure" shutter speed
        The "Mid Exposure" shutter speed
        +1 EV step from the "Mid Exposure" shutter speed
        +2 EV steps from the "Mid Exposure" shutter speed
        +3 EV steps from the "Mid Exposure" shutter speed

12.   Preview the resulting bracketing sequence displayed by the Promote Control in the bottom of the screen. If the entered parameters result in a bracketing sequence that is out of available exposure range, Promote Control will display an "[Invalid Settings]" message.

Generally choosing more steps is better, because you are then able to select the images you need from the resulting sequence. However, taking more than 11 total exposures with a step of 1.0EV or higher is rarely useful, and should only be used for very high contrast scenery, or when there is no time to reliably measure a start exposure. Overall, a combination of 1.0EV step with 7 total exposures works well for most situations.

Configuring the HDR

To configure the HDR settings, access the Promote Control "Setup Menu" by pressing and holding the "Left" button, while simultaneously pressing the "Right" button. You may also enter "Setup Menu" by pressing and holding "Left" and "Right" buttons simultaneously.

The "Setup Menu" is a series of screens each of which controls one setup value. Use the "Left" and "Right" keys to move between setup screens. Use the "Up" and "Down" keys to change values displayed on the screens. When finished making changes, press the "Center" key to return to the mode setting screens. As with all Promote Control settings, the "Setup Menu" values are saved whenever the Promote Control powers down, and are restored each time it powers back on. Most "Setup Menu" values default to settings that work well in standard applications.

OPERATING THE WIDE-FIELD IMAGER IN FLIGHT

Once programmed, charged, and mounted, the PoSSUM Wide Field Imager is ready for flight. Make sure to use a fresh set of batteries in both Promote Control and your camera and use a large memory card. The control system only needs to be engaged prior to flight and the sequence started just prior to launch. Both of these actions are performed through switches on the PoSSUMCam Main Control Panel. Once the Sequence Start switch in engaged, the system should behave autonomously throughout the mission.

POST-FLIGHT IMAGE PROCESSING OF HIGH DYNAMIC RANGE IMAGES

The following Windows/Mac applications may be considered for merging High Dynamic Range bracketed image series:

- Photomatix (commercial)
- HDR Efex Pro (commercial)
- HDR Darkroom (commercial)
- Picturenaut (freeware, Windows only)

Full Documentation

The complete user guide for the Canon EOS 5D mark III can be downloaded here:

http://usa.canon.com/cusa/consumer/products/cameras/slr_cameras/eos_5d_mark_iii#Brochures AndManuals

MESOSPHERE CLEAR AIR TURBULENCE EXPERIMENT (MCAT)

In this section, the Mesosphere Clear Air Turbulence (MCAT) instrument is described, as well as a background on the temperature, pressure and density structure of the mesosphere and how this is used to measure pressures and temperatures in the mesosphere.

In this section, the Mesosphere Clear Air Turbulence (MCAT) instrument is described, as well as a background on the temperature, pressure and density structure of the mesosphere and how this is used to measure pressures and temperatures in the mesosphere.

Figure 70. The Mesosphere Clear Air Turbulence (MCAT) instrument.

Mesospheric Temperature Detection

The mesosphere is the layer in the Earth's atmosphere between the stratopause at ~50 km and the mesopause at ~100 km. Temperatures in the mesosphere decrease with altitude due to less ozone heating and more infrared cooling to space. The atmospheric pressure near 50 km is only one thousandth of the surface pressure (1/1000 bar = 1 mbar = 100 Pa). As in the lower atmosphere, the pressure keeps decreasing exponentially with the e-folding "scale height" ($H = RT/Mg \sim 7$ km). This is equivalent to a factor of 10 every 16 km, so the pressure is only 0.001 mbar near the mesopause. The composition is almost identical as we find near the Earth's surface, mostly nitrogen and oxygen, and the relationship between densities, pressure and temperature can be represented by the ideal gas law.

164

In this section, the Mesosphere Clear Air Turbulence (MCAT) instrument is described, as well as a background on the temperature, pressure and density structure of the mesosphere and how this is used to measure pressures and temperatures in the mesosphere.

Mesospheric Temperature Detection

The mesosphere is the layer in the Earth's atmosphere between the stratopause at ~50 km and the mesopause at ~100 km. Temperatures in the mesosphere decrease with altitude due to less ozone heating and more infrared cooling to space. The atmospheric pressure near 50 km is only one thousandth of the surface pressure (1/1000 bar = 1 mbar = 100 Pa). As in the lower atmosphere, the pressure keeps decreasing exponentially with the e-folding "scale height" ($H = RT/Mg \sim 7$ km). This is equivalent to a factor of 10 every 16 km, so the pressure is only 0.001 mbar near the mesopause. The composition is almost identical as we find near the Earth's surface, mostly nitrogen and oxygen, and the relationship between densities, pressure and temperature can be represented by the ideal gas law.

The temperature of the mesosphere and its gradient vary strongly with season and latitude. The high-latitude stratopause is warmest in summer (~270 K ~ 0 °C) due to stronger ozone heating, and the resulting temperature and pressure gradients drive zonal mesospheric wind jets, easterly in the summer hemisphere and westerly in the winter hemisphere. However, the mesopause is observed to be coldest in summer (~140 K ~ -130 °C) due to a global meridional circulation from the summer hemisphere to the winter hemisphere. Air slowly ascends in the summer mesopause, thereby expanding and cooling adiabatically. This circulation is mechanically driven by a continuous flux of internal atmospheric gravity waves. They mostly originate in the troposphere from flow over mountains or deep convection also responsible for thunderstorms. Many waves will propagate upward, grow in amplitude due to the decreasing density, until they "break". They transfer their energy and momentum to the mean flow, slowing down and even reversing the zonal mesospheric jets. The result is the global meridional circulation in the mesosphere.

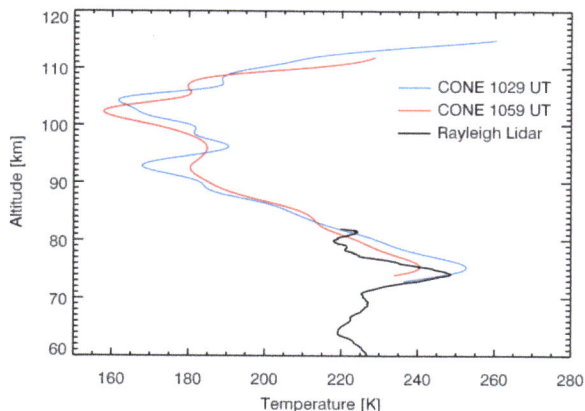

Figure 71. Temperature profiles measured on 17 February 2009 over Poker Flat, Alaska (credit: G. Lehmacher)

While the basics of global mesospheric dynamics has been suggested more than fifty years ago, the detailed processes of gravity wave propagation, interaction with other waves and the background flow, wave instabilities and dissipation, and the generation of turbulence are still intensely studied. They play crucial roles for the accurate physical description of the mesosphere, short-term predictions of "space weather", and long-term trends of temperatures and densities, even ranging up into the upper atmosphere where low-earth satellites orbit.

Temperature modulations due to gravity waves are often 20 K or more over a depth of a few kilometers, giving rise to stable mesospheric inversion layers, where the temperature gradient is positive, and adiabatically unstable layers, when the temperature decreases more than ~10 K/km. Such temperature variations correspond to density variations of 10 %, and can persist for hours or days. Figure 76 shows an example of temperature profiles measured over Alaska with such layers near 70 km and 90 km. They could be called the very largest scales of mesospheric "clear air turbulence" (CAT) in analogy to similar structures and instabilities in the troposphere, which can be dangerous to aircraft. But turbulence is also and more often evident in density and temperature fluctuations at spatial scales much smaller than gravity wave scales. A large Reynolds number ($Re = UL\rho/\mu$) is a characteristic of well-developed turbulence. Since the Reynolds number is proportional to the fluid density ρ, turbulence will eventually become less important with altitude in favor of viscous fluid motion. This transition is called the turbopause, which occurs also near the level of the mesopause (100 km). Therefore, depending on altitude and turbulence strength, turbulent density fluctuations are observed at scales of 100 meters down to meters and centimeters.

Since gravity wave amplitudes increase with altitude, the upper mesosphere is the region of the strongest effects of turbulence, such as mixing of heat and trace gases. Crossing through the region with a sounding rocket or space plane allows for immediate and local observation ("in situ") of the density and temperature structure and is the only way to observe the distribution of small-scale turbulent density fluctuations.

Vacuum Gauges

There are many different techniques available to measure pressure or density at the required pressures of 0.001 to 1 mbar. Examples are ionization gauges, capacitance gauges, and thermal conductivity (Pirani) gauges. The latter are most commonly used in the laboratory for this pressure range. Here we are mainly concerned with hot-cathode ionization gauges, which are fast and sensitive, with a large dynamic range. In these gauges, which are based on the design of vacuum triode, a filament is heated and electrons are emitted and accelerated towards a positive anode (Figure 77). The electrons gain sufficient energy to ionize gas molecules they encounter and the collected positive ion current is proportional to the electron emission current and the neutral gas density. Normally the emission current is held constant by adjusting the filament heating current. Pressures up to 1 mbar are the upper limit that the electron emission current can be maintained and ionization gauges are normally used at much lower pressures (0.001 - 10^{-9} mbar). While the ionization gauge is a standard laboratory technique, matters are more complicated by attaching the probe to a spacecraft, which moves at supersonic speeds and is surrounded by shock waves (Figure 73).

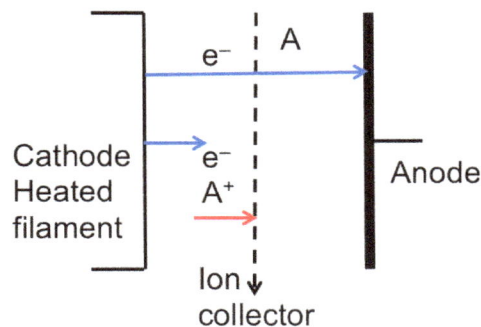

Figure 72. Hot-cathode ionization gauge schematic.

One method is to measure the stagnation pressure by directing some gas into a volume where it comes into thermal equilibrium with the instrument walls. The ratio between ambient and stagnation pressure may be calculated from the supersonic Rayleigh-Pitot equation. Such an instrument is called a Pitot probe or Pitot tube, since a small tube protrudes from the front or the wings of the vehicle connecting it to the pressure sensor. A standard appliance on aircraft, it has been applied to sounding rockets as well. Suborbital spacecraft for the mesosphere travel at speeds of Mach number M ~ 3 (1000 m/s). The Mach number is the ratio between vehicle speed and the speed of sound. The stagnation pressure is roughly proportional to the vehicle speed or Mach number squared, therefore about 10 times larger than the ambient pressure. At low pressures in the upper mesosphere, when the mean free path of air molecules becomes a substantial fraction of the instrument dimension, the Rayleigh-Pitot equation is no longer valid and the ratio between ambient and instrument pressure must be determined for molecular flow conditions or numerical flow simulations.

The temperature profile is calculated by integrating the density profile assuming hydrostatic equilibrium and a start temperature $T(z_0)$,

$$T(z) = \frac{1}{n(z)}\left(T(z_0)n(z_0) - \frac{M}{R}\int_{z_0}^{z} n(z')g(z')dz'\right) \qquad (1)$$

Figure 73. Shock waves in supersonic wind tunnel (credit: XCOR Aerospace).

Since it takes a finite length of time for the gas to pass through the tube to the instrument, small-scale turbulent density fluctuations may be averaged out and lost in this process. If we wish to observe 1-meter scales at a speed of 1000 m/s, a measurement resolution of one millisecond or less is required. Therefore, an open sensor provides better time resolution for atmospheric observations. The CONE ionization gauge (Fig. 1) has been developed for this purpose. Its transparent and spherical construction allows for air to flow around the nested electrodes. The measured pressure is therefore less dependent on the Mach number, only about 2 to 3 times greater at M ~ 3, and less sensitive to the direction of the air flow due to the spherical geometry.

All pressure gauges need direct exposure to the ambient air and therefore need to extend from behind doors or the main body. On sounding rockets instruments are often stowed below a nosecone, which is deployed during the flight.

Drag Acceleration

Besides measuring the density directly, the acceleration due to the drag force on a body is also proportional to the air density. For large velocities the drag acceleration (or rather deceleration) is proportional to the velocity squared,

$$a_D = \frac{C_D A}{2m} \rho |v|^2 \qquad (2)$$

For objects in free fall like satellites, suborbital rockets or rocket payloads, and the absence of other inertial forces (e.g, due to rotation) the drag force is the largest force and can be measured by sensitive accelerometers. It strongly depends on the area over mass ratio, A/m, and for mesospheric densities and typical speeds, it can range from 10 m/s^2 for a light, inflatable, falling sphere, to very small values of 10^{-2} to 10^{-4} m/s^2 for a heavy vehicle.

The principle has been developed in the 1960s and 1970s in form of the passive, inflatable falling sphere as a standard technique to measure densities, temperatures, and winds in the mesosphere. Folded spherical balloons of 1-meter diameter and 150-gram mass are launched onboard of small rockets and deployed and inflated at 100-120 km. The descent is tracked by precision radar and the full trajectory information is used to derive the atmospheric parameters. Smaller spheres have been equipped with accelerometers, GPS receiver, power, and telemetry to obtain winds and densities in the lower thermosphere.

A sphere is the preferred shape, since the drag force is independent from the orientation. Nevertheless, a lot of complexity is hidden behind the drag coefficient C_D, which depends on Mach number and Reynolds number, which also depends upon temperature and density. For the inflatable falling sphere, wind tunnel experiments have been compiled to an extensive drag table. For shapes like a cylindrical sounding rocket payload, the orientation must be known, and the drag coefficient must be determined from other measurements or numerical simulations. Fig. 74 shows the result of a pressure distribution for a supersonic sphere. If we may assume that orientation and drag coefficient change only gradually, the residual variations in acceleration may be representative of density variations due to gravity waves or turbulence.

The PoSSUM MCAT System

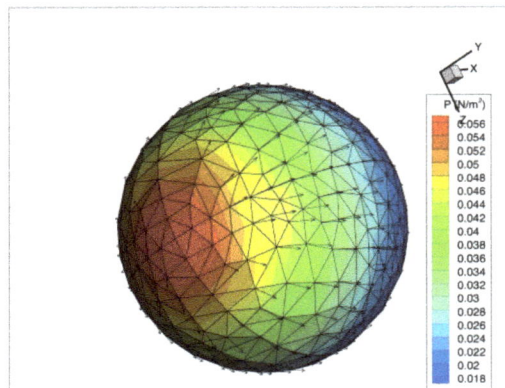

Figure 74. Pressure distribution for a sphere in supersonic flow (credit: G. Lehmacher).

The purpose of the PoSSUM MCAT system is to characterize the variations of atmospheric density and temperature for altitudes of ~60-103 km, which are key data to help interpret other observations and guide the gravity wave models. Ideally, we would use an open ionization gauge like CONE that is extended into the atmosphere beyond the shock wave envelope. An alternative is a Pitot tube experiment that is capable of measuring mesospheric pressures. Such a Pitot tube may already be part of the avionics equipment of reusable space planes, and only a suitable pressure gauge may be required.

The same is true for an accelerometer experiment. Any sounding rocket payload contains accelerometers to monitor vibrations during launch and motor burn. More sensitive accelerometers will provide data about inertial motion and drag. This information is useful for monitoring the environment for onboard micro-g and milli-g experiments, as well as deriving the density and turbulence structure, especially during re-entry. Suitable accelerometers require very little space and power and should be easily integrated into any payload, ideally near the center of gravity.

To derive a temperature profile, according to Equation 1 above, only a relative density profile is required. For both pressure gauge and accelerometer experiments, the variation of the most relevant correction factors, ram pressure correction and drag coefficient, respectively, must be known. Using knowledge of the velocity vector, orientation, and a detailed surface model of the spacecraft, modeling software must be used to calculate the flow environment and drag coefficients for the mission specific parameters. Small-scale variations can be directly used to estimate atmospheric density fluctuations.

Operational Procedures

Operator interface is simple an involves simply 1) turning on power and data collection systems just before takeoff, and 2) extending the probe after Main Engine Cut Off (MECO)

Operations Procedures:

T-30 seconds: MCAT + MASS Power – **ON**

T+180 seconds (MECO): MCAT Probe – **EXTEND**

T+480 seconds (Prior to Re-entry): MCAT Probe – **RETRACT**

On Landing: MCAT + MASS Power - **OFF**

MESOSPHERIC AEROSOL SAMPLING SPECTROMETER (MASS)

MASS is a PoSSUM instrument designed to complement observations on the PoSSUM Noctilucent Cloud Tomography Experiment. Designed to fit on a suborbital Reusable Launch Vehicle (sRLV), MASS is a reduced size version of a similar MASS instrument that has flown four times into the polar mesosphere on sounding rockets from the Andoya Rocket Range in Norway. In this module, the concepts behind the Mesospheric Aerosol Sampling Spectrometer (MASS) instrument will be described, as well as the physical properties and the operational concepts of the instrument.

Operational History of the MASS Instrument

The MASS instrument is an 8-channel mass spectrometer that measures the number density of both positively and negatively charged aerosol particles in 4 ranges of mass spanning 0 – 50,000 atomic mass units (amu). The instrument flew in July 2007 twice, once into noctilucent clouds (NLC) and once into Polar Mesospheric Summer Echoes (PMSE), and returned data on the numbers of icy particles responsible for these phenomena. In 2011, the MASS instrument was flown twice outside the NLC season (Figure 80) and returned data on the number density and charge state of meteoric smoke particles (MSPs) that are probably the condensation nuclei for NLC; it is commonly believed that NLCs form from nucleation on meteoric smoke particles.

Figure 75. The Mesospheric Aerosol Sampling Spectrometer (MASS) instrument.

In 2014, the MASS instrument was miniaturized and is currently undergoing tests on sounding rockets, where it will be mounted at the forward deck of the rocket payload under a deployable nosecone. The rocket flights have employed attitude control so that the air flowed into the slit is nearly parallel to the instrument axis. The shape of the instrument, including the slit opening geometry and air exit windows, were designed for the rocket speed and the forward rocket bulkhead position, and proper aerodynamics flow patterns were ensured by extensive simulations using the Direct Simulation Monte-Carlo method.

The MASS instrument is currently being matured for use on PoSSUM campaigns. For these campaigns, the instrument performance will be enhanced by combining detection capabilities for the nanometer-sized meteoric smoke particles and the larger icy NLC particles. The science objective of the MASS instrument to be flown on PoSSUM flights is to obtain simultaneous measurements of the number densities of NLC particles and of their condensation nuclei for validation of models for the nucleation, growth, and sedimentation of icy particles in the mesosphere.

Instrument Description

The MASS instrument is a tapered box-like structure with a slit opening at the top and with four collecting electrodes on each of two sides. The current MASS instrument to be flown on PoSSUM missions is a reduced size version of the previously designed and flown MASS instruments [Knappmiller et al., 2008]. Figure 81 is a schematic diagram showing how the instrument operates and Fig. 3 is a photograph of the MASS instrument.

The MASS instrument can be divided into the following subsystems: sensor, analog electronics, digital electronics and power supply/battery pack. The sensor is the aerodynamically designed mechanical structure with integrated particle collecting electrodes. The analog electronics subsystem includes the current-to-voltage converters and the battery packs for the bias voltages. The digital subsystem consists of the analog-to-digital converters, on-board data handling microprocessor, memory and communication interface. The flown mass instruments did not have any digital parts or memory, and the analog signals were handed over to the telemetry unit of the rocket payload.

Launched from Andoya Rocket Range
2009 – NLC
2011 – Meteoric Smoke

Figure 76. MASS instrument being launched on a sounding rocket from Andoya, Norway

The MASS instruments used on the sounding rockets were powered by +28V provided by a battery pack on the payload. The low voltages (+/- 9 V) required for the operation of the analog electronics were converted onboard. The MASS instrument modified for PoSSUM flights will be powered from a 28V provided by the craft, or from an internal battery pack, if necessary.

During flight, air enters through the slit at the top of the instrument, which is oriented along the expected flowlines of the atmosphere, and exits through the windows near the bottom. One side of the instrument has electrodes with negative voltages to collect positive charge and the other side has electrodes with positive voltages to collects negative charge. The collectors have bias voltages that increase with distance from the entrance slit, causing the more massive particles to be collected on electrodes further from the slit. Sensitive electrometers at the bottom of the instrument measure the current collected by the electrodes, which is of the order of 1 nA. The current is converted to a number density of charged aerosol particles using the payload velocity, slit area, and results from detailed numerical simulations of the airflow through the instrument. The collected particles are sufficiently small that it is unlikely that they carry more than one charge, which simplifies conversion of current to number density.

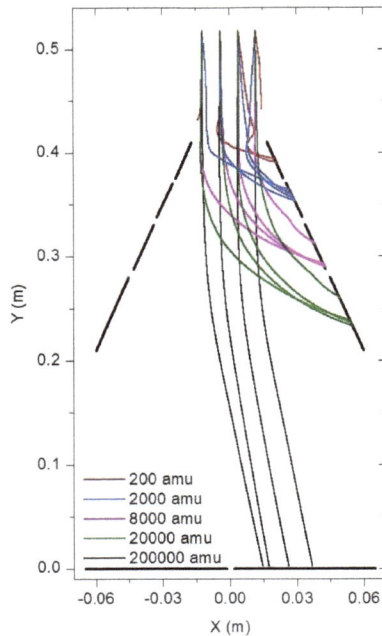

Figure 77. Computer-generated trajectories of singly charged particles of different masses within the instrument. These trajectories are used to determine the range of mass collected by each of the five pairs of electrodes. The fifth pair is on the bottom of the instrument and is used to detect neutrals (credit: Z. Sternovsky)

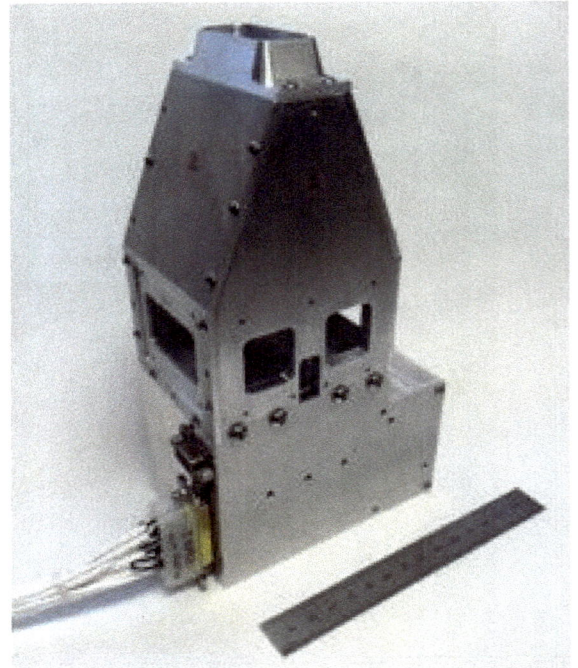

Figure 78. Photo of a partially constructed mini-MASS spectrometer designed to be a part of an instrument cluster on the Norwegian MAXIDUSTY sounding rocket to be launched into NLC in the summer of 2014. The ruler is 15 cm. The air enters the slit at the top and exits via the vents on bottom (credit: Z. Sternovsky)

The miniaturized version of the MASS instrument designed for PoSSUM flights measures approximately 100 mm L x 100 mm W x 150 mm H, and rests upon an electronics box measuring roughly 100 mm x 100 mm x 100 mm. The electronics box accommodates the voltage bias batteries and the analog current-to-voltage converters, and can be separated from the spectrometer by a meter of shielded cable, if necessary. Computer simulations of the instrument were conducted, showing the electrostatic potentials of the instrument. Results of these simulations are shown in Figure 79. In this simulation, charged particles are attracted to the detectors, which are set at a specific electrostatic potential. The first pair (pair closest to the flow intake port) is set at a potential of 4V. The second pair is set to a potential of 10V, the third to 30V, and the fourth and last to 54V. Particle charges are thus binned according to their charge-mass. Figure 5a and 5b show the flow velocities and the particle number densities through the instrument at standard operating conditions.

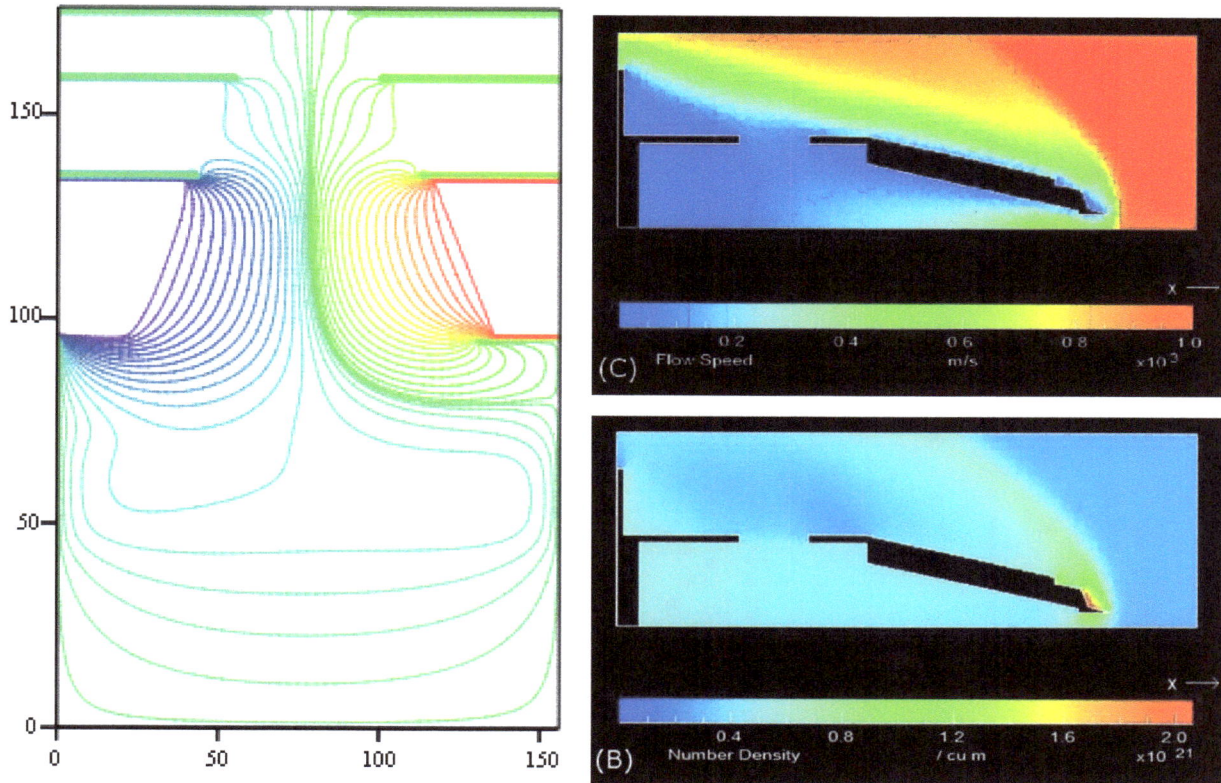

Figure 79. a) Computer model of electrostatic field and gas flow (left) , b) flow speed (m/s) through the MASS instrument at operating velocities (upper right), c) number density of particles passing through the MASS instrument (lower right) (credit: Z. Sternovsky).

Flight Results

The 2007 and 2011 MASS campaigns were successful (all channels returned useful data throughout the desired altitude range) and proved the novel MASS instrument concept to have unique capabilities with great scientific potential. The sounding rocket platform returns data from only a short, nearly vertical path within the mesosphere and having the instrument adapted to suborbital vehicles or dipper satellites would allow more lengthy data sets permitting studies of spatial and temporal variability. Results from the MASS instrument, which was flown through an NLC layer in 2007, are shown below in Figure 80. The charged particles are binned in four groups, drawn by their electrostatic potential. In this way, the number density of particles of the following sizes may be inferred: less than 0.5nm, 0.5-1.0nm, 1.0-2.0nm, and greater than 3.0nm. Number densities are displayed as a function of altitude.

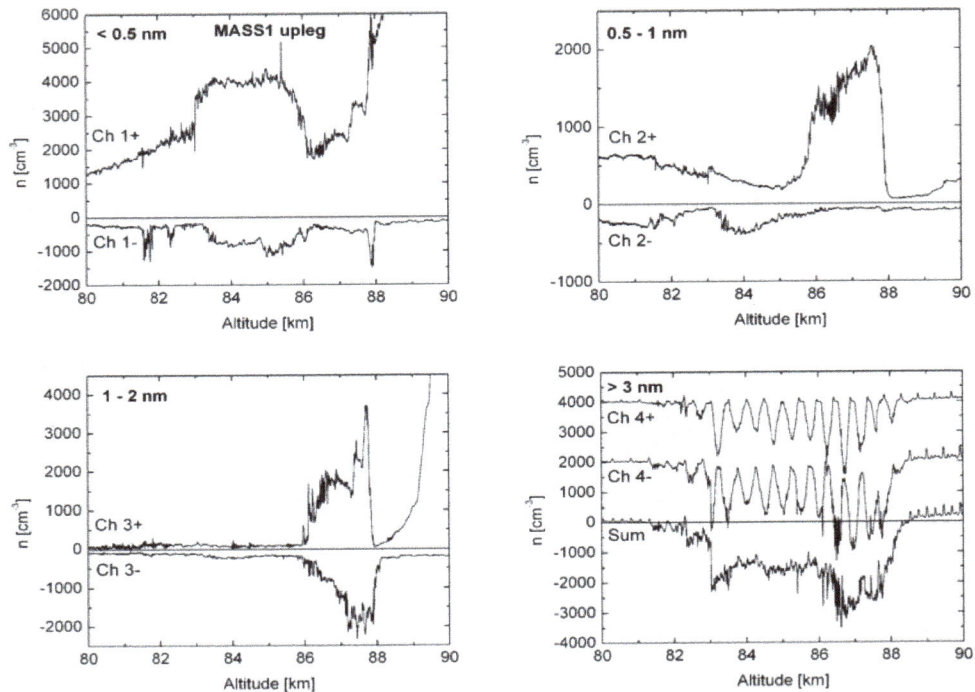

Figure 80. Results from previous NLC Campaigns (credit: Z. Sternovsky)

Benefits of the PoSSUM MASS Instrument

Below is a list of improvements needed to utilize the MASS instrument on PoSSUM flights using rSLVs. This version of the MASS instrument is selected for further maturation due to its smaller size and thus greater potential to fly as a standalone instrument, or as part of suite. There are several unique benefits to implementing the MASS instrument on rSLVs on PoSSUM flights:

1. **There is a need for the horizontal measurement of particle distributions.** A dedicated sounding rocket campaign equipped with an expensive attitude control system allows taking two snapshot measurements of the particles' vertical distributions, upleg and downleg, typically separated by ~70 km. Much of the scientific interest, however, is in the horizontal spatial variability due to gravity waves with long wavelengths. Covering large horizontal distances is possible using 'dipper' spacecraft of future sRLVs. We envision that such enabling capabilities will exist in the near future.

2. **Fast response and high repetition rate measurements.** Launching a sounding rocket requires careful planning and an expensive support crew on ground. Usually, the frequency of lunches is limited to a separation of several days and the trajectory (location of impact) is difficult to change on a short notice. Suborbital Reusable Launch Vehicles, on the other hand, can be used more frequently, on a shorter notice and potentially with an adjustable trajectory. sRLVs that can reach ~100 km altitude are thus an ideal platform for the observations and characterization of NLCs, which show variation on an hourly basis. Further, sRLVs may be able to respond to transient events such as volcanic eruptions that may inject dust to high altitudes.

3. **Measurements from lower altitudes.** The nosecone on the sounding rocket is usually not deployed below 60 km and thus meteoric smoke particle are not characterized below this altitude. sRLVs may offer measurements from lower altitudes.

Operational Procedures

MASS is designed to fit within the CP and CS payload pods onboard the XCOR Lynx suborbital vehicle. Operator interface is simple an involves simply 1) turning on power and data collection systems just before takeoff, and 2) opening the payload bay covers after Main Engine Cut-Off (MECO).

> Operations Procedures:
>
> T-30 seconds: MCAT + MASS Power – **ON**
>
> After Landing: MCAT + MASS Power - **OFF**

MESOSPHERIC SAMPLING OF CO2 and NITROGEN OXIDES

Sampling Method

Accurate monitoring of carbon dioxide (CO_2) and Nitrogen Oxides (NOx) is an objective of PoSSUM flights. CO_2 and NOx are believed to contribute towards the heating of the troposphere while, at the same time, cooling the mesosphere and thermosphere. Their effects on the F2 layer of the ionosphere and wave activity near the mesopause remain uncertain.

CO_2 radiates infrared energy, heating the lower atmosphere and cooling the upper atmosphere, and this selective radiation is isotope selective, meaning that the heating mechanisms are affected by different atmospheric constituents differently. However, only two oxygen isotope measurements have been made in the upper atmosphere and no carbon isotope measurements have been made.

CO_2 exists in abundance above 80 km, and the majority of *in situ* CO_2 abundance measurements of the mesosphere and lower thermosphere took place before 1976 with means that introduced up to 20-30% errors. Meanwhile, CO_2 abundance in the troposphere has increased 21% and the thermosphere is changing in response to climate change.

Furthermore, Nitrogen oxides play an important role in atmospheric chemistry. Nitrogen oxides are known to be hazardous to human health and contribute to ozone production in the troposphere. Nitrogen Oxide (NO), however, helps to cool the upper atmosphere. To summarize, CO_2 and NO are important atmospheric gases with limited understanding of their abundance. This abundance varies due to anthropogenic activity and, in the case of NO, due to the solar cycle.

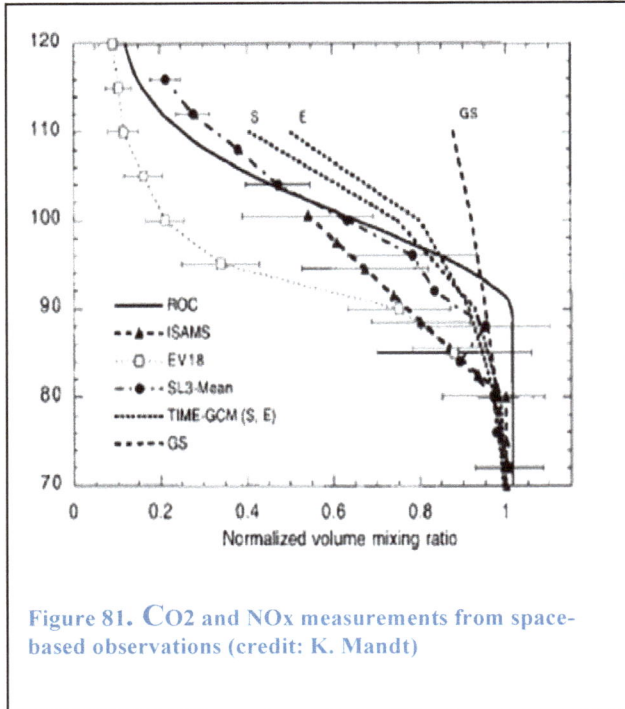

Figure 81. CO_2 and NOx measurements from space-based observations (credit: K. Mandt)

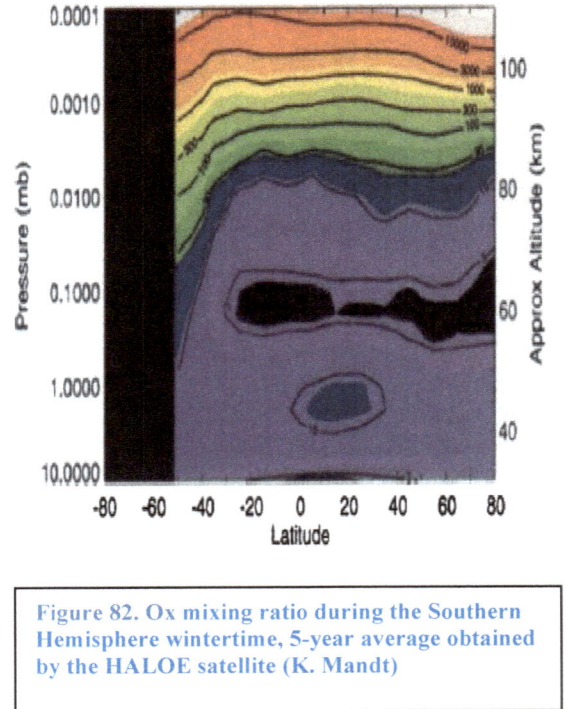

Figure 82. Ox mixing ratio during the Southern Hemisphere wintertime, 5-year average obtained by the HALOE satellite (K. Mandt)

Operational Procedures

Operator interface involves 1) turning on power and data collection systems just before takeoff, and 2) taking a sample at NLC altitudes as the spacecraft penetrates the cloud layer on ascent and on descent. Both of these actions are done by actuating switches on PoSSUMCam's Main Control Panel.

Operations Procedures:

T-30 seconds: SAMPLER Power – **ON**

On Ascending Penetration of NLC Layer: SAMPLE – **ENGAGE**

On Descending Penetration of NLC Layer: SAMPLE - **ENGAGE**

After Landing: SAMPLER Power - **OFF**

PART IV: NOCTILUCENT CLOUD TOMOGRAPHY EXPERIMENT

Noctilucent Cloud Tomography Experiment Concepts

Observing Noctilucent Clouds from Research Aircraft

Observing Noctilucent Clouds from Suborbital Spacecraft

The Suborbital Space Flight Simulator

NOCTILUCENT CLOUD TOMOGRAPHY EXPERIMENT CONCEPTS

Image credit: Virgin Galactic

Module Objectives

➢ Understand the Science Objectives of the NLC Tomography Experiment
➢ Understand the campaign plan and constraints
➢ Successfully fly mission profiles using the Suborbital Space Flight Simulator as both the pilot and scientist

Contributing Author:

Dave Fritts, Ph.D.
Jason Reimuller, Ph.D.
Vasco Ribeiro

Science Objectives

The PoSSUM Noctilucent Cloud Tomography Experiment will address two specific science goals relating to our understanding of the upper mesosphere:

1)   Use noctilucent cloud and OH-layer tomographic imaging methods to identify the dominant gravity wave instability and turbulence dynamics, scales, and intensities, contributing to gravity wave dissipation and momentum deposition, and:

2)   Perform high-resolution numerical modeling of gravity wave instability and turbulence dynamics to interpret noctilucent cloud imaging quantify these dynamical influences in the upper mesosphere.

Campaign Plan

Each PoSSUM Tomography Experiment sortie will employ a manned reusable suborbital vehicle (e.g. Virgin Galactic Spaceship Two) from a high-latitude spaceport. Project PoSSUM noctilucent cloud campaigns require launches from latitudes where noctilucent clouds could be observed, since noctilucent clouds form generally at latitudes higher than 60 degrees. The clouds also form with greater intensity at higher latitudes. The leading candidate of the 2016 Mission is at Eielson AFB, near Fairbanks, Alaska.

Initial calibration flights will occur prior to deployment in order to 1) qualify the payload in the suborbital spaceflight environment, 2) provide added flight crew training, and 3) produce key background imagery from the same cameras.

Mission Profile

The initial design of PoSSUM instrumentation assumed the use of the XCOR Lynx Mark II vehicle, which was a suborbital rocket plane with a targeted apogee of 103 km. The vehicle will launch vertically with enough velocity to ascend through a NLC layer, assumed to stratify at 83 km of altitude. After the discontinuation of the development of the XCOR Lynx vehicle in 2016, PoSSUM instrumentation was reconfigured for use on the Spaceship Two suborbital vehicle, whose profile is shown in Figure 87. The mission profile of Spaceship Two is similar to that of the Lynx in that both vehicles will be decelerating to subsonic velocities as they pass through the altitudes at which NLCs form and then accelerate to supersonic velocities as it penetrates the NLC layer again upon descent.

As the boresight of the imagers will be oriented longitudinally and forward, the spacecraft will assume an attitude where the axis along which the cameras will be oriented would be pointed northward from main engine cut-off to apogee. The pitch of the spacecraft would be adjusted manually so that the payloads would see the limb of the noctilucent cloud layer up through penetration of the cloud layers.

Launch Commit Criteria

When strong cloud formations are observed from the ground or from LiDAR, the spacecraft will be launched to an altitude that transitions the noctilucent cloud layer. The clouds will be under direct illumination from the sun and the attitude of the spacecraft would be oriented north to the presumed region of highest cloud density.

Launch opportunities could not be guaranteed on any specific night as there is no way to predict on what days the noctilucent clouds will be present. However, historical trends indicate one good opportunity every three to five days. A 'Commit to Launch' decision will be made when strong cloud formations are observed visually or from local ground-based LiDAR.

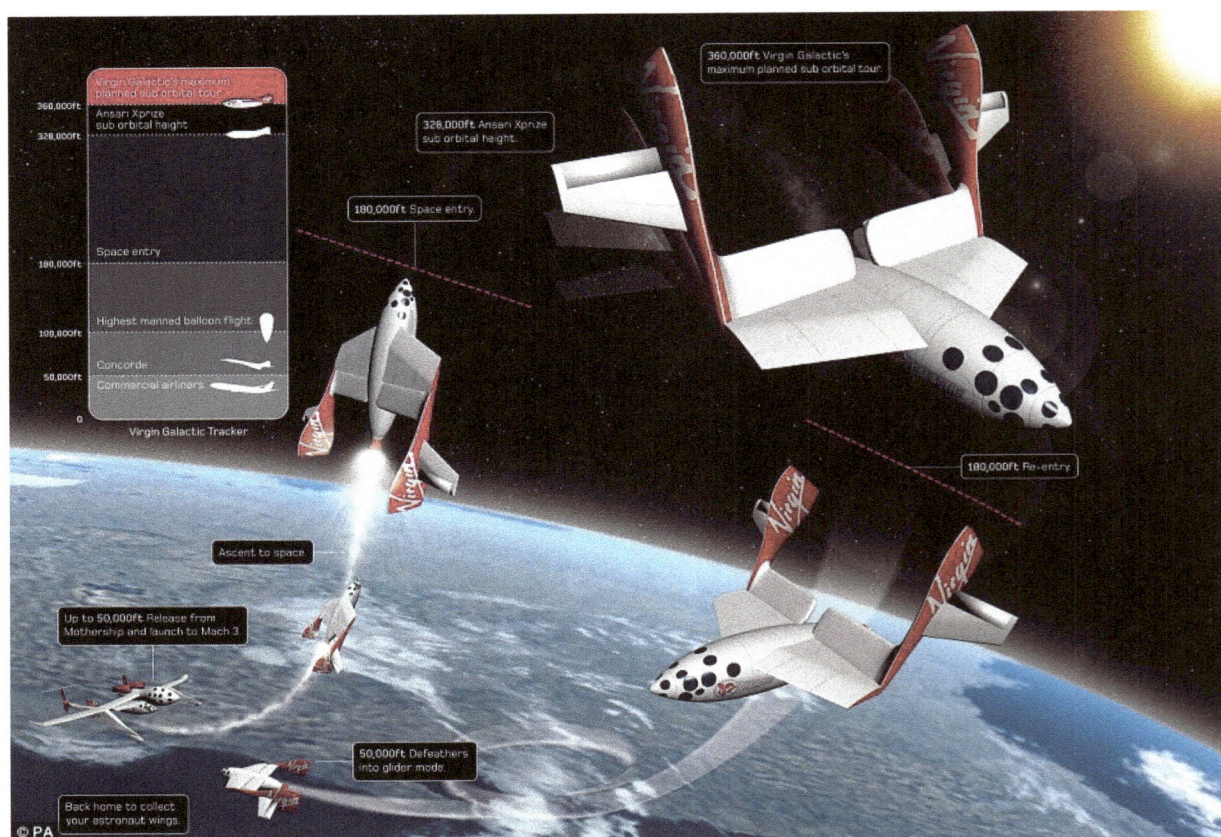

Figure 83. Spaceship Two Mission Profile (Credit: Virgin Galactic)

The Need for Tomographic Imagery of Small-Scale NLC Structures

Transport and deposition of energy and momentum by small-scale gravity waves play central roles in vertical coupling extending from Earth's surface to the mesosphere and lower thermosphere. However, the details of these dynamics, their spatial scales and intermittency, and their consequences for atmospheric structure and variability depend on instabilities and turbulence dynamics at the smallest scales of motion. These dynamics are poorly understood at present because measurement techniques able to quantify them are very limited.

180

Satellite remote-sensing instruments fail to measure the most important small-scale gravity waves; the best current satellite sensors can resolve horizontal wavelengths on the order of 10-100km. Ground-based radars, lidar, and imagers can often describe qualitatively small-scale gravity waves and larger-scale instability structures, but typically fail to quantify these events and their transitions to turbulence at smaller scales. Until recently, only high-resolution in-situ rocket and radar measurements have succeeded in quantifying turbulence intensities, but even these have not defined the character of the turbulence fields or their impacts on the evolving larger-scale environment. The absence of a method to perform systematic measurements of small-scale gravity waves and their instabilities and dissipation, and the implied lack of knowledge of the consequences of gravity wave momentum deposition, represent major deficiencies in our understanding. They also cause gravity wave effects to be poorly described in global models at all altitudes. The PoSSUM Tomography Experiment is designed to provide an improved understanding of the instability and turbulence dynamics accounting for small-scale gravity wave dissipation and momentum deposition, and the spatial and temporal scales of such events.

Gravity wave instabilities and turbulence dynamics in the upper mesosphere was inadvertently observed by star trackers onboard the EBEX cosmology experiment aboard a stratospheric balloon over the Antarctic in Austral summer 2011-12. These images yielded serendipitous images of noctilucent clouds at altitudes of 83km that revealed detailed turbulence structures as small as the "inner scale" of turbulence at ~10-20 m. Comparisons of a number of these images with high-resolution numerical simulations of GW instability dynamics and turbulence reveal a potential to interpret and quantify these dynamics, the turbulence intensities, and their consequences at larger scales [Miller et al., 2015].

The PoSSUM Tomography Experiment will explore these same small-scale observables, now known to exist and contain a wealth of information pertaining to energy and momentum deposition, through tomographic methods. Much like an MRI builds a three-dimensional representation of the human body, PoSSUM tomographic imaging methods will observe dynamical features in our upper mesosphere. The imaging and supporting numerical modeling developed for Project PoSSUM will capture gravity waves, instabilities, and turbulence dynamics at scales from tens of kilometers down to several-meter resolution and will quantify gravity wave instability and turbulence events to a degree not previously possible with any other measurement technique. These structures have been difficult to resolve from previous means of observation from space-based or ground-based imagers but are believed to contain most of the information pertaining to energy and momentum deposition in the upper atmosphere.

Gravity Waves and Associated Energy and Momentum Transport

Gravity waves are now widely understood to play key roles in controlling the large-scale circulation, thermal structure, and variability of the upper mesosphere. As gravity waves propagate upwards, forcing from below by large and small-scale gravity waves, their modulation by background filtering conditions, and their transport and deposition of energy and momentum are the dominant influences on these upper mesospheric dynamics at the small spatial and temporal scales that are not visible from previous means of observation. On global scales, gravity waves force closure of the mesospheric jets in summer and winter and drive a residual circulation from the summer to the winter hemisphere that results in strong cooling of the summer mesopause and warming of the winter mesopause at high latitudes [Fritts and Alexander, 2003]. Gravity waves also appear to have potentially large influences extending to much higher altitudes than initially believed. These include instabilities that may occur at altitudes well above the usual "turbopause" altitudes of 105 – 110km. (~105-110 km) [Lund and Fritts, 2012; Fritts et al., 2015a] and localized gravity wave packets and momentum deposition that yield secondary gravity waves that may have strong influences extending well above the mesosphere [Yamada et al., 2001; Vadas and Fritts, 2002, 2004; Vadas, 2007; Vadas and Liu, 2009; Fritts et al., 2014a; Laughman et al., 2015].

Importantly, the major dynamical contributions in this region of the atmosphere accompany gravity waves having small horizontal wavelengths and packet scales of approximately 30-200 km, and these events are very frequent. However, there has been very sparse observational guidance on the character, scales, and consequences of local gravity wave packets and their dissipation and momentum deposition to date. This is the focus of PoSSUM tomography flights.

Instabilities, modulation of gravity wave momentum deposition, and turbulence

From global atmospheric models, we now understand the qualitative aspects of gravity wave forcing of the upper atmosphere, but we still know very little about the gravity wave interactions and instability dynamics and their quantitative effects in the mesosphere and above. Specific roles of gravity wave and instability dynamics include energy and momentum fluxes, flux divergence and flow accelerations accompanying gravity wave dissipation, energy transfers via wave-wave interactions, constraints on gravity wave amplitudes, turbulent mixing, and evolution of the gravity wave spectrum in altitude and time

To date, the various roles of gravity waves and associated instabilities are poorly quantified because the small scales on which they often

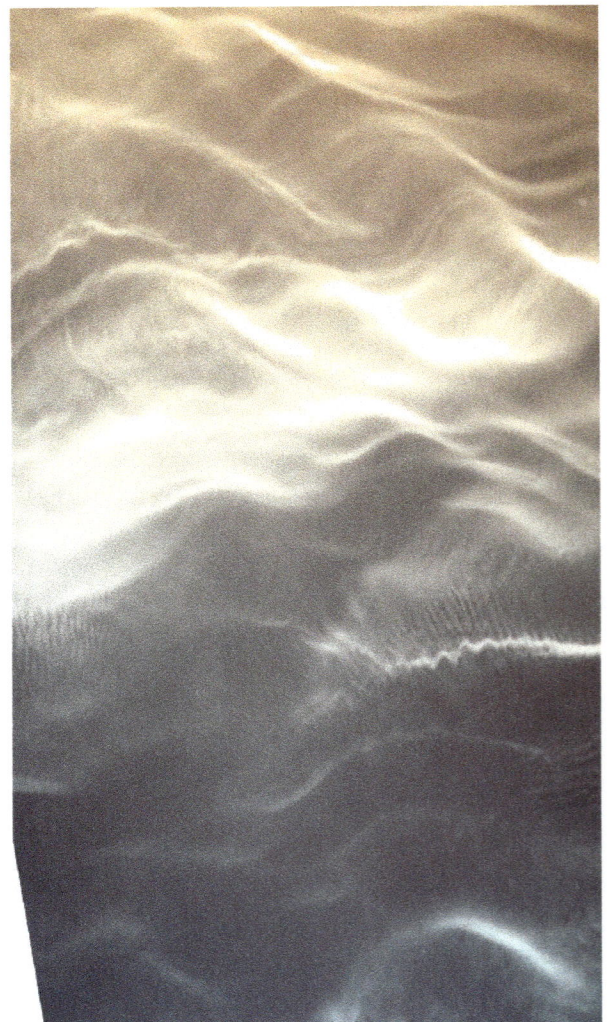

Figure 84. NLC image showing gravity wave and instability structures extending from ~59-61.3°N (~250 km, bottom to top) projected assuming an 82 km altitude, so vertical displacements yield apparent horizontal structures [Baumgarten and Fritts, 2014].

occur are not discernable from ground-based observations or from satellites. Thus, the complexity of instability dynamics, and our inability, until recently, to characterize such processes observationally spanning the relevant scales have remained largely unknown [e.g., Yamada et al., 2001; Hecht, 2005; Lehmacher et al., 2007; Pfrommer et al., 2009].

New theoretical, modeling, and measurement capabilities, however, are enabling a dramatically more quantified view of these complex dynamics. PoSSUM balloon observations will seek to focus more on the temporal variability of these gravity wave signatures while the PoSSUM tomographic images obtained through suborbital rSLV flights will model these features at high spatial resolution.

More general forms of these instabilities, and others, also readily occur through *superpositions* of various gravity waves and mean motions at larger and smaller spatial scales. These interactions occur in the upper atmosphere at most times and locations due to the ubiquity of gravity waves from multiple sources at lower altitudes and gravity generation by interactions and instabilities within the upper atmosphere. An NLC image from the ground in Germany showing multiple superposed gravity waves and embedded instabilities is shown in Figure 88.

The implications of these gravity wave and instability dynamics for local momentum deposition, turbulence, mixing, and gravity wave spectral evolution with altitude, are largely unknown. However, these are also the most important processes in understanding the variable responses to gravity wave forcing in the mesosphere and throughout the atmosphere. Significant progress will require observations spanning the major gravity wave and instability scales, and high-resolution numerical modeling to guide their interpretation.

Secondary gravity waves arise from both nonlinear interactions and instabilities (and dissipation) within the gravity wave field. Despite the apparent importance of this source of larger-scale gravity waves that would impact the thermosphere to much higher altitudes, observational guidance on the scales of these events is almost nonexistent at present.

PoSSUM Imagery and Tomography

Specifically needed are observations that will quantify the spatial and temporal scales of gravity wave packets and their instabilities and dissipation at an altitude (or altitudes) in the upper atmosphere where their propagation and instability dynamics can be quantified as fully as possible. The PoSSUM Noctilucent Cloud Imagery and Tomography experiments will quantify for the first time the finest details of small-scale gravity wave, instability, and turbulence dynamics occurring in the upper atmosphere [e.g., Miller et al., 2015]. PoSSUM high-resolution modeling of gravity wave and instability and turbulence dynamics will allow analyses that:
    1) quantify gravity wave scales, amplitudes, and intrinsic properties,
    2) characterize the dominant instability dynamics, spatial and temporal scales,
    3) identify turbulence onset dynamics, evolutions, intensities, and intermittency, and
    4) estimate the implications of instability and turbulence dynamics for gravity wave

momentum deposition and secondary gravity wave generation.

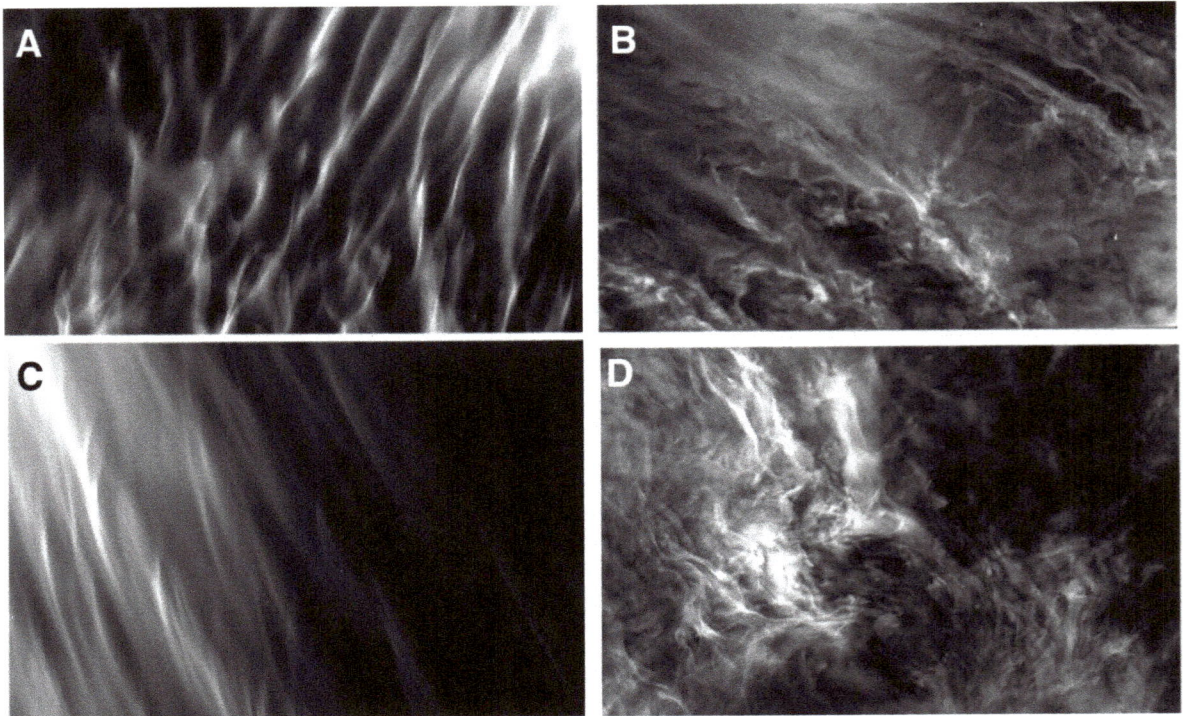

Figure 85. Example of EBEX images showing laminar intertwining vortices (left) and clear turbulent features that exhibit coherent vortices (right). In each case, high sensitivity to small spatial scales apparently arises due to the occurrence of initial instabilities on thin vortex sheets

To illustrate these capabilities, four NLC images obtained during the EBEX measurement program are shown in Figure 90. PoSSUM numerical models agree reasonably with the smallest scales seen in these images and these models are able to achieve realistic descriptions of gravity wave, instability, and turbulence dynamics that would enable identification and quantification of similar dynamics observed NLC layers. The more laminar features at left show what appear to be intertwining vortex structures suggestive of a Kelvin-Helmholtz Instability (KHI). The more turbulent fields at right illustrate some of the diversity in the NLC images observed to date. Extensive and complex vortex clusters occur frequently, are often localized and have different character than adjacent dynamics, but also exhibit dynamical linkages over large distances. In many cases, especially when turbulence appears to be more "energetic", these dynamics reveal elongated, small-scale features on the order of ~10m. In other cases, the flows often appear to be largely laminar, with little or no indications of significant turbulence or small-scale structures.

The ability to observe turbulence and instability dynamics at the smallest scales in the upper mesosphere arises where peak NLC layer brightness coincides with the formation of sheets of high stratification and wind shear (or vorticity). This is because these sheets are the sources of a large fraction of instabilities and turbulence in the stratified and sheared atmosphere [e.g., Fritts et al., 2009b, 2013, 2015b]. The formation of these sheets causes increased NLC brightness gradients that are further enhanced in the presence of instability and turbulence advection and enable them to dominate the brightness gradients averaged along the viewing direction. *Thus, NLC brightness variations represent highly sensitive tracers of small-scale instability and turbulence dynamics.*

PoSSUM Numerical Modeling of Gravity Waves and Instabilities

Imagery data obtained through PoSSUM missions will be analyzed and modeled through numerical simulation which will include general superpositions of higher- and lower-frequency gravity waves in varying mean flows. NLC images and those arising in our modeling of idealized flows often exhibit common features that are indicative of specific instability dynamics and gravity wave characteristics, including amplitudes, intrinsic properties, and momentum fluxes. However, we also want to achieve more quantitative estimates of these dynamics than are possible to infer from similarities of NLC and model images alone. This requires that PoSSUM imaging spans scales extending from >100 km down to ~10-20 m. Models developed to support previous PoSSUM research demonstrate gravity wave breaking and illustrate how vortex ring dynamics and orientations change as gravity wave amplitudes increase. The evolution of a breaking gravity wave is shown in Figure 90.

Previous numerical simulations of KHI viewed from the ground reveal that these KHI dynamics can likewise be quantified. PoSSUM's NLC imaging and modeling capabilities will be able to quantify KHI dynamics at the NLC altitude extending to the smallest turbulence scales, thus characterizing KHI dynamics and influences where KHI is the dominant instability and in MS flows where KHI often participates as an initial instability accompanying larger-scale gravity wave breaking (see left and right panels of Figure 91).

PoSSUM numerical modeling techniques have already yielded multiple instability events having surprising similarities of simulated NLCs to EBEX observations. Such comparisons suggest that even simple multi-scale flows are able to capture some of the diversity in real multi-scale dynamics and key instability characteristics in such flows in the upper mesosphere [e.g., Miller et al., 2015]. As examples, four realizations of the dynamics seen in the multi-scale numerical simulations described by Fritts et al. [2015c] are shown in Figure 92. These panels show the following dynamics (top to bottom):

1) turbulent KHI evolving on a strong vortex sheet above a gravity wave breaking event beginning somewhat earlier (see the coherent spanwise structures with embedded strong 3D vortices at top right in the top left panel),

2) the decaying phase of the KHI event showing larger-scale 3D structures in the turbulence field (left of center in the second left panel),

3) an evolving gravity wave breaking event (at lower right in the third left panel) exhibiting streamwise vortices trailing the breaking "front" at right, and

4) an intrusion event advecting fluid to the left with a leading edge at the domain center of the lower left panel and trailing turbulence dynamics extending to the right through the periodic domain boundary [see Fritts et al., 2013].

Figure 93 shows numerically-modeled structures at mean altitudes of 80, 82, and 84 km illustrating the tendency for instabilities to form when laminar flow thins, much like waves breaking on a beach. The MCAT and MASS instruments, combined with in-situ knowledge of oxygen and NOx concentrations, will be invaluable in the maturation of these mesospheric models constructed from tomographic imagery obtained from PoSSUM imagery systems.

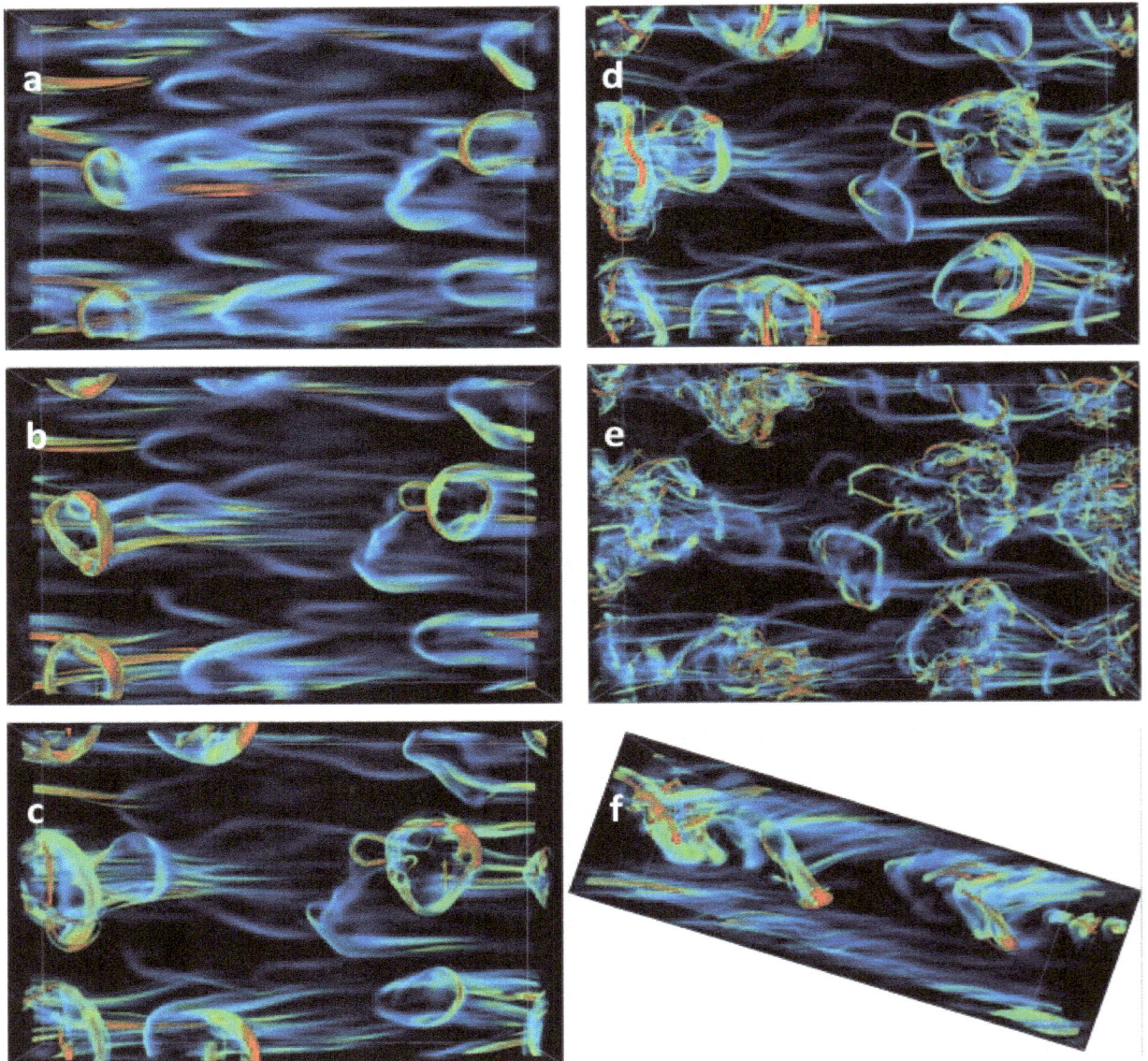

Figure 86. 3D views showing the formation of streamwise-aligned vortices and their initiation of vortex rings that trigger the transition to turbulence (a-e). [Fritts et al., 2009b].

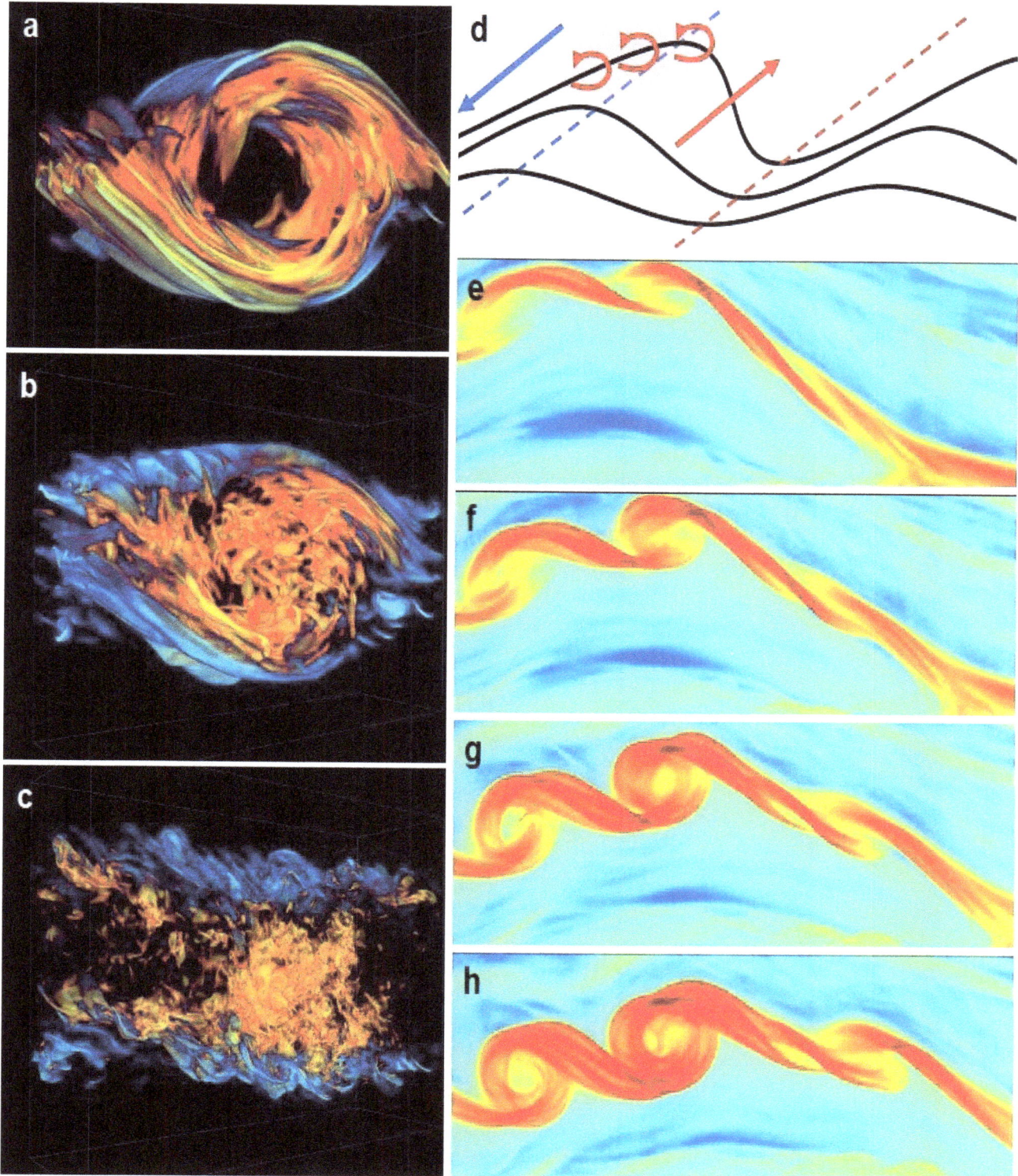

Figure 87. Numerical simulations of KHI in idealized (left) and multi-scale (right) flows used to quantify KHI in ground-based imaging of NLCs and airglow [Fritts et al., 2014b, c]. Panel d shows the location of small-scale KHI within a local gravity wave, and panels e-h show a local KHI event initiated by a local gravity wave [Fritts et al., 2013, 2014c].

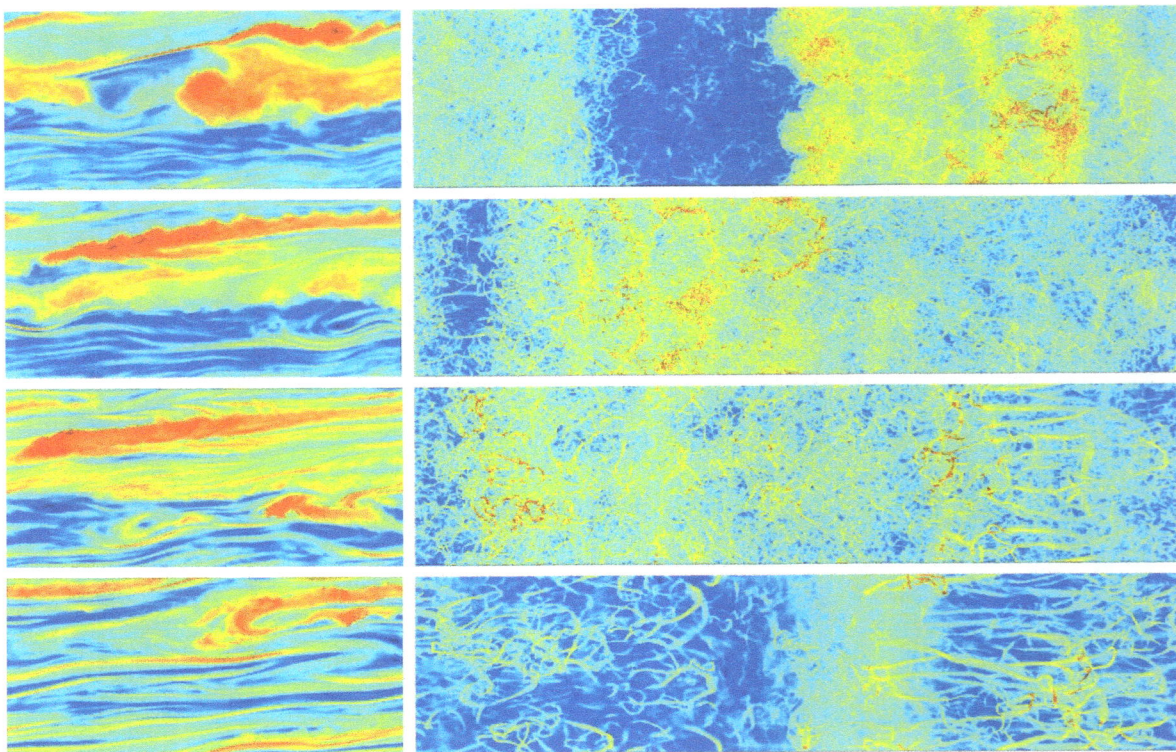

Figure 88. Vertical cross sections (left) and 3D views from above (right) showing various dynamics in a numerically-modeled flow described by Fritts et al. [2015b].

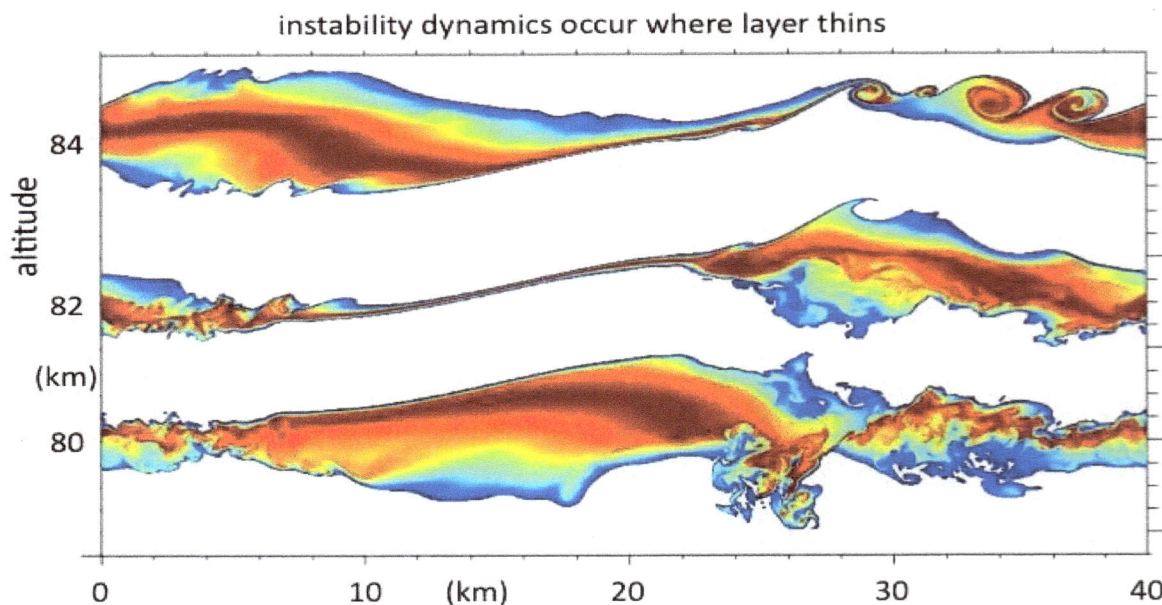

Figure 89. Noctilucent cloud dynamical models constructed by PoSSUM team researchers that will be improved by imagery obtained through PoSSUM flights (credit: GATS)

OBSERVING NOCTILUCENT CLOUDS FROM SUBORBITAL SPACECRAFT

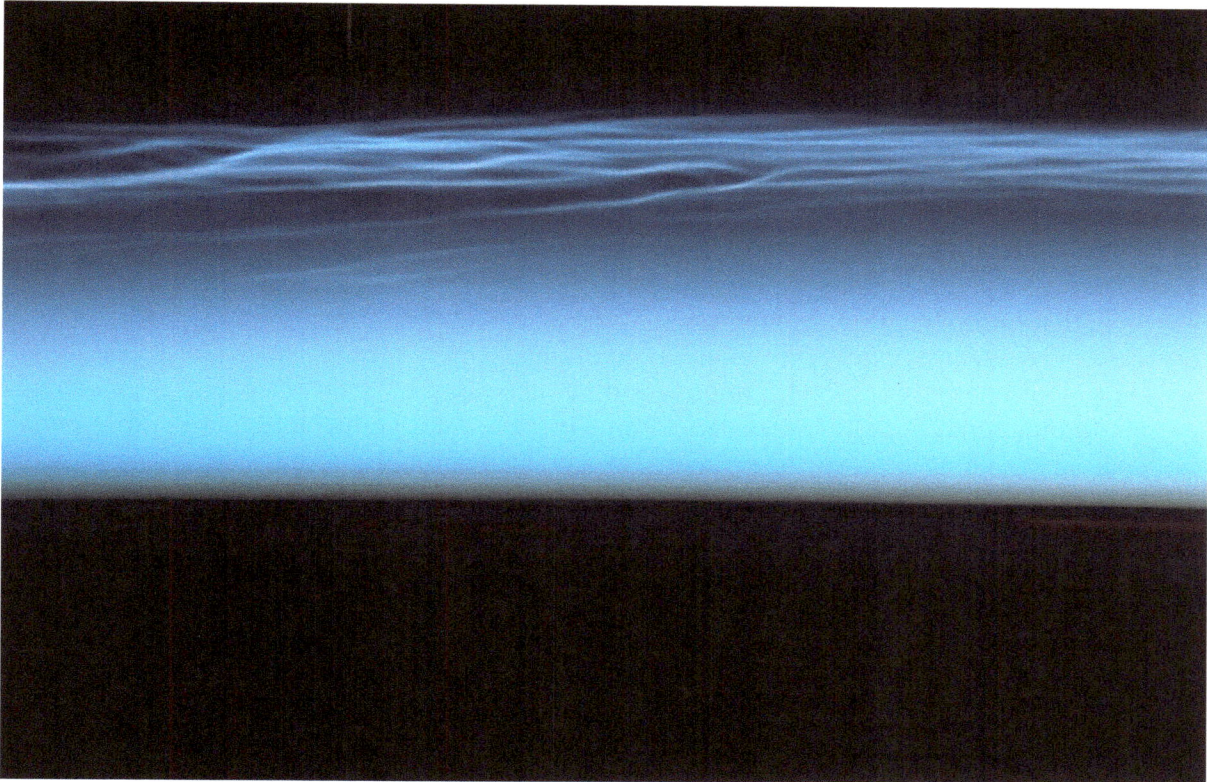

Module Objectives

- Become familiar with previous Noctilucent Cloud photographic campaigns from Space
- Understand the geometry to best observe Noctilucent Clouds from Suborbital missions
- Understand the principles behind the modeling Noctilucent Cloud data

Contributing Author:

Jason Reimuller, Ph.D.

PREVIOUS PHOTOGRAPHY FROM SPACE

Imagery of noctilucent cloud structures has successfully been obtained from orbital space stations. The observation geometry for space observations was solved by Professor Willman of the Estonian Institute of Astrophysics and Atmospheric Physics in 1977. Professor Willmann determined that the multilayered structures of a noctilucent cloud field may be observed from space. His calculations assumed a station height of 350 km. From this altitude, NLCs are seen at an essentially constant height corresponding to approximately two degrees above the horizon.

Following from Professor Willman's calculations, observations carried out on 'Salyut 6' by Y.V. Romanenko and G.M. Grechko in the period from 23 December 1977 to 2 February 1978 showed that NLCs were recorded during 164 orbits. Onboard 'Salyut 6', Soviet cosmonauts used film Aviaphot pan30 (Agfa Gavaert). The normal exposure time was set to 1/30 second and depended on the ISO speed of the film used. Lens aperture was chosen according to the brightness of visible noctilucent clouds and the twilight sector in the range from 4 to 22, and focal apertures of 11 to 16 were determined to produce the best images.

Images of noctilucent clouds continued to be taken by both astronauts and cosmonauts onboard the International Space Station (ISS). Notably, Don Petit gained a reputation for contributing to noctilucent cloud science through images obtained through his 6-month say on the ISS. It is important to realize that, although the images obtained from the ISS have no background scattering, the farthest latitude that the groundtrack of the ISS reaches is 51.6 degrees. As noctilucent cloud structures generally form polewards of 62 degrees, observations taken from the ISS are at a distance of at least 600 nautical miles. At these distances, even the most state-of-the-art camera systems will only be able to resolve the most prominent structures on the clouds. Further, the geometry of space stations relative to the clouds allow for viewing at only one scattering angle. Resolution obtained from PoSSUM flights will be greatly improved and be able to resolve structures on the order of meters, exhibiting the small scale structures that will contain information on turbulence and wave interactions in the upper mesosphere. Furthermore, the trajectory of PoSSUM slights will allow observations of noctilucent cloud structures through a wide range of viewing angles, enabling tomographic reconstruction of image sequences obtained through suborbital flight.

The following sections detail how suborbital mission profiles may be planned so that the most useful imagery may be obtained.

VIEWING GEOMETRY

Approximating Solar Elevation

In the northern hemisphere, the following relationships may be used to estimate the highest and lowest points that the Sun appears at any given day at any given location. In polar environments, it is possible to have positive solar elevation at midnight (for locations above the polar circle near the summer solstice) or nights that never fully get dark, as one would see in sub-polar regions near the summer solstice.

- Colatitude = 90 – Latitude

- Solar Elevation at Midnight = Solar Declination (δ) – Colatitude   (Sun due north)

- Solar Elevation at Local Noon = Solar Declination (δ) + Colatitude   (Sun due south)

Solar declination can be obtained through the table provided in Appendix A1 for each day of the year or approximated using the following equation:

$$\delta = sin^{-1}[sin\ sin\ (23.45^o)\ sin\ sin\ \left(\frac{360}{365}(d-81)\right)]$$

where d is the day of the year with January 1 as d = 1.

Approximating Solar Azimuth

The Sun traces a path in the sky that places it due north at local midnight (when the sun is lowest below the horizon), rising and extending eastward until it reaches a position of due south at local noon (where the sun is highest in the sky). Since the day has 24 hours, the sun will appear to move 15 degrees each hour. Therefore, to know the time and longitude is to know the exact bearing of the Sun relative to the observer. At any given time, the solar azimuth may be approximated by the following relationship:

> Solar Azimuth (degrees) = Time of Solar Midnight
> + 15 degrees * (Time elapsed in hours since Solar Midnight)

For example:

Solar Midnight at Fairbanks is at 1:55AM. The time is now 5:30AM. At what azimuth is the sun?

Answer: (5h30m – 1h55m) = 3h35m = 3.583hours

3.583hours * 15 = 53.7 degrees east of north.

Computing Local Midnight from time zone and longitude

It is important to know the value of local midnight at the location where observations will be made. This value is the time, expressed in local time, when the sun is due north and has the greatest value for the solar depression angle. If the observation location lies directly on a meridian (e.g., a longitude of 105 degrees, 120 degrees, 135 degrees, or other multiple of 15), then local midnight will correspond to 00:00 local, though if the region observes Daylight Savings Time, that value will be 01:00 local.

To compute the time of local midnight at any longitude, follow these steps:

1. Compute your longitude in decimal form: degrees + (minutes/60) + (seconds/3600). If you're in the Eastern Hemisphere, make this a negative number.

2. Compute your "standard meridian" (the longitude of the center of your time zone): difference in hours between your time and "Universal time" (Greenwich or Zulu time) times 15 (for example, the US Central time zone is six hours different, times 15 means that the "standard meridian" for the Central Time Zone is 90 degrees). Make this a negative number if you're in the Eastern Hemisphere.

3. Subtract the standard meridian from your longitude.

4. Divide that difference by 15 to get the longitude deviation time (the amount that your local time is ahead or behind the clock time for your time zone)

5. Add 0.123 to that time (the correction for the deviation between the way our clocks/calendars work and the way the earth/sun work: this is found in the current Almanac for a given date - in this case, 0.123 for March 20th)

6. Add the resulting number to 12 (or subtract it from 12 if it is negative) to get the time of your local noon.

7. Convert the decimal number to hours, minutes, and seconds.
   a. The whole number is the hours.
   b. Multiply the fraction by 60. The whole number is the minutes.
   c. Multiply the fraction by 60. The result is the seconds.

Scale

It is important to take the frame so that there is a frame of reference, such as a planet or a star field, in the picture. This is important for determining the spatial scale of the noctilucent cloud field.

Calculating Dip: Observations at Altitude

Dip is the effect of an observer that is elevated off the surface. Even standing *on* the surface doesn't mean our eyes are *at* the surface, so dip must always be calculated, even for an observer at sea where the horizon is very pronounced. In this case, an observer's eyes might be 1.8 meters above the ground, and this produces a calculable effect. For airborne applications, especially those involving spaceflight, these effects can become quite significant. Fortunately, for space-based observation, the effects of atmospheric refraction are less significant, so we will make an initial approximation neglecting its effects.

The diagram of Figure X shows a vertical plane through the center of the Earth (at C) and the observer (at O). The radius of the Earth is R, and the observer's eye is a height h above the point S on the surface. Because we are assuming that there is no refraction, the observer's apparent horizon coincides with the geometric horizon, indicated by the dashed line OG, tangent to the surface of the Earth. We can now apply elementary trigonometry to calculate the dip:

$$\cos(d_g) = CG/OC = R/(R+h)$$

But even at suborbital altitudes, h is very small compared to R, so we can apply the small-angle approximation $\cos x = 1 - x^2/2$ to the left side of the equation, leaving the approximate result:

$$(d_g)^2/2 = h/R \text{ , or } d_g = \text{sqrt} (2h/R)$$

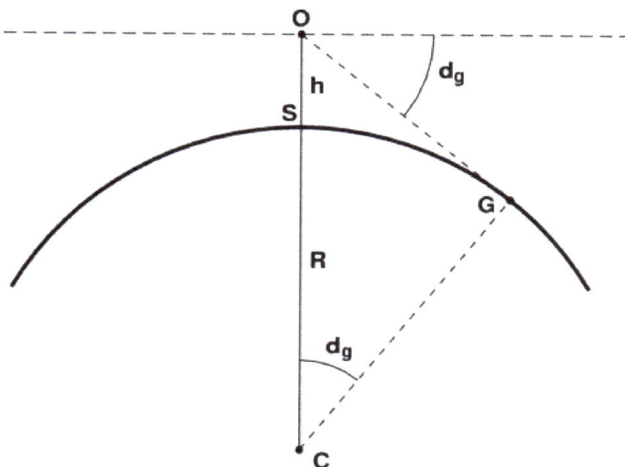

Calculating for significant altitudes on a PoSSUM Flight in Table 10, we calculate the dip at NLC altitudes to be approximately 9.23 degrees.

Altitude	Altitude (m)	Dip (deg)
Commercial Jet	10000	3.203
MECO	48000	7.017
NLC Altitude	83000	9.228
Apogee	103000	10.279

Table 10. Effects of Dip at significant altitudes of a PoSSUM flight

CALCULATING OBSERVATION GEOMETRY FOR SUBORBITAL NLC OBSERVATIONS

To plan a noctilucent cloud flight, we want to make sure that the signal to noise (SNR) ratio received by the imagers is ideal, so we carefully plan location and time when the clouds will be strongest and the scattering angles favor the best imagery. We first introduce some background on observations of NLC from the ground.

Observing NLCs from the Ground

For ground observations, as shown in Figure 90, NLCs are observed best when the sun is about 6-9 degrees below the horizon. This is because the column of air below the NLC is obscured by the shadow of the Earth, so it does not directly scatter light from the sun. When the sun in less than six degrees below the horizon, an observer receives too much scattered light from the atmosphere (the 'noise') to view the NLC (the 'signal'). When the sun is greater than 10 degrees below the horizon, NLCs are not visible since no direct sunlight reaches them.

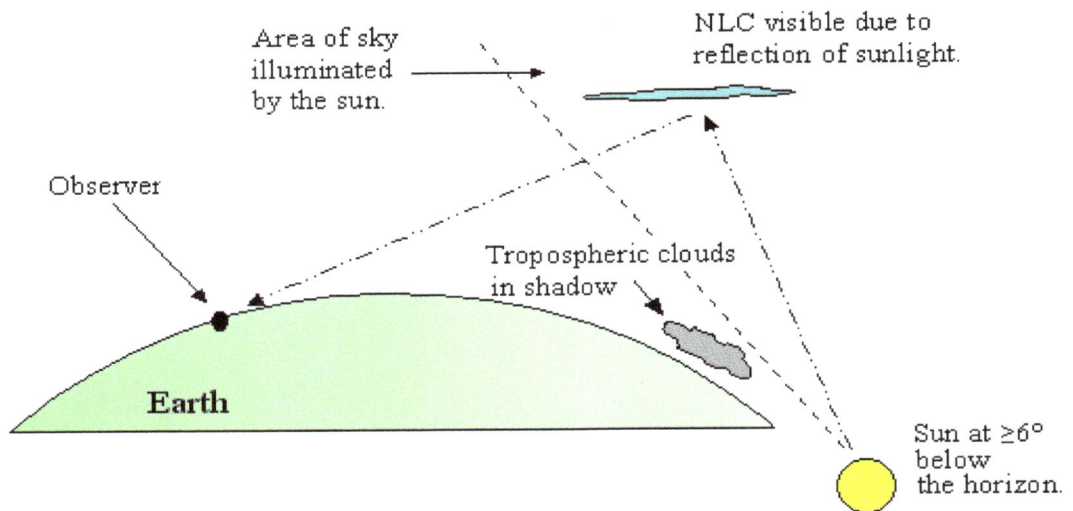

Figure 90. NLCs observed from the ground

Observing NLCs from Suborbital Altitudes

When we can rise above 30 km in altitude, there is very little atmosphere that will scatter sunlight. The background is the blackness of space, and the NLC scatters light without interference from the atmosphere below it. We establish several assumptions and requirements here:

Assumptions:
- Observations are taken towards the north
- Assume no atmospheric scattering
- Minimize background effects of the earth

Requirements:
- Do not look directly at the sun
- Observe the cloud with incident light producing forward scattering within +/- 60 degrees in azimuth.
- Observe the cloud with the sun at a Solar Depression Angle between zero and two degrees for solar azimuth +/- 25 degrees from North.
- Observe the NLC layer with the Solar Elevation less than 5 degrees for solar azimuth between 25 and 60 degrees from north.
- Minimize background of the Earth

We want to calculate the Solar Depression Angles for the time +/- 4 hours from local midnight on the day of flight. From the ground, the sun reaches its highest Solar Depression Angle at local midnight, when the Sun is due North. Figure 91 shows two solar tracks for an observer looking North from a high latitude at a time near the summer solstice. For an observer on the ground (the lower arc), the Sun appears to dip below the horizon in the Northwest. The sky dims until it is darkest at Local Midnight. The sky then lightens as the Sun approaches sunrise in the northeast.

Now, if we ascend to NLC altitudes by means of a suborbital rocket, the dip effect causes the solar depression angle to lessen. As we rise, the sun will also appear to rise, but we are also rising past the absorbing atmosphere, so that even though we are reducing the Solar Depression Angle, the images of the NLCs will become sharper.

The 'Exclusion Zone' is the region within which the Sun will flood an imager looking directly to the North. For the PoSSUM Wide-Field Imager, this angle is 44 degrees, or +/- 22 degrees from North (1.5 hours).

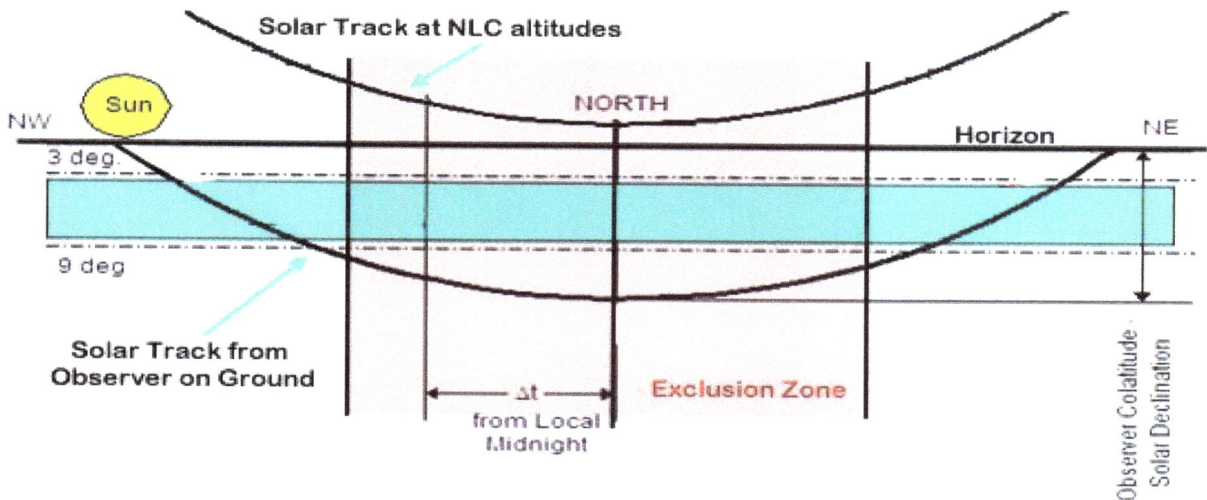

Figure 91. Solar position from observers on the ground and observers from NLC altitudes

Using the NOAA Solar Position Calculator

An effective solar position calculator is available online through the National Oceanic and Atmospheric Administration (NOAA): http://www.esrl.noaa.gov/gmd/grad/solcalc/

The results of two iterations, one on July 23 (the start of the planned NLC campaign) and the second on July 31 (the end of the planned NLC campaign) for Eielson AFB, Alaska, are displayed below. We see that local midnight remains constant at 01:57AM but the Solar Depression Angle at Local Midnight increases from 5 degrees on July 23rd to 7 degrees on July 31st.

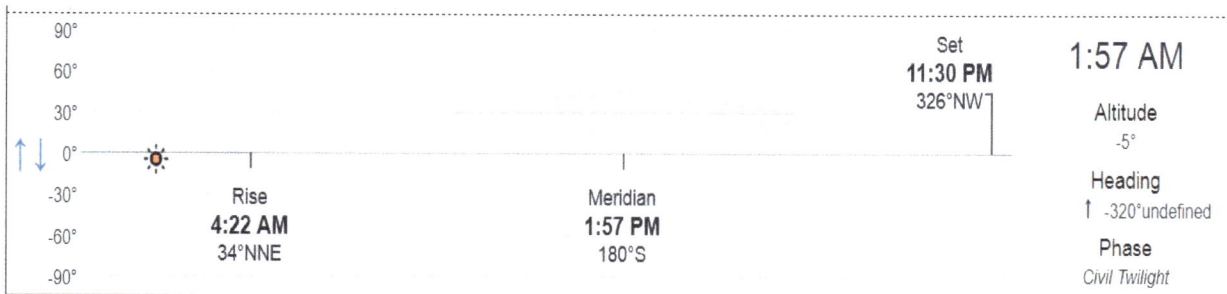

Figure 92. Solar Track at Eielson AFB for July 23rd (credit: NOAA)

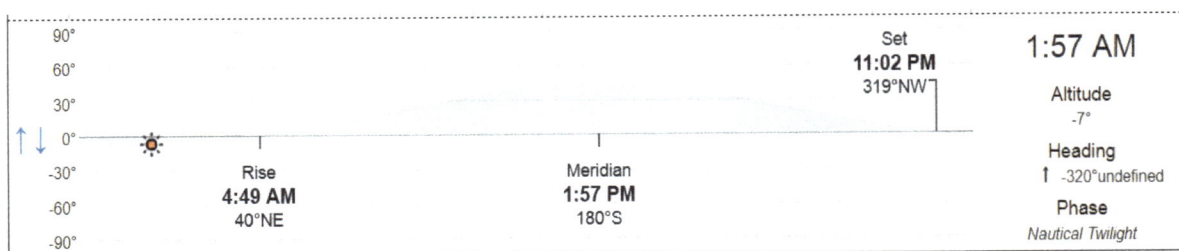

Figure 93. Solar Track at Eielson AFB for July 31ˢᵗ (credit: NOAA)

2016	Sunrise/set		Daylength		Astro. Twilight		Naut. Twilight		Civil Twilight		Solar noon	
Jul	Sunrise	Sunset	Length	Diff.	Start	End	Start	End	Start	End	Time	Mil. mi
23	4:22 AM ↑ (34°)	11:30 PM ↑ (326°)	19:08:10	−6:51	-	-			Rest of night		1:57 PM (45.0°)	94.431
24	4:25 AM ↑ (35°)	11:27 PM ↑ (325°)	19:01:18	−6:52	-	-			Rest of night		1:57 PM (44.8°)	94.422
25	4:29 AM ↑ (35°)	11:23 PM ↑ (324°)	18:54:24	−6:53	-	-			Rest of night		1:57 PM (44.6°)	94.414
26	4:32 AM ↑ (36°)	11:20 PM ↑ (323°)	18:47:30	−6:54	-	-			Rest of night		1:57 PM (44.4°)	94.405
27	4:36 AM ↑ (37°)	11:16 PM ↑ (322°)	18:40:35	−6:55	-		Rest of night 2:19 AM	-	-	1:36 AM	1:57 PM (44.1°)	94.396
28	4:39 AM ↑ (38°)	11:13 PM ↑ (321°)	18:33:39	−6:55	-		Rest of night 2:37 AM	-	-	1:19 AM	1:57 PM (43.9°)	94.386
29	4:43 AM ↑ (39°)	11:09 PM ↑ (321°)	18:26:42	−6:56	-		Rest of night 2:49 AM	-	-	1:07 AM	1:57 PM (43.6°)	94.376
30	4:46 AM ↑ (40°)	11:06 PM ↑ (320°)	18:19:45	−6:56	-		Rest of night 2:58 AM	-	-	12:57 AM	1:57 PM (43.4°)	94.365
31	4:49 AM ↑ (40°)	11:02 PM ↑ (319°)	18:12:48	−6:57	-		Rest of night 3:07 AM	-	-	12:48 AM	1:57 PM (43.2°)	94.354

Table 11. Sunrise and Sunset times at Eielson AFB throughout projected campaign

Determining Launch Windows

We want to pick a launch window that meets the following constraints:

1. Clouds must be present, so we assume we can launch between 1 July and 15 August from latitudes of at least 60 degrees.
2. We want to image the cloud with forward scattering (see Mie Scattering discussion) but we do not want to aim the imagers directly towards the sun.
3. As cloud density is highest towards the pole, we want to take images preferable to the North.
4. We also don't want the sun to be too high in the sky at cloud altitudes.

196

So, we want to launch when the sun is at least 22 degrees off North so that the imagers may be pointed towards the North, but at a time when the sun will have an elevation of between 0 and 5 degrees elevation. As the wide-field imager has a field of view of 44 degrees, we would want the sun to be 22 degrees of north, or 1.5 hours from local midnight. Thus, if local midnight is 1:55AM, then we would want to launch near 12:20AM or near 3:20AM. Let us consider two examples:

EXAMPLE 1: Launch at 12:20AM on 1 August form Eielson. (Apogee at 12:25AM)
  Solar Declination = 18 degrees.
  Solar Depression Angle at Local Midnight = (90-64) – 18 = 7.4 degrees
  Dip (adjustment for NLC altitude) = 9.1 degrees
  Solar Elevation at NLC Altitude at local midnight = 1.7 degrees (just over horizon)
  Correction for 1.5 hours before or after local midnight: 7.4 - 5.7 degrees = 1.7 degrees
  Solar Elevation Angle = 0
  Solar Azimuth Angle = 23 degrees West of North

EXAMPLE 2: Launch at 1:50AM on 10 August form Eielson. (Apogee at 1:55AM)
  Solar Declination = 15.8 degrees.
  Solar Depression Angle at Local Midnight = (90-64) – 18 = 9.9 degrees
  Dip (adjustment for NLC altitude) = 9.1 degrees
  Solar Elevation at NLC Altitude at local midnight = -0.8 degrees (just under horizon)
  Solar Elevation Angle = -7 degrees
  Solar Azimuth Angle = 0 (North)

Note: Example 2 will place the sun below the horizon, so there will be no direct scattering that will reach the imagers during NLC transit.

Ground Observation Support

Imagery obtained on the rSLV will be combined with imagery from two ground stations to support tomography development. Tomography has been used successfully to investigate structures in the ionosphere, in the aurora and airglow emissions and more recently in noctilucent clouds. Unlike conventional tomography applications (e.g. medical imaging), suitable atmospheric data for tomographic applications of mesospheric phenomena have been difficult to obtain due to sparseness of suitable ground-based data and restricted viewing angles, nevertheless tomography is proving to be a powerful tool for investigating a variety of structure and dynamics in the upper atmosphere.

The PoSSUM campaign will augment imagery obtained on the rSLV by imagery obtained through two ground-based observation stations, located to the south of the launch site and mutually orthogonal to each other, for the purpose of developing enhanced tomography. For the PoSSUM experiment we plan to broaden our tomographic capabilities and their application to the unique viewing geometry that space-borne measurements that the rSLV will provide. The relatively close proximity of the measurements to the noctilucent cloud layer and the viewing from below through and then above the layer will provide unprecedented wealth of high-resolution information removing many of the limits imposed using sparse data and enabling a more complete tomographic study of the internal structuring of the layer.

AERIAL GROUND IMAGING ANALOG EXERCISE

Module Objectives

- ➢ Become familiar with setup and operation of the imaging camera and Gyro-Stabilizer
- ➢ Understand the effects of Sun angle on scene illumination
- ➢ Develop an aerial imaging plan for the target, based on planned take-off time
- ➢ Become familiar with basic navigation techniques and methods
- ➢ Gain an appreciation of the planning required for airborne imaging
- ➢ Successfully fly the imaging plan, collecting target photographs and corresponding Geotracker data

Contributing Author:

Elizabeth Balga
Wai Yin (Wilson) Cheung
Jeska Clark
Kim Ellis
Ken Ernandes
Chris Vanacore

Introduction

The Aerial Ground Imaging Ground Imaging Exercise is an analog flight, providing practical familiarity of the airborne imaging process, including the navigation skills needed to support the image collection.

Each flight mission team has two members: an Instrument Operator for capturing the study area images and a Navigator for ensuring the acute flight route as planned by communicating with the pilot proactively.

To achieve the goal of this exercise, the team members must have a basic knowledge of camera operation and aviation navigation.

Analog Exercise Objectives

The objectives for this exercise are to provide practical experience with airborne imaging activities, including skills that would be useful for a noctilucent cloud imaging campaign.

- Become familiar with setup and operation of the imaging camera and Gyro-Stabilizer. These are the basic IIAS instruments for aerial imaging.
- Develop an understanding of the effects of Sun angle on scene illumination.
- Develop an aerial imaging plan for the target, based on planned take-off time and the practical considerations involved.
- Become familiar with basic navigation techniques and methods. Aeronautical charts generally have too large a scale to plan a mission such as this. Thus, developing basic charting and navigational skills facilitates the planning process for situations in which an existing computer-based application or a commercial chart is not available.
- Gain an appreciation of the planning required for airborne imaging. Airborne imaging is not as simple as you might think. In addition to real-time skills, planning flexibility is needed to react to significant changes in things such as takeoff time.
- Successfully fly the imaging plan, collecting target photographs and corresponding Geotracker data. Classroom training is not a substitute for real world experience. The flight provides the opportunity to put your plan in action and collect actual data.

Target and Imaging Area Familiarization

The imaging target is Lake Hell 'N Blazes (a.k.a. Lake Hellen Blazes), located approximately 9 nautical miles (NM) (or 16 km) south-west of Melbourne, Florida. The lake, which is the source of the St. Johns River has a 260-acre area (ref. 1). It is approximately 1 NM across its longest dimension and 0.3 NM wide perpendicular to the long direction. The geographic coordinates of the lake's center are approximately:

Latitude:	**28.02°N**
Longitude:	**80.79°W**

Figure 94 provides a (Google Earth – ref 2) overhead aerial view of Lake Hell 'N Blazes, plus several other nearby bodies of water. Familiarization is important for both the Instrument Operator and the Navigator to ensure correct target identification.

The image collection mission begins from Melbourne Orlando International Airport (KMLB), depicted in Figure 2 in the upper right corner (image also from Google Earth). The airport center coordinates (ref. 3) are approximately:

Latitude: **28.10°N**
Longitude: **80.65°W**

The distance from KMLB to the target is approximately 9.6 NM. Given that, you can use the "60xD=STreet" formula below to estimate the time to the target area once the aircraft is departing westbound out of KMLB. In the "60xD=STreet" formula, the number 60 is an hour to minutes conversion factor, D is a distance in NM, S is the aircraft speed in knots (KT – or NM/hour), and T is time in minutes.

$$60 \times D = S \times T$$

Example 1 is used to compute the time from a westbound departure from KMLB, to the arrival time in the target area.

EXAMPLE 1: Determine the time (T) in minutes to reach the target area, using distance as ($D = 9.6\ NM$) and aircraft speed presumed at ($S = 120\ KT$).

Using algebra: $T = \frac{60 \times D}{S} = \frac{60 \times 9.6}{120} = 4.8\ minutes$

This result can be rounded to the nearest minute (i.e., 5 minutes). You should consider including at least three additional minutes for the takeoff, climb out, and airport traffic pattern flight, if calculations presume a specific takeoff time.

This result can be rounded to the nearest minute (i.e., 5 minutes). You should consider including at least three additional minutes for the takeoff, climb out, and airport traffic pattern flight, if calculations presume a specific takeoff time.

Lake Hell 'N Blazes are depicted and labeled in figure 2's lower left corner. Approximately 6 NM to the east is an abandoned military flight training airfield, outlined with a red square. This landmark is aptly named "*Abandon.*" It is a Visual Flight Rules (VFR) checkpoint for contacting Melbourne Tower for permission to re-enter KMLB's "Class D" airspace. This should be the last waypoint in your mission profile. The geographic coordinates for Abandon are approximately:

Latitude: **28.02°N**
Longitude: **80.68°W**

It is the Navigator's responsibility to ensure the pilot is briefed that Abandon will be the checkpoint to contact Melbourne Tower for re-entering KMLB's airspace.

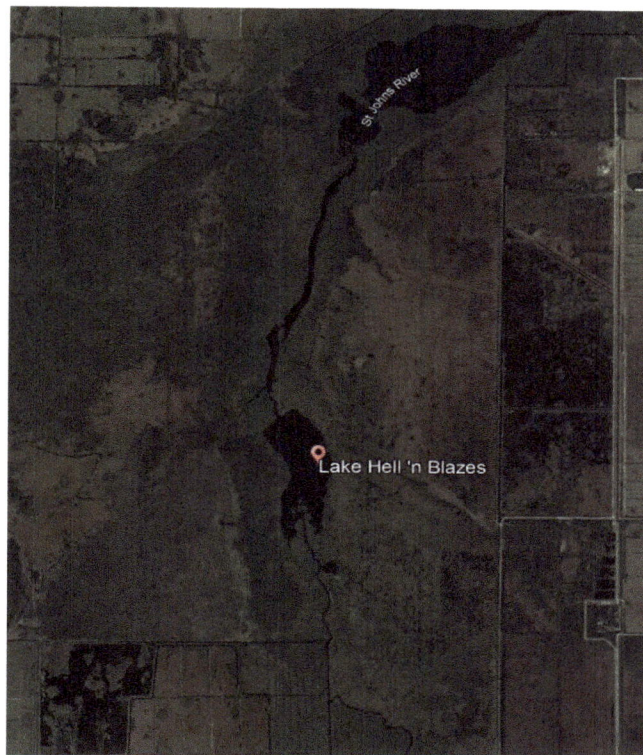

Figure 94. Lake Hell 'N Blazes and Surrounding Bodies of Water

Camera and Geotracker Setup

Details on the camera and geotracker setup and operation will be presented on site. The camera is mounted on the Pilot's (i.e., left) side of the aircraft. Therefore, the planning and imaging legs develop a flight plan for imaging the target from the aircraft's left side. Operation and restrictions relative to the local Class D airspace must be considered, particularly for graceful entries and exits into the target area.

The Class D airspace around Melbourne Orlando International Airport generally extends 5 statute miles (~4.3 nautical mile / 8.1 km) from the airport center to 1900 feet above the airport's 33 ft field elevation. A 2500 ft Mean Sea Level imaging altitude is recommended to operate well above Class D airspace. This altitude facilitates imaging from any sun angle at the recommended 6 nautical mile center-of leg standoff range. This altitude and standoff range facilitate a slightly less than 10° imaging [depression] angle below the local horizon.

The camera and attached Geotracker with stabilizer are mounted on the Pilot's (i.e., left) side of the aircraft, as shown in figure 3, for obtaining the images of the study area (in this case Lake Hell 'N Blazes). Make sure all batteries are fully charged and carry extra batteries or a power bank in order to ensure adequate power. It is highly recommended that both mission teammates run a pretest on the ground before the airborne mission.

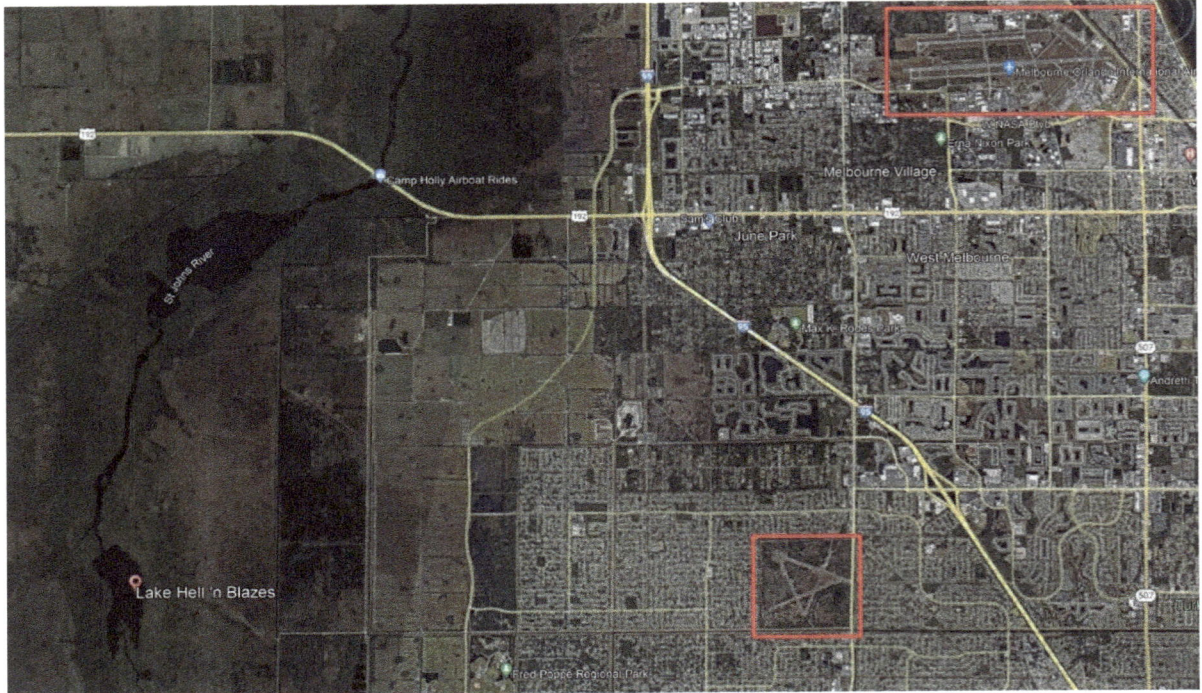

Figure 95. Imaging Area Key Landmarks

Figure 96. Instrument Operator (Wilson Cheung) in charge of capturing the study area images during the flight mission.

Imaging Range and Camera Field-Of-View (FOV)

Selecting the camera FOV is somewhat subjective for this exercise. Important considerations and tradeoffs include the target size, the aircraft speed, and the planned standoff range from the target. The camera FOV will depends directly on the target size (s) and standoff range (r) by the following relationship:

$$FOV = 2\tan^{-1}\left(\frac{s}{2r}\right)$$

You should consider adding a few additional degrees to ensure the target is well-framed in the image. Then, given a desired FOV angle, the corresponding standoff range is:

$$r = \frac{s}{2}\cot\left(\frac{FOV}{2}\right)$$

But there are additional factors that must be considered when selecting a FOV, which are related to the standoff range (r) and the aircraft speed (v). The angular rate of change of the target direction ($\dot{\theta}$) may be computed by:

$$\dot{\theta} = \frac{v}{r}$$

The closer the range from the target, the more quickly you must slew the camera horizontally to keep the target centered in the imaging FOV. Conversely, longer standoff ranges allow for slower slewing, and make it easier to keep the target centered in the image.

Target Sun Angle Profile

This exercise has three distinct imaging legs, each of which has its center abeam the target at a specific Sun angle. Directions are referenced by standard angular lines of azimuth. Figure 98 relates the four cardinal directions – North (N), East (E), South (S), and West (W) – plus the four intercardinal (or ordinal) directions – Northeast (NE), Southeast (SE), Southwest (SW), and Northwest (NW) – to their corresponding azimuth angles.

The angles are such that the middle leg is viewing directly "down Sun" (i.e. the Sun behind the camera, with the earlier and later legs viewing ±45° to the down Sun direction, as shown in figure 97.

The imaging profile has four waypoints (WPs), followed by the "Abandon" WP. The triangles are situated such that the "Down Sun" line (between WP2 and WP3) is

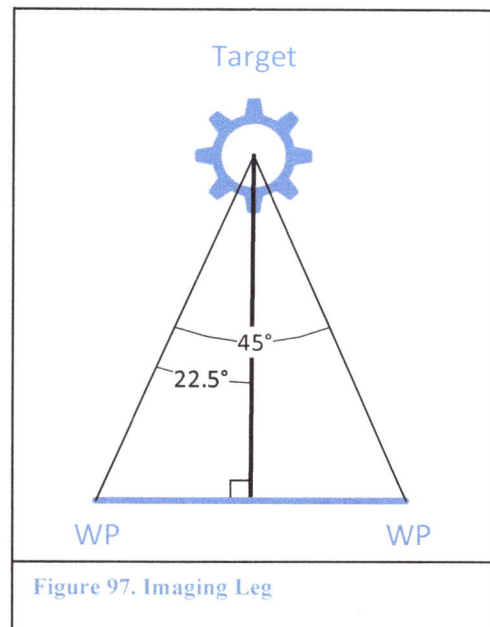

Figure 97. Imaging Leg

aligned with the Sun's direction during the imaging sequence. The line midway between WP1 and WP2 is 45° west of the Sun's direction and the line midway between WP3 and WP4 is 45° east of the Sun's direction.

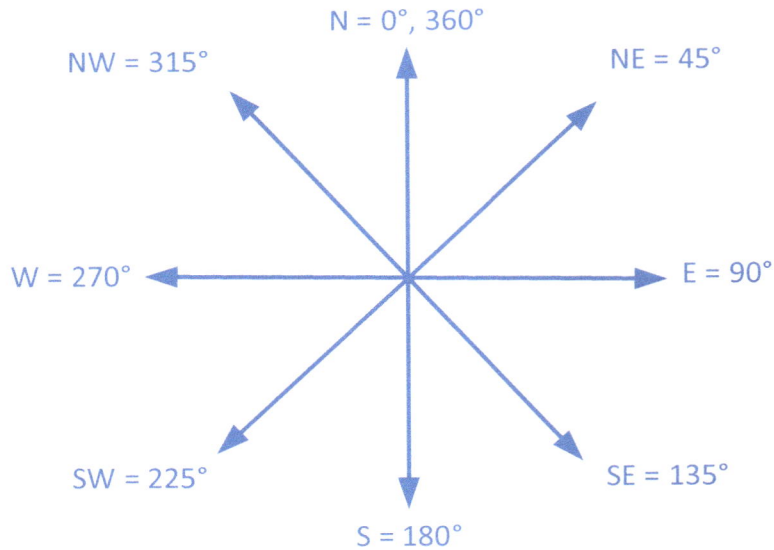

N = 0°, 360°
NW = 315°
NE = 45°
W = 270°
E = 90°
SW = 225°
SE = 135°
S = 180°

Figure 98. Azimuth Angles Related to Cardinal and Ordinal Directions

If the imaging sequence occurred at local noon, the Down Sun line would be on a North-South azimuth (i.e., 180°), relative to the target. In such a scenario, the approximate azimuths of the WPs would be as presented in Table 12.

Waypoint	Azimuth
WP1	248°
WP2	203°
WP3	158°
WP4	113°

Table 12. Example WP Azimuths for 180° Sun Azimuth

Leg	Center Azimuth
WP1 to WP2	226°
WP2 to WP3	181°
WP3 to WP4	136°

Table 13. Central Leg Azimuths for a 117° Sun Azimuth

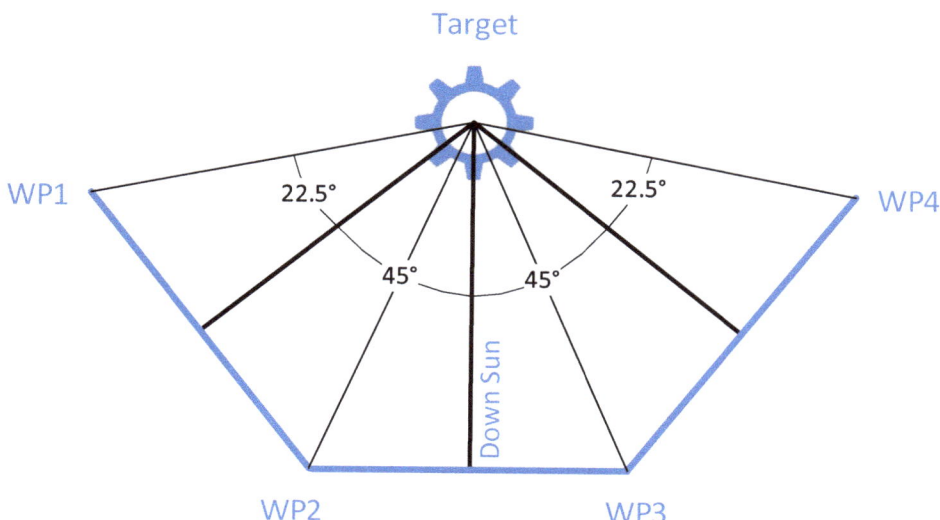

Table 14. Target Imaging Profile

Profile Alignment to Planned Takeoff Time

The target imaging profile needs to be aligned to the local Sun line azimuth at the time the aircraft is flying through the target area. Put simply, the target imaging profile in Figure 14 is rotated so the Down Sun line is along the local Sun azimuth, at the time estimated for the middle of your imaging sequence.

The first step in the alignment process is determining the time of local noon at the target's longitude. Since the Earth rotates at 15°/hour, central longitude for each time zone is a multiple of 15. Thus, the central meridian for the target is 75°W. The second step is determining the time offset of the target from the central meridian. Since the target is located at 80.68°W:

$$dlon = 75°W - 80.68°W = -5.68°$$

$$dT = -\frac{5.68°}{15°/hour} = -0.3867 \ hours \times 60 \ min/hour \approx -23 \ min$$

The negative sign indicates the local time is ~23min earlier at the target. Thus, local noon occurs 23 minutes later at the target or at 12:23 zone time.

If your mission takeoff time is, for example 08:00 zone time, you might estimate a total of 10 minutes after takeoff to reach the Down Sun location (i.e., you would reach this location at 08:10 zone time). The third step is determining the angle to rotate the Target Imaging Profile in figure 5 to align the Down Sun line with the Sun's azimuth at 08:10 zone time. Using the Earth's rotation rate, the Sun azimuth is computed as follows:

$$ZoneNoon = 12 \ h + \frac{23 \ m}{60 \ m/h} = 12.3833 \ h$$

$$DownSunTime = 8\,h + \frac{10\,m}{60\,m/h} = 8.1667\,h$$

$$dAZ = (DownSunTime - ZoneNoon) \times 15°/h \approx -63°$$

$$AZ = 180° - 63° = 117°$$

Using the $-63°$ azimuthal offset, the WP azimuths in table 1 are adjusted in Table 15 to align the Target Imaging Profile with the Sun azimuth for the example takeoff time.

Waypoint	Azimuth
WP1	185°
WP2	140°
WP3	095°
WP4	050°

Table 15. Rotated WP Azimuths for a 117° Sun Azimuth

The imaging lines at the center of the three imaging legs were 225°, 180°, and 135°. These are likewise adjusted by -63° in Table 16 for the rotated Target Imaging Profile.

Leg	Center Azimuth
WP1 to WP2	162°
WP2 to WP3	117°
WP3 to WP4	072°

Table 16. Rotated Central Leg Azimuths for a 117° Sun Azimuth

Figure 99 represents the rotated Target Imaging Profile for a 117° Solar Azimuth.

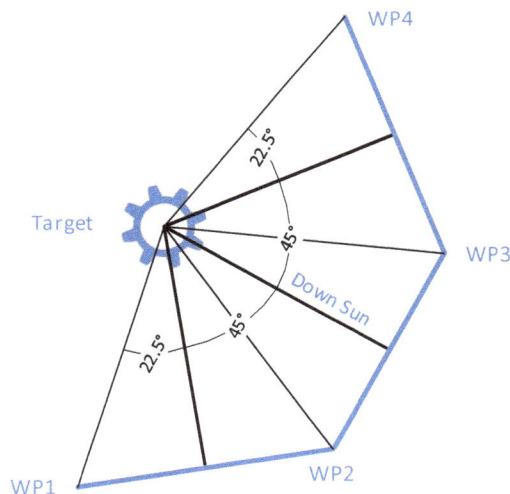

Figure 99. Target Imaging Profile Rotated for a 117° Solar Azimuth

Note that, for this particular profile rotation, a heading from KMLB directly to WP1 would require an unreasonably sharp turn to enter the first imaging leg. (Turns should generally be less than 90°.) For these cases, it is prudent to add a setup "WP0" somewhat North and West of the target to facilitate a smooth entry to the first imaging leg.

Geographic Coordinates

Geographic coordinates include parallels of latitude and great circle meridians that converge at the North and South Poles as illustrated in figure 8 (ref. 4). Parallels of latitude are referenced to the Equator (0 degrees) and range from -90 degrees (South Pole) to +90 degrees (North Pole). Positive latitudes are in the northern hemisphere, and thus conventionally given an "N" label. Likewise, negative latitudes are conventionally given an "S" label.

Longitudinal meridians are referenced to the Prime Meridian, which passes through the Greenwich Observatory outside London, England. Longitude ranges from -180 degrees to +180 degrees. Positive longitudes are East of Greenwich and are conventionally given an "E" label, while negative longitudes are West of Greenwich and given a "W" label.

Latitude and longitude are measures of angular arc on what we approximate as a spherical surface for the Earth. Latitude and longitude are further subdivided into 60 minutes of arc (symbolized by an apostrophe) but may also be denoted as decimal fractions of degrees.

The Earth is divided into time zones ranging from -12 to +12 hours, relative to Greenwich time. The central meridian for each time zone is evenly divisible by 15°, since the Earth rotates at 15° per hour. Since the imagining area is between 80° and 81°W, the time zone's central meridian is at 75°W. Dividing this central longitude by 15 shows the zone time to be 5 hours earlier than Greenwich time.

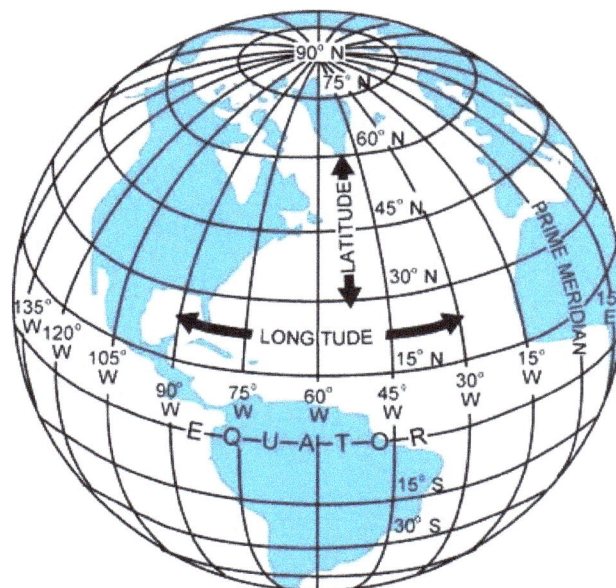

Figure 100. Meridians of Longitude and Parallels of Latitude

Imaging Plan Development

The imaging plan uses the azimuths determined in the Sun alignment process to determine the imaging WP coordinates and corresponding aircraft ground tracks. One method of doing this is creating a paper geographic grid chart to render the target area. This chart approximates the imaging area as rectangular. This approximation must, however, have the correct latitude-to-longitude aspect ratio for the angular and distance relationships to be correct.

While circles of latitude are always parallel, longitudinal meridians are great circles defined by the intersection of planes with the Earth's surface that passes through Earth's center. Figure 7 (ref. 4) shows how the meridians all converge, meeting at the North and South poles. Thus, the further North you go, the closer together are the meridians.

This rendering of a small portion of a spherical surface can be adequately approximated on a flat chart by spacing meridians closer together than the latitudinal parallels. The ratio of meridional spacing to that of the parallels is the cosine of the chart's central latitude. A convenient central latitude for our target is 28.0°N. Thus 0.883:1 is a good approximate spacing ratio for a rectangular grid of meridians and parallels.

If, for example, one was to choose an $x = 30\ cm$ spacing between meridians, the above ratio dictates that the parallels would need to be spaced $y = 34\ cm$ apart, which was determined as follows:

$$\frac{x}{y} = \cos lat$$

Using the above aspect ratio, prepare a latitude/longitude grid on a sheet of paper such that you can draw the rotated Target Imaging Profile corresponding to your takeoff time. You will need meridians for 80°W and 81°W to include KMLB, the Target Area, and the Abandon waypoint. You may [optionally] include a half-degree meridian at 80°30'W. A finer meridional grid is not recommended, since it is likely to introduce clutter that would be more confusing than helpful. Similarly, you will need parallels at 27°30'N, 28°0'N, and 28°30'N. Figure 8 shows a typical latitude/longitude grid for this exercise.

Once the latitude/longitude grid is prepared, you will need to plot dots for the three key landmarks: KMLB, the Target, and the Abandon check point. These may be located on the grid using standard navigational instruments such as: dividers, right triangles, straight edges, rulers, and parallel rules. A ruler (as a scale) and dividers are generally the best way to determine longitude and latitude offsets from your reference grid lines. Be sure to use the correct proportion (consistent with the aspect ratio) when determining latitude versus longitude offsets.

After you plot KMLB, the Target, and Abandon, you can then plot the rotated Imaging Profile. This is best accomplished using a special protractor called a plotter. Light bearing lines should be drawn, originating from the target along the seven azimuth directions from tables 2 and 3. The lengths of these azimuth lines should be longer than the scaled WP distances computed when determining the Imaging Range and Camera FOV.

Plotting Tools

The plotting tools and instruments used are those traditionally used for maritime piloting and navigation. These were chosen over aeronautical plotting tools since the process in this exercise requires custom chart scaling. The primary tools used are:

- Dividers - looks like a compass used for drawing circles, except that it has two sharp points, instead of a pencil lead as one point.
- Parallel Rules - are two straight edges, separated by four pivot points. These may be extended by a desired amount to create a line parallel to another line or reference direction.
- Rolling Rules (optional) - have a similar function to parallel rules, but a straight edge may be translated on a chart parallel to a line or reference direction by rolling wheels.
- Right Triangle - is a simple tool for creating a direction or line perpendicular to another line or reference direction.
- Plotter (optional) - has a protractor for establishing angles. Most plotters also have straight edges.

Plotting Basics

Plotting techniques are summarized below and consider the conventions used for geographic coordinates. These methods facilitate easy and accurate chart and plot generation.

1. All measurements are made using dividers.
   a. Longitude measurements are made from the longitude scale for your central latitude, which is established in the lower right corner of the Universal Plotting sheet.
   b. Latitude and distance measurements are made from the vertical scale. When measuring distance, one should note that 1 degree of latitude equals 60 nautical miles. One degree of latitude is also divided into 60' (i.e., minutes) of arc. Thus, a nautical mile, by definition, equals one minute of latitude.
2. Lines parallel to meridians may be established using either the parallel rules or the rolling rule.
3. Perpendicular lines, relative to meridians, parallels of latitude, bearings, or course lines may be established using a right triangle or plotter.
4. Courses and bearings may be transferred from the central compass rose using the parallel rules or rolling rule. Alternatively, the plotter may be used.

For the aerial imaging exercise, the scale is reduced by a factor of eight. Therefore, all measurements in this exercise must consider this scaling factor.

Preparing the Universal Plotting Sheet

Universal Plotting Sheets (UPSs) have five parallels of latitude reference lines and one central meridian. The following YouTube video is a quick tutorial for preparing the UPS:

https://www.youtube.com/watch?v=VXJI0-HRp9k

The central parallel should be labeled **28°00'N**. The topmost and bottommost parallels should be labeled **28°15'N** and **27°45'N**, respectively. Thus, the chart's scale is being decreased by a factor of eight.

The graphic in the lower right corner provides the means to create a longitude scale. Draw a line corresponding to the 28° central latitude to establish the longitude scale. As with the latitude scale, this longitude scale also decreased by a factor of eight.

Label the central meridian as **80°45'W**. Locate 28° on the outer scale of the compass rose. The distance from the 28° mark to the central meridian equals the compass rose radius, reduced by the cosine of 28°. Adjust the dividers such that when they are rotated about the 28° location, they come in tangency to the central meridian. Now (maintaining this spacing) place one of the dividers points on the center of the compass rose and "walk" it out two steps to the right along the central parallel to establish the **80°30'W** meridian distance. Repeat this process to the left to establish the **81°00'W** meridian distance. Walk the parallel rules from the central meridian, out to each of these meridional distances, to draw the meridians.

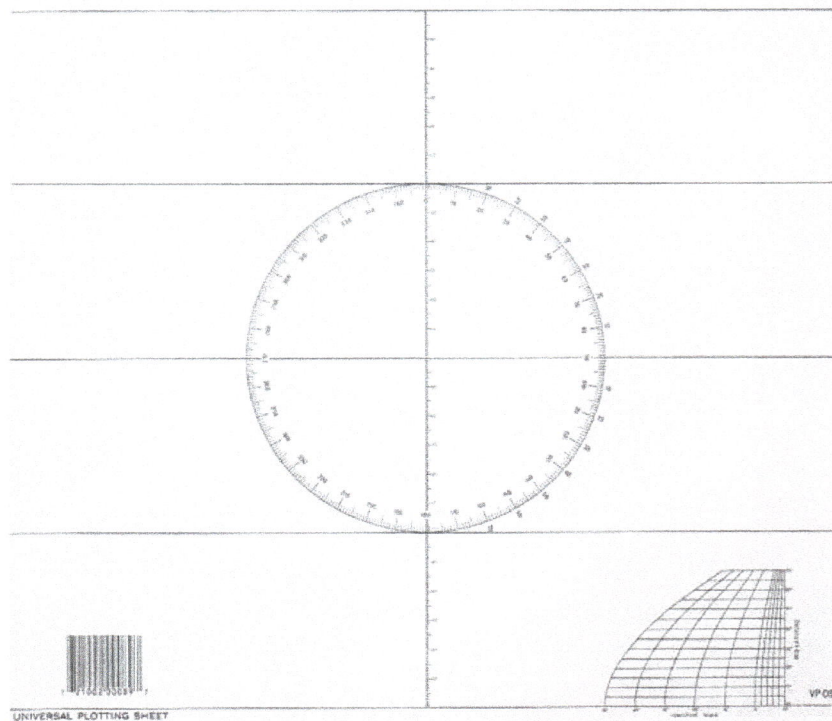

Figure 101. Latitude / Longitude Grid on Universal Plotting Sheet (UPS)

After you plot KMLB, the Target, and Abandon, you can then plot the rotated Imaging Profile. This is best accomplished using a special protractor called a plotter. Light bearing lines should be drawn, originating from the target along the seven azimuth directions from tables 2 and 3. The lengths of these azimuth lines should be longer than the scaled WP distances computed when determining the Imaging Range and Camera FOV.

Plotting Key Reference Points

The first three key references to locate on your chart are **KMLB (28°01.2'N 80°39.0'W)**, **Lake Hell 'N Blazes (28°01.2'W 80°47.4'W)** and **Abandon (28°01.2'N 80°40.8'W)**. Their locations on the chart can be established by measuring off their latitudes on the vertical scale (considering the "8" scaling factor) from the compass rose center. Likewise, their longitudes can be measured relative to any of the meridians, using the longitude scale and the "8" scaling factor.

The systematic process to establish a location is as follows (using the basic plotting techniques):

1. Locate the reference point latitude on the central meridian.
2. Use the parallel rules (or rolling rule) to set a horizontal straight edge at the reference point latitude.
3. Measure the reference point longitude, relative to one of the meridians, using the longitude scale with the dividers.
4. Use the dividers measurement to set the longitude (relative to the reference meridian), along the horizontal straight edge.
5. Mark your reference point with a dot and give it an abbreviated label.

Marking Bearing Lines from the Target

You will need to draw seven (light) bearing lines from the target. The bearing azimuths are those in tables 1 and 2, adjusted (by 15°/hr or 1°/4 min) based on the scheduled takeoff time. These bearings should be of a generous length. Their direction should be transferred from the compass rose to the target, using a parallel instrument, consistent with the plotting techniques.

Creating Imaging Leg Triangles

The imaging legs are the base of isosceles triangles represented in figure 4. Thus, there are three of these triangles, corresponding to the three imaging legs. The centers of the imaging legs correspond to the azimuths in table 2, rotated for the scheduled takeoff time. The range to standoff from the target should be measured along these three azimuths, using the vertical scale and the "8" scaling factor.

The imaging legs are perpendicular to the three central azimuths and pass through the standoff range on the central azimuths. Thus, lines perpendicular to the central azimuths are drawn to establish each leg, intersecting the [rotated] azimuth lines corresponding to table 1. This establishes the three imaging leg triangles, consistent with what is seen in figures 6 and 7.

Locating Waypoints

The waypoints are vertices on the triangles, along the imaging legs. Their coordinates are measured from the chart by a process that's essentially the reverse of how KMLB, Lake Hell 'N Blazes, and Abandon were plotted on the chart. The process to measure the coordinates is:

1. Use the dividers to measure the waypoint latitude relative to a reference parallel of longitude. The dividers are adjusted so they come in tangency with the reference parallel.
2. The divider spacing is measured as an offset against the latitude scale, considering the "8" scaling factor.
3. The latitude offset is applied (added or subtracted) to/from the latitude of the reference parallel against which it was measured.
4. Use the dividers to measure the waypoint longitude relative to a reference meridian. The dividers are adjusted so they come in tangency with the reference meridian.
5. The divider spacing is measured as an offset against the longitude scale, considering the "8" scaling factor.
6. The longitude offset is applied (added or subtracted) to/from the longitude of the reference meridian against which it was measured.

Depending on the scheduled takeoff time, WP1's location may be such that a direct leg from KMLB would require a sharp course change to enter imaging leg #1. (As a general rule, course changes should be no more than 90°.) Given the common scenario in which a direct leg from KMLB requires a sharp turn, it is recommended that a WP0 be created two or three miles North of WP1, in order to allow a gentle transition to the imaging legs.

Measuring Course Lines

Course lines, where not already existing, should be established from KMLB to the first WP, and from the final WP to Abandon. The headings for each of these legs need to be measured and listed as "true" headings on the flight card.

The easiest way to measure the true headings is using a parallel instrument (i.e., parallel rules or rolling rule). The process is to align the straight edge with the course line and then bring it up, so it is centered on the compass rose. At that point the course is read on the compass rose.

True versus Magnetic Headings

Up to this point, azimuth angles have been related to the geographic (or true) cardinal directions. True directions are referenced to the Earth's rotational North and South poles. However, an aircraft navigates using a magnetic and/or gyroscopic compass, which use the magnetic poles as a reference. Thus, the final step is applying an adjustment (called magnetic variation) to your true aircraft headings for your pilot's benefit.

Since the geomagnetic poles do not coincide with the geographic poles, plus that their orientations shift slowly, The International Association of Geomagnetism and Aeronomy (IAGA) continually maintains updates to the International Geomagnetic Reference Field (IGRF – ref. 4). Figure 9 illustrates the declinations (or azimuthal contours) for the year 2020. These contours (a.k.a. isogonic lines) are lines of equal magnetic variation. Note the location of the North Magnetic Pole over eastern Siberia and the South Magnetic Pole near Antarctica, south of Australia.

The figure 9 isogonic map provides a full world view of magnetic declination, and is provided for conceptual insight. Figure 10 is a "close up" of isogonic lines over the 48 contiguous United States (ref. 6). This illustration, while dated (the magnetic poles have shifted noticeably) allows one to infer that eastern Florida has a westward shift from true North to the direction of the magnetic pole. Because of the dynamic shifting in the isogonic lines, a current Sectional Aeronautical Chart (or updated information in ForeFlight) is needed to determine an authoritative value for the target area magnetic variation. For the purpose of the examples, a 7°W magnetic variations with be used. (Note that actual magnetic variation for the target area is subject to drift over time, so current references need to be considered.)

Correcting for Magnetic Declination

A logical question is, given a 7°W magnetic variation (for example), how are headings between waypoints converted from true to magnetic, so they are useful for the pilot to fly the intended ground tracks between your WPs?

Since the magnetic pole is to the West for westward variations, there is an additional 7° of angle in our example. Thus, to get the magnetic heading, the 7° *westward variation is added* to the true heading. Likewise, *eastward variations are subtracted* from true headings to get magnetic headings.

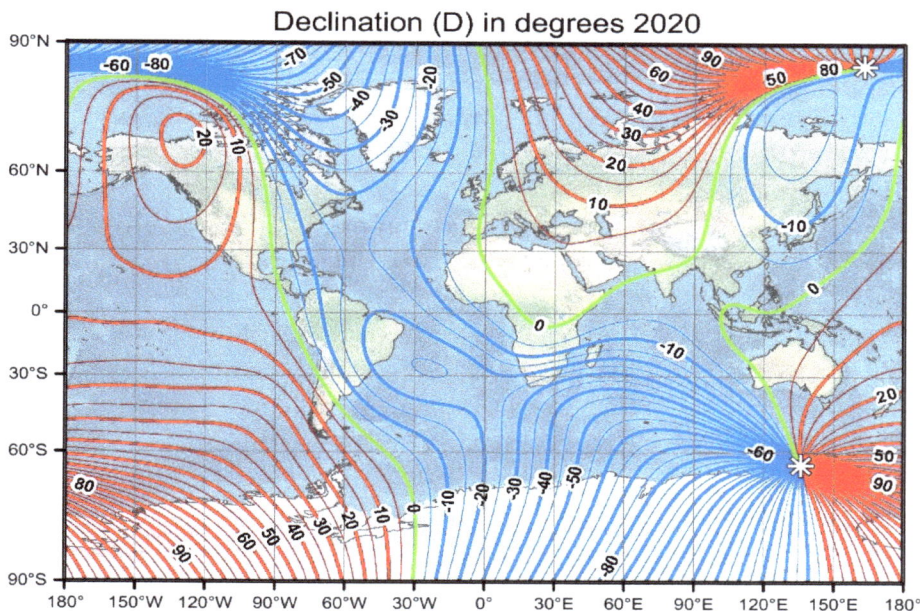

Figure 102. International Geomagnetic Reference Field Declination Map

Isogonic Lines Show The Pattern of Magnetic Declination

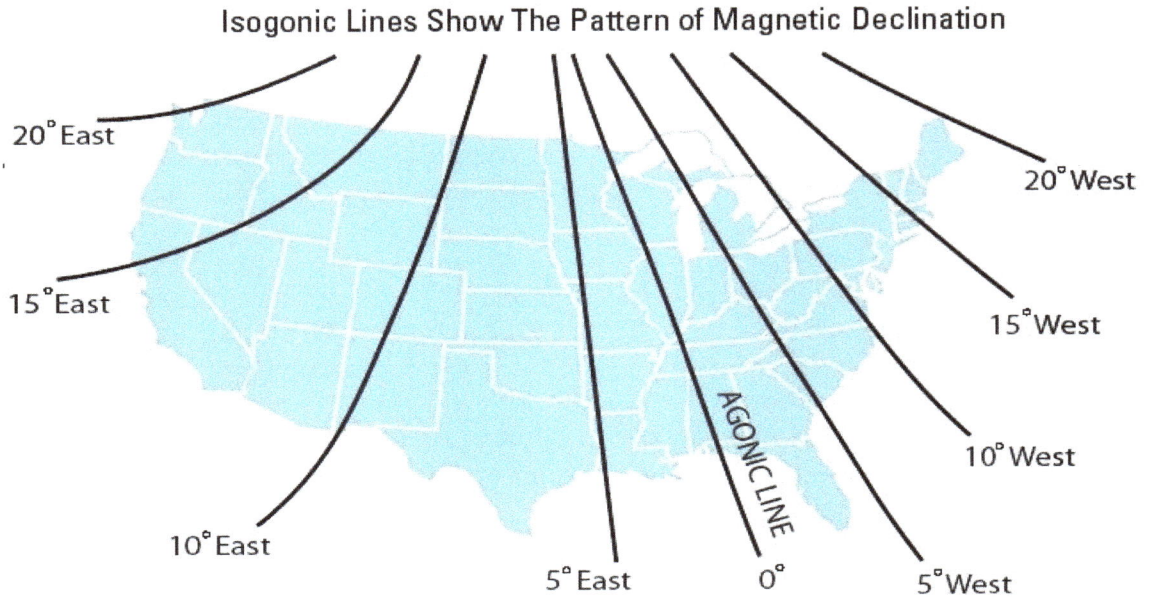

Figure 103. Isogonic Lines over the 48 Contiguous United States

Flight Card and ForeFlight Preparation

The Flight Card provides information useful during the conduct of the flight. Figure 12 shows the card's Waypoint Plan portion. Each flight segment should have its applicable information added as follows:

1. The first column is the sequential segment number.
2. The second column is the time estimated for the beginning of the segment. Time estimates begin with the planned takeoff time from KMLB. Each subsequent time computes the time for the previous leg (determined from its length in NM and the 120 KIAS aircraft speed using the 60D = ST formula) and adds this to the time for the previous leg. Note that ~3 minutes should be added to the time of the first leg to account for takeoff and climb out.
3. The third column is the reference waypoint and its geographic coordinates.
4. The fourth column is the true heading (0°-359°) along the segment. This is the true heading to get from the current waypoint to the next waypoint.
5. The fifth column is the magnetic heading along the segment. For this exercise, the magnetic variation is 7° West. Thus, enter 7 for variation above the Waypoint Plan table and circle the "+W." The entries in this column should thus be the true heading, plus 7 degrees. Note that negative magnetic headings, or headings greater than 359° should be corrected to the 0°-359° range. For negative headings, add 360°; subtract 360° from any heading greater than 360°.
6. The Sixth column is the segment's length in NM.

Variation: ________ deg -E / +W

Waypoint Plan					
#	HH:MM	Reference	Course (deg)		(NM)
Seg.	Time	WP / Lat / Lon	True	Mag.	Dist.

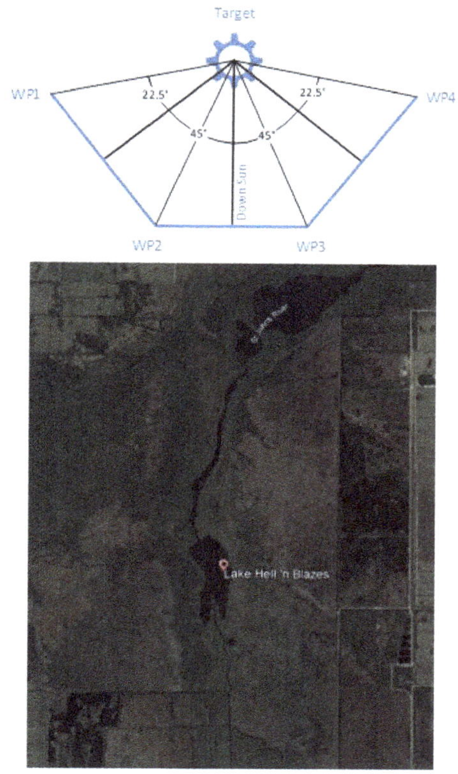

Figure 104. Waypoint Plan Portion of the Flight Card

Consider the following example: The takeoff time from KMLB is 09:30, the distance to the first Waypoint is 9.4 NM, and the true course to the first Waypoint is 242°. The Flight Card's first line in the Waypoint plan should appear as in figure 13.

The final segment begins at Abandon with the true and magnetic courses and distance set for a return to KMLB. (Hint: the true course is 19° and the distance is 6.0 NM.) While the pilot doesn't need this information is entered to the flight card for completeness (and as a check on your plotting accuracy).

Navigation Utility in Wilderness Applications

In certain situations, a GPS is not available or reliable. A resourceful way to triangulate an unknown location with limited resources is by utilizing natural landmarks, a topographic map, and a magnetic compass. Before using a map and compass, it is important to have a topographic map that is up-to-date. In addition, having knowledge of the declination in the region is imperative. Declination is taken into consideration when establishing an azimuth from the map to the compass (subtract the declination) or compass to map (add the declination), however, special considerations need to be considered when navigating in polar regions.

Waypoint Plan					
#	HH:MM	Reference	Course (deg)		(NM)
Seg.	Time	WP / Lat / Lon	True	Mag.	Dist.
1	09:30	KMLB 28°01.2'N / 80°39.0'W	242	249	9.4

Figure 105. Sample Flight Card Segment Entry

Using the Compass

The compass with a mirror is a valuable tool for any navigation situation. There are numerous techniques when navigating in the wilderness, and most techniques can be learned online, at your local outdoor shop, and with practice. However, a technique used to find an unknown location on a map with two known landmarks is called resection.

The process for re-sectioning is first to find at least two landmarks on the map that can be also identified on the ground, for instance, the highest seen point. With a compass, measure the azimuth of each of the landmarks on the ground. Adjust for declination for the physical landmarks to the map, and convert to back azimuths (add the declination, from the compass to the map). Using the two known landmarks on the map, draw the lines of the back azimuths until the lines cross. Thus, the crossed lines will be the current location.

Special Considerations for Polar Navigation

The continent of Antarctica is our last wilderness frontier on Earth. Remote landscape, extreme environmental conditions, enormous ice shelves, and myriad exotic life forms epitomize the public's perceptions of this southernmost wilderness. Navigating across the featureless expanse of Antarctica is a challenging task with the Global Positioning System (GPS). Before GPS, Solar and magnetic compasses were used to set bearings or check a course and dead reckoning allowed quicker travel between readings. A simple way to set a rough bearing was to use shadows – a technique still used today. Knowing the longitude and, therefore, the local time, one can estimate in which direction a shadow will fall at any given hour and maintain a bearing relative to the direction of the shadow. Mountains, nunataks and other landmarks provided additional navigational aids and, where possible, caches were established near some noticeable feature and large snow cairns were built to pinpoint the cache location. In some cases, this means making a dog leg to retrieve a cache, before continuing in the desired direction. Some expeditions used a sledge odometer to measure distance traveled and regular sightings using a sextant and artificial horizon were taken to calculate an exact position.

Figure 106. Sun in the polar regions is a reliable 'GPS' for navigation in the wilderness. Halo is an optical phenomenon produced by light (typically from the Sun or Moon) interacting with ice crystals suspended in the atmosphere. Ice crystals can also float near the ground under cold weather conditions, in which case they are referred to as diamond dust. Thus, the halo is always found in the polar regions very often. (photograph by Wilson (Wai-Yin) Cheung)

These methods relied on having an accurate chronometer, and overcast days presented a challenge, as there were no shadows to follow and no way to take sightings. Navigation within 100 miles of the Pole is particularly challenging, as the lines of longitude converge and the slightest mistake in a bearing – due to misreading a watch, the sun, or the sextant – could send a team far off course. With the advent of GPS, expedition members no longer needed to master the art of celestial navigation. The first expedition to use GPS was the 1989-90 Würth-Antarktis-Transversale, a two-man Antarctic traverse by Reinhold Messner and Arved Fuchs. At the same time Geoff Somers, chief navigator on the International Transantarctic Expedition 1989-90, an epic dog-sled crossing led by Will Steger, navigated by sextant and watch.

However, due to the map projection, limited GNSS navigation satellites and variation between magnetic and true 90° S, navigating the featureless polar regions is still a trying task. As figure 12 shows, a standard GPS receiver cannot accurately navigate after the aircraft passes beyond 85 degrees South. This is the reason why all the flight missions are by VFR in Antarctica.

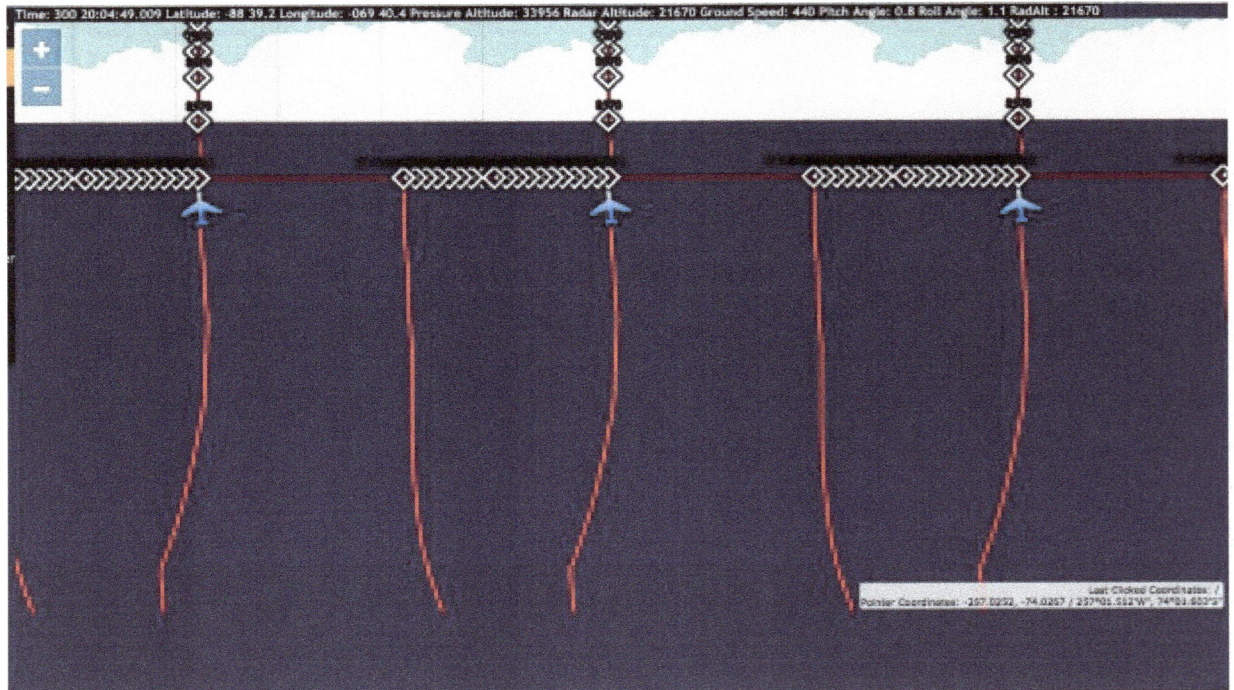

Figure 107. The GPS has difficulty reading how the plane will head south and then north after the 85th parallel. Note how we will "Fly off the map" at 85° S (NASA image)

Finding your way in Polar regions

Using the following simple principle, the sun is the best and easiest way to navigate as the Earth rotates around its axis once in 24 hours, which means that it rotates 360 degrees in 24 hours, or 15 degrees per hour (as previously mentioned). While you compare these degrees with the hours on your watch, you will know that (in the Northern Hemisphere), the sun will be precisely in the East (90 degrees) at 9 pm, and at noon in the South (180 degrees). If you would like to ski to the North Pole and you start at 9 am. The sun will be at 135 degrees (9 am *15 degrees). Your own shadow will thus be to the left of your azimuth, at 315 degrees (135 degrees + 180 degrees). The same technique also applies to the South Pole (shown in Figure 13).

Be aware that the sun moves from left to right in the Arctic and in the opposite direction in Antarctica from right to left.

The tempo of the interior Antarctic expedition is mainly constructed with the elemental influence of the weather at hand. Once you begin the expedition in the wilderness setting, the daily practices must conform to the cyclical progression of sunlight, bodily circadian rhythms, and weather. For instance, there is no official time zone in Antarctica. The time setting then usually follows the time zone of the departure country or closer national station to accommodate logistics and communications. However, once entering latitudes greater than 66 degrees South in the Antarctic summer, the sun will never set below the horizon. Experience in several skiing expeditions to the Geographic South Pole as a guide indicates the South direction, once one knows he/she is facing the Sun. One can turn off the GPS and watches to save battery in the cold condition as one has a reliable reference on this wilderness landscape. Thus, the position of the Sun serves as a reliable time reference for the rhythm of everyday life in Antarctica, while this natural timing information brings substantial relief for navigation.

Summary

This chapter provides an introduction to navigation, with emphasis on the practical knowledge for the aerial ground imaging exercise. The procedures for planning and conducting the exercise were initially developed by the PoSSUM 2001 class, and further refined from lessons learned there and from subsequent classes. The techniques embody planning and navigation methods used in terrestrial, maritime, and airborne activities.

This chapter thus provides the background knowledge of Earth geodesy, geomagnetism, and navigational planning techniques necessary for successful execution. The exercise details provide the imaging methods over the three target legs and the necessary information to plot the flight plan and prepare the flight card, with methods to tailor the plan to the solar angles consistent with the imaging date and time.

The primary goals of this chapter and imaging exercise are teaching general navigation skills as could be applied to PoSSUM aeronomy campaigns, such as are conducted in the follow-on AER 103 Remote Sensing of Noctilucent Clouds course. However, information is also provided for general navigation knowledge for other exploration activities.

References

1. Wikipedia, n.d., downloaded 2022-February-02, https://en.wikipedia.org/wiki/Lake_Hell_%27n_Blazes.
2. Google Earth, © 2021 by Google LLC.
3. AirNav.com, 2022-January-27, https://www.airnav.com/airport/KMLB.
4. Dauntless, n.d., downloaded 2022-February-03, Latitude and Longitude (Meridians and Parallels), https://www.dauntless-soft.com/PRODUCTS/Freebies/Library/books/AK/8-2.htm.
5. SpringerOpen, International Geomagnetic Reference Field: the thirteenth generation, 2021-February-11, https://earth-planets-space.springeropen.com/articles/10.1186/s40623-020-01288-x.
6. U.S. Geological Survey (USGS), n.d., downloaded 2022-February-03, https://www.usgs.gov/educational-resources/magnetic-declination-varies-considerably-across-united-states.
7. MyOpenCountry, 2019-December-22, What Is Compass Declination & How To Adjust For It?, https://www.usgs.gov/educational-resources/magnetic-declination-varies-considerably-across-united-states.

THE IIAS SUBORBITAL SPACE FLIGHT SIMULATOR

Module Objectives

- Learn the Suborbital Spaceflight Simulator's main characteristics and systems
- Operate the simulator through different flight regimes
- Learn how to perform a complete mission as both pilot and scientist.
- Installing and setting up the simulator

Contributing Author:

Vasco Ribeiro

This section explains how to perform a complete simulated PoSSUM tomography mission with the IIAS Suborbital Space Flight Simulator (SSS). The SSS was designed exclusively for IIAS's Project PoSSUM and simulates both profiles similar to the XCOR Lynx (now defunct) and Virgin Galactic Spaceship Two profiles.

This simulator will help you to practice a mission in a small commercial, two-seat, piloted space transport vehicle in a half-hour suborbital flight to 350,000 feet and then return safely to a landing at the take-off runway. The SSS has Horizontal Takeoff Horizontal Landing (HTHL) capability, designed for suborbital flights up to 105km (350,000ft) allowing it to go higher than conventional high-altitude aircraft or balloons and lower than orbital spacecraft. It uses its own rocket propulsion system to depart a runway and return safely.

THE SUBORBITAL SPACE FLIGHT SIMULATOR

The simulated Virgin Galactic Space Ship 2 spacecraft is based around several assumptions:

- The wing area is sized for a gliding ratio of 8 and landing at moderate touchdown speeds near 110 knots.
- The SSS is about 18 meters (60 feet) in length with a double-delta wing that spans about 8 meters (27 feet).
- Two large rudders give it good directional stability. The flight control surfaces are mixed elevator and ailerons controls (elevons) and rudder. Wing speed brakes help on re-entry to give almost optimal pitch.

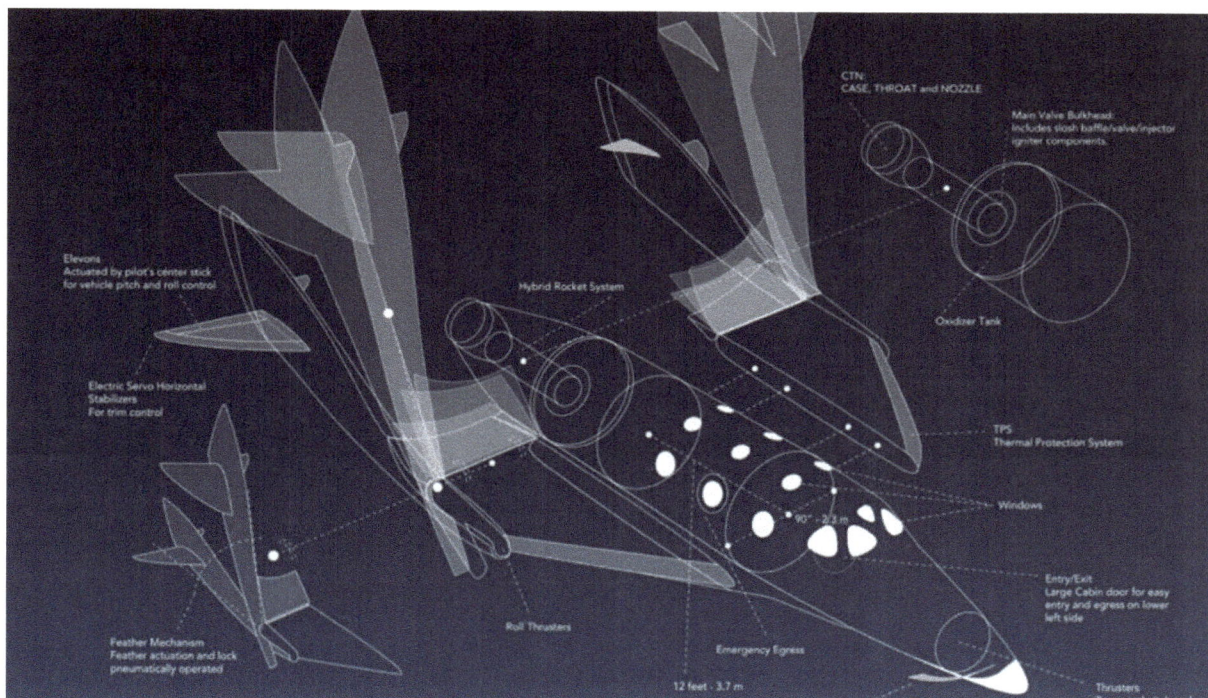

Figure 108. Space Ship 2 Control surfaces

- The empty weight is 5,000lbs and the maximum take-off weight (TOW) 13,600lbs. Target altitude is 361,000 ft.
- The propulsion is by one hybrid rocket engine producing a maximum of 60,000 lbf for take-off and climbing to 190,000ft at Mach 2.9 (Delta-V).
- The model has hydraulic flight controls and landing gear by means of an electric pump. In case of failure, it can still be control manually. An artificial stability system helps to control the vehicle.
- A Reaction Control System (RCS) controls the vehicle in pitch, roll and yaw when in the mesosphere since flight control surfaces are inoperative at such high altitudes.

Table 1 A Comparison of the Rocket-Propelled Planes				
	SS2	SS1	X-15	Space Shuttle
Body Shape	Blunt, bullet	Blunt, Bullet	Long bullet	Blunt, Boxy, Delta wing
Control System	Manual	Manual	Manual	Fly-by-wire
Length	60 ft	28 ft	50 ft 9 in	122.17 ft
Height	18 ft	8 ft 9 in	13 ft 6 in	56.58 ft
Wing Span	27 ft	16 ft 5 in	22 ft 4 in	78.06 ft
Wing Area	507 ft^2*	161.4 ft^2	200 ft^2	2960 ft^2
Aspect Ratio	1.62	1.6	2.5	2.06
Wing loading		49.07 lbs / ft^2	170 lbs / ft^2	120 lbs/ft^2
Empty Weight	5000 lbs **	2640 lbs	14600 lbs	172000 lbs
Loaded Weight		7920 lbs	34000 lbs	295000 Lbs
Thrust to weight ratio	2.08	2.08	2.07	1.5
Mission	Suborbital, Space tourism	Suborbital, Winning X Prize	Alt/Spd record	Orbital, docking with ISS
Target Altitude	361000 ft	367360 ft	354330 ft	3168000 ft
Rate of climb		1367 ft/s	1000 ft/s	770 ft/s
Target Speed	3813.33 ft/s	2336.905 ft/s	7502.46 ft/s	25403 ft/s
Re-entry environment	6G	5.4G/M3.25	8G	5G/M23
Engine Type	Hybrid	Hybrid	Liquid	Solid/liquid
Engine Thrust	60000 lbf	16600 lbf	70400 lbf	2 x 2800000 lbf (0 stage)/1225704 (1st stage)/12000lbf (2nd Stage)
Specific Impulse	250s	250s	276s	269s/455s/316s
Burn Time	87s	87s	86s	124s/480s/1250s

Figure 109. Rocket Plane Comparison

Touchscreen panels

The IIAS suborbital simulator is designed to refine crew resource management techniques and skills in the manipulation of PoSSUM science instrumentation in real-time in a simulated environment. The simulator itself has three projection screens mounted outside of a cockpit that houses two touch screens that simulate the cockpit interior. The two bucket seats simulate spacecraft seats and are designed to secure persons wearing a Final Frontier Launch, Entry, and Ascent (LEA) spacesuit via a 5-point harness.

The two touchscreen panels are present in separated screens, the pilot and co-pilot ones. The pilot panel has the main flight instruments and an Avidyne Primary Flight Display (PFD). The co-pilot panel has an Electronic Centralized Aircraft Monitor (ECAM), the radios, GPS and fuel management buttons. The PFD artificial horizon was modified to give information on pitch angles for climbing, re-entry and gliding. A Head-Up Display makes things easier for the pilot giving basic information about the flight parameters without taking the eyes from the flight path.

Figure 110. Pilot panel with the central PFD and redundant dials around it.

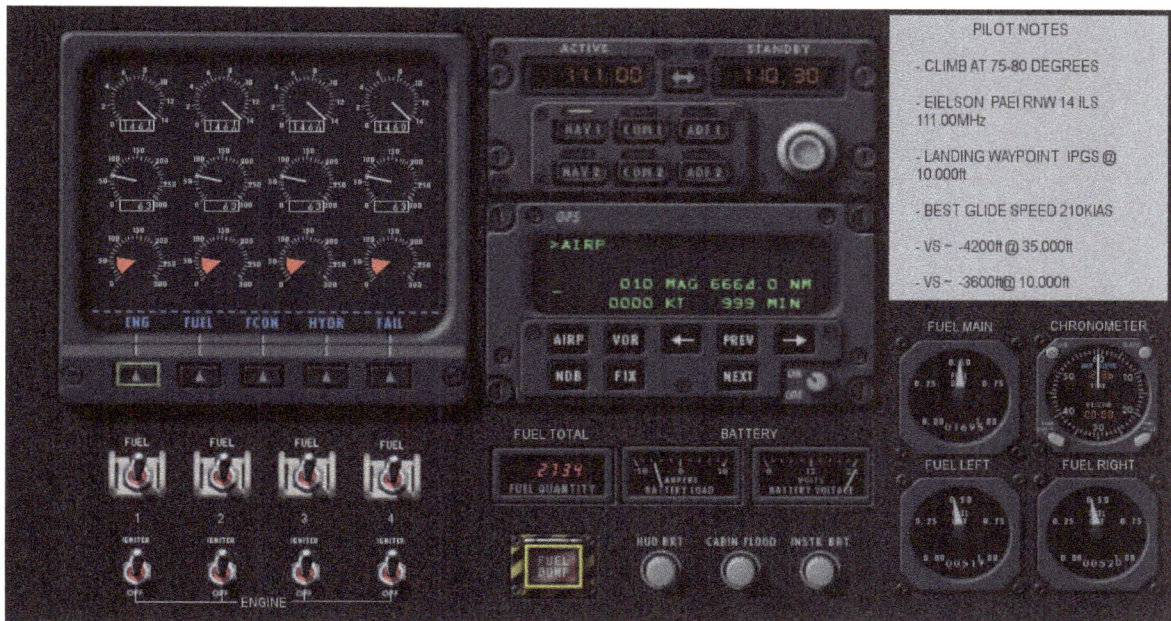

Figure 111. Co-pilot panel with ECAM, radios and GPS, engine and fuel management buttons.

THE BASELINE PoSSUM TOMOGRAPHY EXPERIMENT

The baseline PoSSUM tomography experiment flight will depart from an air-drop location 100nm south of Eielson AFB (ICAO code PAEI) near Fairbanks, Alaska. The initial heading of the spacecraft will be to the northwest. After drop, set the pitch attitude to degrees while making a coordinated turn to a bearing of 360 degrees (True North). The vehicle will be under full thrust from takeoff until about 180,000ft or more, depending on initial weight. After the Main Engine Cut Off (MECO) event at +183s, the vehicle will follow a parabolic trajectory for about 3 minutes. Apogee will occur at about +4m30s. Vehicle re-entry will occur at about +5m56s, this being the most delicate part of the flight. After regaining positive control of aerodynamic flight controls, you will perform a left turn to align with waypoint IPGS (Initial Point for Glide Slope).

The final approach should be at a speed of around 200KIAS and an attitude of -8 degrees glideslope. Touchdown airspeed should be at 110KIAS (+17m40s).

Figure 112. Head-Up Display with speed on the left, altitude on the right, flight path vector and ILS markers on the center.

Setting up the vehicle

Some settings need to be adjusted before flight:
1. Fill the 'tanks' to 'full' (simulating the hybrid propulsion solid-fuel system). By default, the 'fuel tanks' are half-full. The payload weight bar should be set between 300 to 400lbs. Remember that the Takeoff Weight (TOW) will determine the maximum altitude of your flight.
2. Make sure the Noctilucent Cloud Plugin is activated.
3. Set the Date and Time to 1 August, 2017 and midnight (0000LT), respectively.

Flight regimes

Pre-Flight Configuration for Scientist-Astronaut (MS1)

Before flight, the pilot will check all systems using the checklist, setting all radio frequencies for NAV and GPS waypoint navigation. The flight is short and the workload high to set up during flight.

Prior to the closing of the cockpit door, the scientist-astronaut will confirm that all mission payloads are ready for flight. Support crew will power systems that the astronaut will not be able to manipulate once in a spacesuit.

    Support Crew: PoSSUMCam MAIN POWER – ON
                          WIDE-FIELD IMAGER – Configured
                          RED EPIC – Configured

Prior to the decision to launch, MS1 will engage all systems and make sure they are functional.

    MS1:   WIDE FIELD IMAGER POWER - ON
                          SAMPLER POWER – ON
                          MCAT/MASS POWER - ON
                          MISSION ELAPSED TIME – Set to Zero
                          WIDE FIELD IMAGER – SEQ START
                          RED EPIC – ON
                          RED EPIC - REC

Lastly, just before launch, MS1 will start the chronometer:

    MS1:   Mission Elapsed Time - ON

Take off

The SSS has three fuel tanks, a main tank in the fuselage plus two in the wings. As the wings have a lower Centre of Gravity (CG) the resultant CG will be lower, meaning that in first stage of flight a down pitch moment will be compensated by lift of the wing. Set ¼ lower trim (nose up) for a soft take off. Give ¾ throttle. The rotation will occur at 170 KIAS. Keep in mind that the wheel gear creates resistance and a pitch down momentum.

Climbing

After take-off, increase throttle to 100% and build speed. The climbout will be as the vehicle is pitched to an attitude of 75 to 80 degrees at an Indicated Air Speed (IAS) of around 350 KIAS. The vehicle should never exceed 400 KIAS. The wing fuel will burn first. After the wing fuel has been burned the CG will be aligned with the resulting vector of thrust, which would lead the pilot to do some elevator trim adjustments. Turn the vehicle gently to a bearing of 360 degrees while on the climb. Note that the IAS will decrease as air density gets lower with altitude. Remember that your climbing pitch will give you the final base distance after re-entry. **NOTE:** Climbing pitch below 75 degrees will send you far away from base resulting in an emergency landing.

Above 100,000ft

As the atmosphere begins to get thinner, the vehicle accelerates to Mach 2+ and flying surfaces control remains less effective, keeping the flight path to the north and climbing pitch by means of trim. The vehicle is very stable, but avoid sharp movements. At this altitude, engage the RCS system by touching the control on the touchscreen monitors.

Main Engine Cut Off (MECO)

The fuel will burn to around 180,000ft or more depending on initial weight. About three minutes will have passed since take-off. At this point, the speed will be around Mach 2.9 and the vehicle will follow a parabolic trajectory, with the apogee at around 330,000ft. To overcome the ineffectiveness of the control surfaces, the RCS helps to control the vehicle. Under these controls, the vehicle may be turned in any direction using the same control. The control stick will control the RCS system just as it controls the aerodynamic surfaces at lower altitudes.

Parabolic Flight and Data Acquisition Phase

After MECO, two events need to happen: 1) initiate the pitch-down maneuver, and 2) deploy the MCAT pressure probe.

PILOT: **PITCHDOWN** to +10 degrees, Confirm Azimuth North
MS1:   DEPLOY MCAT Probe Switch - UP

The first maneuver will be a pitch down maneuver so that the pitch axis is approximately +10 degrees. In this orientation, the vehicle will be ascending with a heading to the North. At this time, MS1 will deploy the MCAT probe and adjust the iris and zoom of the RED EPIC system.

Ascending Noctilucent Cloud Penetration

On cloud penetration, the pilot will pitch down to capture the wake of the cloud. MS1 will engage the sampler to take the first of two samples.
PILOT: **PITCHDOWN** to -90 degrees,
MS1:   SAMPLE Switch - ENGAGE

Apogee

At this stage of flight, which takes another 3 minutes, the pilot would perform a sort of maneuver pointing the nose where he wants to, without changing his flight path. The NCL experiments and camera operation will be at this stage. Compare it with the 20s to 30s of a parabolic flight with a conventional aircraft. Ground speed is around 380-400 Knots. Below 250,000ft the vehicle must be aligned with horizontal flight path (vector on Map).

Descending Noctilucent Cloud Penetration

Pitch is set to +10 degrees in preparation of the second cloud penetration. Again, MS1 will engage the sampler to take the second of two samples.

> PILOT: **PITCHDOWN** to +10 degrees,
>
> MS1:   SAMPLE Switch - ENGAGE

Re-entry

After the second cloud penetration, the pilot must immediately configure for re-entry attitude.

> PILOT: **PITCHDOWN** to -40 degrees,
>
>   Confirm Azimuth North
>
> MS1:   DEPLOY MCAT Probe Switch - DOWN

The most difficult stage of the flight occurs at speeds around Mach 3+ and very steep dive angle. While the Space Shuttle re-entry was at speeds around Mach 25 and a very flat profile while slowing down, the SSS will fall literally to Earth, slowing down when below 120,000ft with a 4g deceleration and pull up for a brief period. The pitch angle should be -40 degrees to keep good directional stability giving some 40 degrees of Angle of Attack (AOA). Less, it will yaw side-to-side, backflip, and ultimately will be destroyed after losing control. Moreover, you will not slow down as required resulting in a higher speed and, consequently, a lower altitude before gliding which in turn will affect your gliding radius. A drag brake in the upper wing surface should be deployed prior re-entry and retrieved immediately after the vehicle pulls up. The Stall Warning Light (SWL) will pop up at 180,000ft and then, while keeping the dive angle and slowing down, at 90,000-80,000ft it starts to pull out. Turn off the RCS and then start a left turn after the vehicle has pulled out.

Gliding

Gliding starts at around 60,000ft when the SWL is off. Continue turning left to waypoint IPGS bearing. At this point, calculate if you have excess altitude (energy) using the rule of thumb of one and a half mile for each 1,000ft of altitude considering the distance to base or IPGS (+8000ft). The speed is in the range of 200-220KIAS and -7.5 degrees glideslope. Use common manoeuvers to lose altitude if needed, like corkscrew or S-turn. For corkscrew keep in mind that SSS could lose as much as 10,000ft in one complete turn. Do not slow down below 200 KIAS to lose altitude since you will sink rapidly and you may need to recover speed back again.

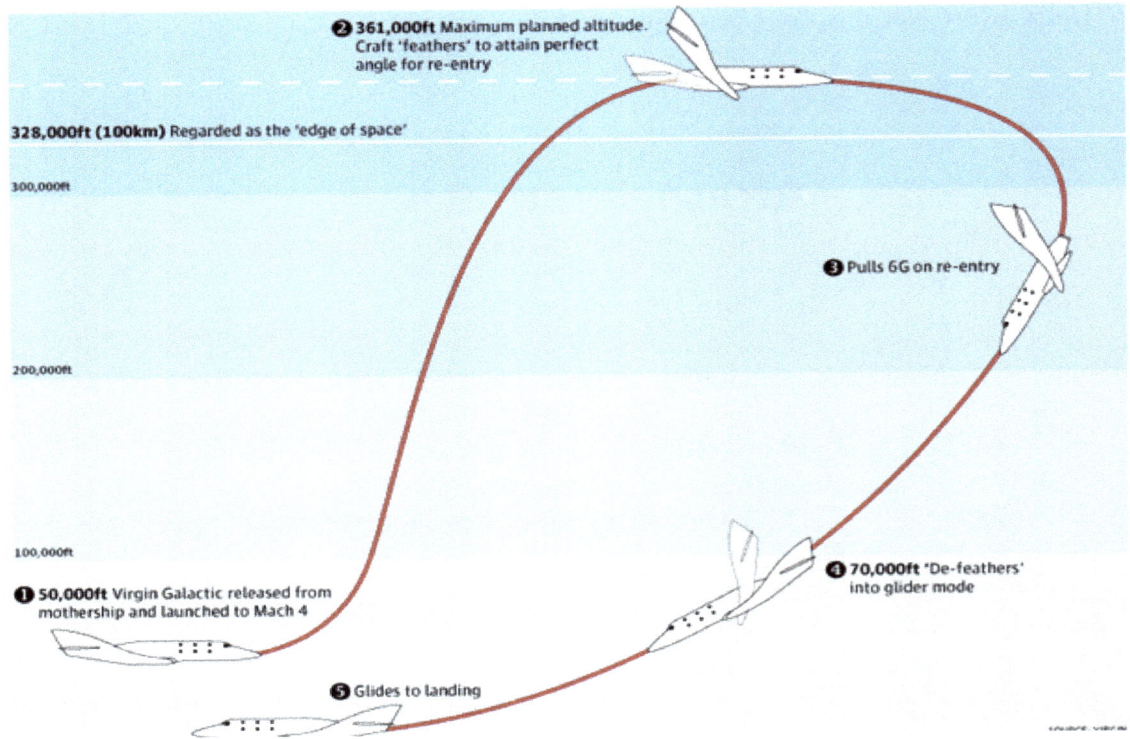

2 361,000ft Maximum planned altitude.
Craft 'feathers' to attain perfect
angle for re-entry

328,000ft (100km) Regarded as the 'edge of space'

300,000ft

3 Pulls 6G on re-entry

200,000ft

100,000ft

1 50,000ft Virgin Galactic released from
mothership and launched to Mach 4

4 70,000ft 'De-feathers'
into glider mode

5 Glides to landing

Figure 113. Space Ship 2 Flight Profile

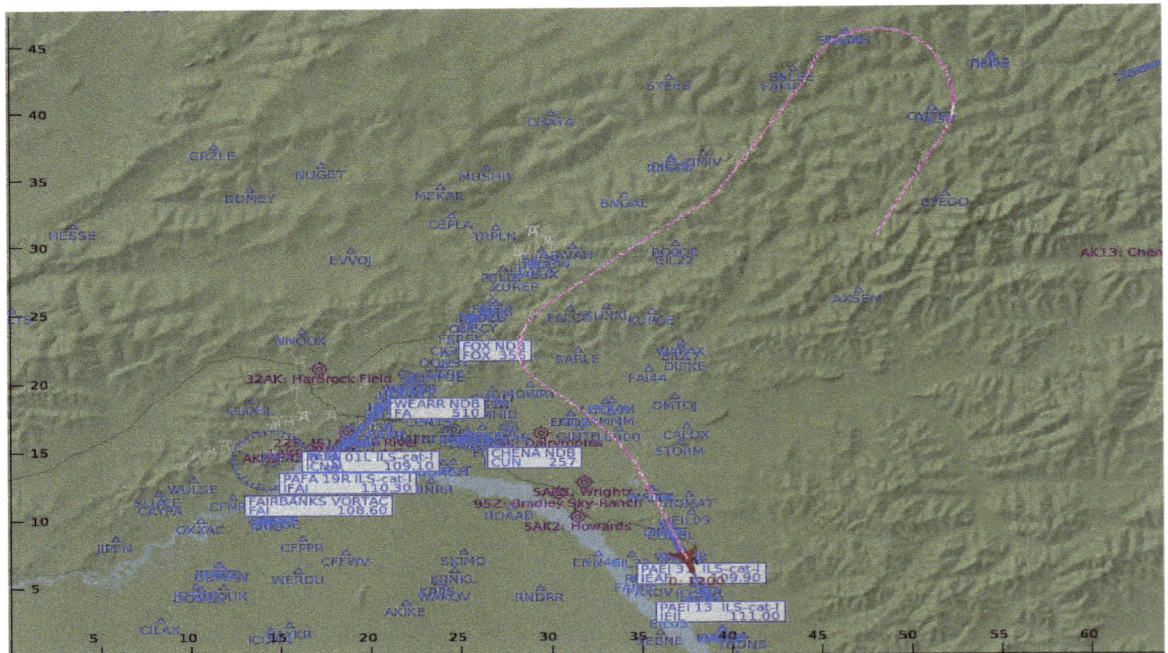

Figure 114. Flight path after apogee. Note that the left turn was maiden at a distance from base at some 45 miles. This would require 30.000ft altitude when starting the glide. The map shows a wide trajectory to correct altitude by extending the gliding distance.

228

Landing

Get to waypoint IPGS with at least 8,000ft and 220 KIAS. Use the flight path vector on HUD to point to the start of the runway so you will keep a good glideslope. Once you got the ILS marker, use it to be sure of your relative position to the ILS glideslope, which is of -8 degrees. This will help you to manage your energy.

When crossing the threshold of the runway, you may use the landing gear as a speed brake. A pitch down moment occurs, so you might have to trim up. Flare it before touching down and float if necessary for contact at 110 KIAS.

Setting up Navigational aids (Radio and GPS)

The SSS Simulator is equipped with radios and GPS to aid navigation albeit the fact that it operates under Visual Flight Rules (VFR). The radio can hold four frequencies of each type: navigational (NAV), communications (COM) and Automatic Directional Finder (ADF). To set them first select the type then, with the selector on the right, set the frequency on Standby Display. With the outer knob, select frequency units, the inner decimal/centesimal in steps of 0.05. To make it active, and in use, just swap position between Active/Standby. The frequencies can be obtained in the local map of X-Plane or in any airport chart.

The GPS has four selection types of entries: Airports, VOR's, NDB and FIX (Waypoints). The entry is an up to five-character code. Airports, AFB, and other airfields have four-character codes, as PAEI for Eielson AFB or the initial point for landing (IPGS). To set just choose which type is the point on the left buttons; select the character position with left/right arrow on the left, setting character with PREV/NEXT buttons. The bearing, distance, ground speed and Estimated time Arrival (ETA) to the waypoint will show up.

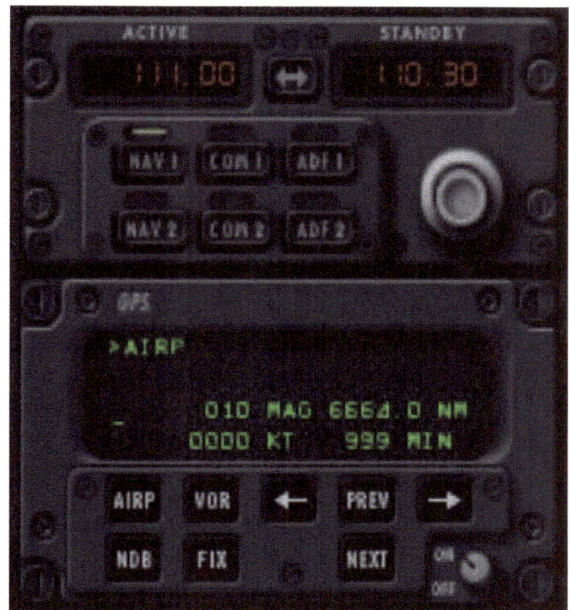

Figure 115. Radio and GPS

Before taking off, set the radio at 111.0Mhz (PAEI runway 14 ILS frequency) at NAV1 to provide custom ILS aid on approach for landing. Make sure that the selector SOURCE is set to NAV1. Set GPS to waypoint IPGS so when gliding starts you can turn directly to it. You also may use the autopilot, by setting heading (HDG) and vertical speed (VCS). As the vehicle operates under VFR, Eielsen AFB runway will be seen from waypoint IPGS and beyond. Again, on pilot's PFD all the information can be displayed on the left buttons cycling trough NAV, GPS and ADF.

Customizing hardware sensitiveness

In order to provide a better control sensation, you will need to set up the joystick sensitivity to assign commands to buttons. Follow the X-Plane menus at Settings\Joystick and Equipment\Axis and Null Zone Tabs. The joystick sensitivity is particularly important to give the proper control feeling. Read more at: http://www.x-plane.com/?article=configuring-flight-controls

Communication

Effective communication techniques must be planned and practiced between the pilot and scientist. The pilot is always responsible for the safe operation of the vehicle; failure for the pilot to accomplish this could lead to a Loss of Crew (LOC) event. The scientist is responsible to meet the science objectives of the mission; failure for the scientist to accomplish this could lead to a Loss of Mission (LOM) event. Obviously, mission safety under the pilot's judgement takes priority; however, the scientist must provide direction to the pilot as needed when changes of vehicle attitude are needed to better meet science objectives.

Launch and Ascent

During launch and ascent, the pilot will follow a pre-determined mission profile. In the case of the PoSSUM Tomography Experiment, the pilot will direct the vehicle to a northward heading after launch and pitch up to an 80-85 degree climb angle. The heading and climb angle are pre-determined and communication will be maintained between the pilot, Air Traffic Control, and Mission Control.

During Parabolic Flight

After Main Engine Cut-off, the scientist will be responsible for providing small attitude correction requests to the pilot, who will comply as long as mission safety would not be compromised. In order to image particular microfeatures of interest, small adjustments might be necessary. The pilot will always try to maintain level roll attitude; however, yaw and pitch may be adjusted as needed. Remember that, under control of the Reaction Control System, roll and yaw are decoupled.

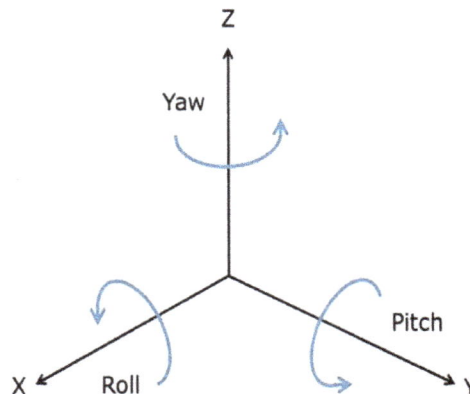

Figure 116. Roll, Pitch, and Yaw. Vehicle will be ascending along the +z axis. X-axis will initially be oriented due north

230

Pitch "up" refers to 'positive pitch' and Yaw "right" refers to 'positive yaw'. Heading is relayed in three-digit True Headings (e.g. "zero-two-zero" for a heading of 20 degrees east of True North). Project PoSSUM uses the following communication protocol. In PoSSUMSim, practice using the following commands:

- "PITCH UP XX DEGREES"
- "PITCH DOWN XX DEGREES"
- "YAW RIGHT XX DEGREES"
- "YAW LEFT XX DEGREES"

As an example, in-cockpit communication a typical parabolic flight might proceed as:

PLT: "MECO"
MS1: "MECO Confirmed, Pitch Down 70-degrees"
PLT: "Roger. Pitch Down 70 degrees"

If a micro-feature of interest is right of the vehicle

MS1: "Yaw Right 20-degrees"
PLT: "Roger. Yaw Right 20-degrees"

Ascending cloud penetration

MS1: "Cloud Penetration. Pitch Down 20-degrees"
PLT: "Roger. Pitch Down 20-degrees"

At MECO

PLT: "Apogee"
MS1: "Confirmed"

While descending cloud penetration

MS1: "Cloud Penetration. Pitch Up 20-degrees"
PLT: "Roger. Pitch Up 20-degrees"

At minimum safe altitude to configure for re-entry

PLT: "Re-entry Attitde"
MS1: "Roger."

Re-entry and Landing
The pilot will configure the vehicle to re-entry attitude once the vehicle descends to the necessary altitude, at which point no further attitude requests will be considered. Again, the re-entry angle and glide approach path are pre-determined and communication will be maintained between the pilot, Air Traffic Control, and Mission Control.

PoSSUM PILOT-ASTRONAUT MISSION CHECKLIST

BEFORE FLIGHT

Battery................................ON
Avionics..............................ON
HUD...................................ON
Landing lights....................OFF
Windshield heat.................ON
AOA sensor heat.................ON
Pitot tube heat....................ON
Hydraulic pump...................ON
HUD bright....................AS REQ
Instruments bright........AS REQ
Cabin flood....................AS REQ
RCS..................................OFF
Art stab...............................ON
Trim................................CHECK
Brakes...............................ON
Brake Chute.......................OFF
CG display.......................CHECK
Altimeter...CHECK BARO PRESS
Selector SOURCE NAV1..CHECK
GPS....................SET WPT IPGS
Radio.........................SET NAV 1
Radio.......STANDBY 111.00Mhz
Radio..........ACTIVE 111.00Mhz
Avidyne PFD..............SET NAV 1
Avidyne PFD.................SET GPS

ENGINES START

Fuel Tank selector...............ALL
Throttle lever......................IDLE
Turn fuel engine 1................ON
Turn fuel engine 2................ON
Turn fuel engine 3................ON
Turn fuel engine 4................ON
Start engine 1.............ENGAGE
Start engine 2.............ENGAGE
Start engine 4.............ENGAGE
Start engine 4.............ENGAGE
Engine lights.........................ON
MFD systems.................CHECK
Elevator trim..................1/4 UP

TAKE OFF

Brakes................................OFF
Throttle...........................100%
Rotate....................170 KNOTS
Landing Gear........................UP

CLIMBING

Pitch 75+ degrees...CHECK TRIM
Bearing 360 degrees........CHECK
MECO at 190,000 ft..........CHECK
Speed Mach 3...................CHECK
RCS....................................ON

PARABOLA LESS 250,000 ft

Map dir. vector forward..CHECK
Drag Brakes...................DEPLOY
Elevator trim..................1/4 UP
Descent Pitch -40.............CHECK

RE-ENTRY LESS 200,000 ft

Stall Warning Light ON.....CHECK
Descent Pitch -40..............CHECK
AOA 40 degrees...............CHECK
Speed Mach 3+.................CHECK
G accelerometer 4g...........CHECK

GLIDING

Stall Warning Light OFF....CHECK
RCS...................................OFF
Drag Brakes...................RETRACT
Elevator trim....................AS REQ
Speed below 45,000 ft..220 KIAS
Glide Slope 7-8 degrees....CHECK
Waypoint bearing..............CHECK
Altitude............................CHECK
Waypoint distance............CHECK
Glide Slope radius.............CHECK

WAYPOINT IPGS

Altitude > 8000ft...............CHECK
Speed > 200KIAS................CHECK
ILS markers.......................CHECK

FINALS

ILS markers.......................CHECK
Radio altimeter (HUD)......CHECK
Drag brake........................AS REQ
Landing gear.....................DOWN
Elevator trim....................AS REQ

LANDING

Speed 160KIAS..................CHECK
Flare.................................CHECK
Touch down 110KIAS........CHECK
Drag chute...........................ON
Brakes.................................ON

PoSSUM SCIENTIST-ASTRONAUT MISSION CHECKLIST

Prior to Cockpit Door Closing:

PoSSUMCam MAIN POWER	**ON**
WIDE-FIELD IMAGER	**Configured**
RED EPIC	**Configured**

Prior to Launch Commit Decision:

MCAT + MASS POWER	**ON**
SAMPLER POWER	**ON**
WIDE FIELD IMAGER POWER	**ON**
RED EPIC	**ON**
Mission Elapsed Time	**Set to Zero**

Prior to Takeoff:

WIDE FIELD IMAGER	**SEQ START**
RED EPIC	**REC**
Mission Elapsed Time	**ON**

Take off

RED EPIC	**ADJUST IRIS AND ZOOM**

Climbing

RED EPIC	**ADJUST IRIS AND ZOOM**

Main Engine Cut Off (MECO)

MCAT Probe Switch	**EXTEND**
RED EPIC	**ADJUST IRIS AND ZOOM**

Ascending Noctilucent Cloud Penetration:

SAMPLE Switch	**ENGAGE**
RED EPIC	**ADJUST IRIS AND ZOOM**

Apogee:

Mission Elapsed Time	**NOTE**
RED EPIC	**ADJUST IRIS AND ZOOM**

Descending Noctilucent Cloud Penetration:

SAMPLE Switch	**ENGAGE**
RED EPIC	**ADJUST IRIS AND ZOOM**

Re-entry

MCAT Probe Switch	**RETRACT**

After Landing:

Mission Elapsed Time	**NOTE**
MCAT + MASS POWER	**OFF**
SAMPLER POWER	**OFF**
WIDE FIELD IMAGER	**OFF**
RED EPIC	**OFF**

PART V: SCIENCE OUTREACH

Science Outreach

Going Forward with IIAS

SCIENCE OUTREACH

Module Objectives

- ➤ Define public education and community outreach
- ➤ Define science communication
- ➤ List the benefits of a public and educational outreach program
- ➤ List the primary steps involved in developing an educational and public outreach (EPO) program
- ➤ Learn the methods of delivery for science education

This section will explain the concepts of educational and public outreach (EPO) and science communication. You will also learn why educational and public outreach and science communication are important to the success of any research program, how to identify the consumers of science, learn the methods of delivering a science outreach program, and how relationships are the key to selling science.

Introduction

When you engage in any type of outreach, you are, for all intents and purposes, engaging in personal selling. The major differences in selling science and selling a mass-produced consumer product are the end-users (consumers) and the product(s) you are offering. The stark reality is that most scientists are not trained in the art and science of selling, marketing, or public relations, and as a result, their research programs are likely missing out on valuable opportunities from lack of exposure and support. Scientific and technical information is complex, and is not easily understood by the general public. In order to gain support for your research program, it is not enough to simply publish your results in scientific journals with the hope that it will filter down to the greater public or lead to new opportunities to expand your program and bolster your budget. As a scientist, it is imperative that you also learn to assume the roles of science communicator and science educator, the mediators between the world of science and the general public.

Public Education & Community Outreach

Public education plays an important and integral part in any publicly- or community-funded, *non-profit organization* (NPO) or non-governmental organization (NGO) research program. Well-planned education and outreach activities can help generate understanding of and support for your research program or project within your local community and beyond. These types of activities can also be used to involve and gain the support of stakeholder (sponsors, partners, and collaborators) and target audiences.

What is Public Education?

Public education is a method of providing information to the general public on issues of interest and/or concern within your target audience or consumer base. In the case of IIAS, public education provides information on the history and background of the program, the overall mission and vision of the program, its research objectives and activities, and how it benefits the greater community.

Information developed and distributed as part of a public education program can take a number of different forms. Lectures and presentations, media coverage, newsletters, and special events can all be used to educate the public about IIAS. The type of public education or outreach program, and its supporting materials, will be dictated by your outreach goals (e.g., solicit donations and sponsorships, prepare an after-school program for your local school district, etc.) and other special considerations that are unique to your target audience.

Why is Public Education and Community Outreach Important?

To maintain long-term program support, it is crucial that you begin by establishing partnerships within your local community in such a way as to provide them with a feeling of "ownership" in your mission with IIAS. Involving people in the "how" and "why" of your mission with IIAS requires significant effort on your part. Ineffective or half-hearted public education programs may confuse community members, diminish support for your mission and the program, or cause people to dismiss your efforts and the efforts of the program altogether. Successful public education programs must be consistent and ongoing. You can't implement a program one week and then forget about it and do it again a year later; it is a never-ending effort.

Public education has the power to generate interest in your mission, as well as in IIAS. When the public becomes interested in your personal story and the Project PoSSUM mission, it can open doors for both you and the program. Think of creative ways that you can involve the public in your mission, whether it be through citizen science initiatives, or involving your alma mater as an educational outreach partner, public participation brings with it the added benefit of building community support for your mission and the IIAS. This point is extremely important. A program that has community support is more likely to be successful than one that faces public opposition.

Public education and community outreach provides the following benefits:

•   An opportunity to disseminate information about your mission and the much larger mission of IIAS;

•   Provide an avenue for soliciting feedback on IIAS objectives, also known as *crowdsourcing*, for the purposes of continuous quality improvement; and

•   A source of information regarding opportunities that exist within the industry, not only for you, but also for IIAS.

All of these aspects are critical to the success of IIAS. The following section will provide you with the information you need to implement your own public education program.

Designing an Effective Public Education & Community Outreach Program

Designing effective educational and public outreach programs requires both funding and expertise. You can often find creative, low-cost ways to accomplish your educational and public outreach goals, even on a shoestring budget. The primary factors to consider when designing an effective educational and public outreach program are:

- Identifying your goals and target audiences

- Crafting a clear and effective message that is appropriate for your audience and in support of your outreach goals

- Selecting an appropriate (for your audience and your event) outreach method

- Developing a way to evaluate the effectiveness of your program for the purpose of continuous quality improvement

Working through these steps does not have to be time or resource intensive. In fact, you probably have considered some of these issues already. But thinking through these steps in a methodical way can help ensure that your resources are well spent and that your outreach program yields the results you're looking for.

There are four major steps involved in implementing a public education and community outreach program:

• Understanding different audiences that exist within your community (target audience/customer base) and how they receive and process information;

• Preparing a formal educational and public outreach (EPO) plan;

• Implementing your outreach program; and

• Evaluating and analyzing your outreach program's activities to determine if your methods are bringing you the results you are looking for, or are helping you to achieve your goals

These four steps are described below.

STEP 1: Identifying Your Audience

Effective public education and community outreach begins with identifying your target audience. This allows you to customize your public education and outreach activities to meet situation-specific needs for your audience as well as for your outreach program.

Understanding your target audience and how they receive and process information is an important part of developing an effective public and educational outreach program. By focusing your efforts on specific target groups and identifying the information best suited to their needs, you can develop a program that maximizes effectiveness while minimizing unnecessary efforts and costs.

The following is a list of questions for you to consider as you design your public and educational outreach program?

• What different age groups can be identified as target audiences (e.g., schoolchildren, high school or college students, professionals, potential sponsors or partners)?

• What components of IIAS are people in your target audience most interested in?

• To what extent are people already educated about IIAS, commercial spaceflight, noctilucent cloud research, other research elements of the IIAS, or the specific topic that you are presenting?

• What communication delivery methods are most appropriate for reaching your target audience (social media, email newsletters, blog posts, brochures, presentations, etc.)?

- Are members of your target audience interested in participating in informational events such as workshops, community dinners and other special events?

- Are there groups within your target audience that are interested in becoming actively involved in supporting your involvement with IIAS? For example, your alma mater as a promotional partner?

After identifying your target audience for a particular outreach event, expand your efforts to include members of your target audience who are respected and influential within the audience/community you are targeting. Be sure to educate and fully inform influential people who are willing to pass on your message to their own networks in order maximize the reach and impact of your message.

Building broad community and industry support through outreach and education can help sustain program funding and momentum.

STEP 2: Preparing Your Educational and Public Outreach Plan

Once you have collected information on your audience and the types of information best suited for your audience, the next step is to develop a formal plan for implementing your public education program and community outreach. Your plan, no matter how simple or complex, should include the following components:

Identify Goals, Objectives, and Obstacles (GOO)

To be effective, your educational and public outreach plan must clearly identify specific goals and objectives that you want to achieve through the implementation of an outreach program. The goals and objectives that you choose to focus on could include securing sponsorships and private donor funding to cover your mission expenses, soliciting new partners or collaborators for a research experiment you are designing, or a community science outreach service project. Keep in mind that your outreach goals are closely linked, and often define, your target audience. Remember to define your goals carefully, making sure to set goals that are achievable given your available resources, timeframes and other constraints.

Your EPO plan should also address any challenges or obstacles that must be overcome to effectively implement your outreach project. Common challenges include successfully delivering educational messages, effectively engaging your audience (getting them to take the desired action you seek), and providing adequate funding to cover your outreach activities. Identifying these challenges, and any potential obstacles, before implementing your outreach plan is crucial to the success of your program.

Once you have identified potential challenges and obstacles that could negatively impact your outreach program, you should determine potential solutions or ways of mitigating their impact. Identifying solutions in advance will help you to develop a more focused and effective program.

Crafting Your Message

Once you have defined your goals and target audience, consider the specific messages and/or information you plan to deliver in order to achieve your outreach objectives. The most effective method of delivery is a simple message to which your target audience can relate. Generally speaking, you want to avoid overwhelming, or even alienating, your audience with information overload. When developing your outreach material, keep in mind the background of your audience, and use language that they can easily understand. For example, the language you would use when delivering a presentation at a scientific conference where the attendees are likely to have a greater understanding for your area of expertise and will be versed in the jargon of the industry, will be different from the language you would use to deliver a presentation to your local Rotary Club. Always keep in mind the needs of the receiver.

Questions to ask yourself as you develop your outreach message could include:

- Why are you interested in educating your chosen target audience?

- What do those in the target audience need to know about your mission or the IIAS?

- What would audience members like to do or be willing to do, to help support your mission?

- What other information can you provide that might act as an incentive or deterrent that will encourage audience members to engage in the desired behavior (what action(s) do you want them to take)?

Keep in mind that different audiences of people typically require different methods of delivery or approaches; methods that work for effectively reaching senior scientists will most likely not work with schoolchildren. Be sure to adapt your message so it resonates with each audience you are trying to engage. Effective messages relate to people on their own terms and in their own language.

Selecting an Outreach Method

The next step is to determine the best method for delivering your message to your target audience. You should consider using more than one method to deliver your outreach material, keeping in mind that there are several different styles of learning, and that learning is not a one-size-fits-all approach. Your outreach and educational messages are likely to have more impact if they are heard more than once. Keeping that in mind, it is a good idea to follow up with your target audience within the first two weeks after your outreach event, before the excitement and interest begins to wane. The goal is to develop a network of actively engaged supporters and allies.

As you consider the numerous options for delivering your outreach material, you will need to consider key factors that might influence your medium of choice, such as your budget, technological requirements, and venue. Also consider using low-cost creative ways of producing your outreach materials. For example, if you want a patch for your research project, and your outreach program is designed to reach and engage schoolchildren and their teachers, you can host a contest and have schoolchildren design and produce their version of the patch, and have a group of friends or professionals from your network vote on and select a winning patch design. When designing any outreach products for your EPO program, always consider ways to incorporate participation from your target audience. This will help your audience members feel a connection to your program. People are more willing to support what they care about.

Public Education & Community Outreach Materials

Public education and community outreach materials can take a variety of forms, including written materials (fact sheets, newsletters, articles, flyers, pamphlets, and brochures); visual materials (signs, posters, charts, pictures, and videos); and events (workshops, school events and programs, public briefings and presentations, and media events). The following list provides ideas on the types of tools, activities and events that you can use to educate and engage your target audience. Remember that this is just a list of ideas. It is not a comprehensive list of things you should do. The number of possible activities and events is only limited by your imagination.

- *Newsletters.* Newsletters provide members of your target audience and supporters with specific information about new developments and progress of Project PoSSUM and your mission. Newsletters are also relatively inexpensive to produce and distribute. Keep the information and communication needs of your audience members in mind as you develop your newsletters.

- *Brochures, Flyers, Articles and Other Written Materials.* Brochures, fact sheets, and flyers are another low-cost method of spreading information. By developing a simple message and distributing it to members of your target audience, you can quickly and easily disseminate timely information and progress updates. To further maximize your reach, you can make written materials available for download on your website, and share across social media networks.

- *Briefings and Presentations.* Briefings and presentations can be used throughout your planning and EPO efforts to keep your target audience and supporters up-to-date on your progress and new developments within the IIAS. These types of methods can be delivered in-person, via online webinar (using tools such as Google Hangout and Fuze Meeting), as a voice-over-PowerPoint that is available for download on your website, or even a combination of these methods to further expand your reach.

- *Media Coverage.* Media coverage of your involvement with IIAS and closely related projects can take a number of different forms. Feature stories in a local or regional newspaper provide the most visible media coverage. This form of outreach can be achieved by sending a news release or making personal contact with a reporter who has an interest in your local community (hometown newspapers are always keen on stories like this) or covers science, technology, and education. News conferences, radio talk shows, podcasts, YouTube, and local news outlets are also effective methods of securing coverage and exposure for your Project PoSSUM outreach program and mission preparations.

- *Workshops and Training.* Workshops and training sessions are valuable ways to reach and engage members of your target audience. These methods can provide a hands-on learning experience for participants, as well as provide you with the exposure and support you're looking for.

● *School Activities and Events.* Educating students about upper atmospheric research, climate science, and space flight offers a rewarding way to engage students and their parents, while simultaneously giving back to your community and accomplishing your program's reach and exposure.

● *Conferences, Industry Meetings, and Special Events.* Another effective way to get your message out is by attending and participating in industry conferences, meetings, and special events. Participating in these types of events enables you to interact directly with your target audience and expand your professional and social network. Meetings and industry conferences are a great way to get your message and information out, while also having the opportunity to learn from others in the industry. Tip: High-profile events can quickly and dramatically increase awareness of your program.

Your EPO plan and program should include several different activities and events. Different forms of information are more effective with some audiences than with others, so a plan that combines several activities and events as part of a comprehensive strategic program will reach more people and is likely to be more effective. Different activities and events can also be used to complement one another.

Identify Available Resources

The type of EPO activities that you select will depend largely upon your available resources. Remember that public education activities do not have to be costly to be effective. Understanding who you are trying to reach will help you use limited resources more effectively.

Examining the EPO activities of other projects and programs in the industry can save you a significant amount of time and money. Don't restrict your scope to examining programs that focus only on the commercial space industry, atmospheric research, or climate science. For instance, EPO activities and materials developed for public health initiatives can also provide you with valuable ideas.

Establish a Timeline

Finally, include a timeline in your EPO plan. Your timeline should coordinate your schedule for implementing EPO activities alongside your IIAS milestones, or any special events you have planned.

STEP 3: Implementing Your EPO Program

Educational and public outreach programs are never fully completed. They continue to evolve as the science upon which they are focused continues to evolve and change. For this reason, it is important that you stay abreast of new developments and pertinent research that are relevant to IIAS, as well as to your own area of expertise.

Once you have an outreach program in place, it is important to take steps to:

- Create meaningful and synergistic partnerships
- Incorporate feedback received from members of your target audience and/or advisors
- Adjust and maintain your education and outreach program materials based on the most current available research
- Implement a program analysis tool to measure your effectiveness (this can include something as simple as a survey)

Partnerships and Outreach Programs Go Hand-in-Hand

Partnerships and outreach programs complement one another. In fact, outreach is required to generate support and create partnerships. In turn, partnerships are crucial to conducting further outreach and sustaining your program. So you need to begin your outreach efforts by first educating your potential partners. Solid partnerships often open the door to new avenues of funding and an expanded resource base.

STEP 4: Evaluate Your EPO Program

Sometimes it is difficult to tell whether your EPO program is effective and reaching the right people. The final step in implementing your EPO program is to develop a method for evaluating each of your program activities and initiatives.

To determine the effectiveness of community education efforts, you can carry out measurement activities. First, consider what would be helpful to measure. For example, you might want to measure:

- Changes in the number of social media followers (as your reach increases, so too should your following).
- Are you seeing an increase in donations or sponsorships?
- Are you receiving new or an increased number of offers from potential partners or collaborators?
- Are your site's Google search rankings improving?
- Are the number of visits to your site or blog (or both) increasing?

By measuring the results of your EPO program, you will be better able to decide whether you need to carry out more efforts or change your education and outreach approaches. Your evaluation of each activity will be determined by comparing results from each activity with your established goals.

The information that you receive as part of your evaluation efforts will assist you in modifying your EPO program to increase its effectiveness. Program measurement and evaluation is important because it can help you identify potential problems with your approach, which affords you with the ability to course correct along the way.

Measuring the success of your outreach and educational program can be difficult, but extremely useful as an assessment and planning tool. Measurement can also assist you in identifying the most effective outreach and education methods to help you plan future activities.

The Science Communicator

Ever heard the saying that "communication is key"? In the world of science, nothing could be further from the truth. One of the most important skills for a scientist to develop is the skill of effective communication. In today's political and economic climate, funding for scientific research continues to dwindle, making the competition for those coveted grant dollars even fiercer. How can scientists give themselves an advantage over their competition? By learning to effectively communicate the value they bring to the marketplace. In other words, as a scientist, are you able to effectively promote your science to a broad audience ranging from the layperson to the seasoned professional? Can you effectively "sell" yourself as a credible and go-to expert in your field of area of expertise? If not, then you are likely missing out on valuable opportunities.

Today's scientist must also learn to assume the role of science communicator. They must be comfortable behind the lab bench, as well as behind the podium. Therefore, a science communicator is someone who is effective in bridging the gap between the complex world of science and the general public.

The Art of Selling Science

To be truly successful in science, especially when it also involves the thrill of exploration, scientists must learn the *art* of *selling* science. Two universal questions that apply to ALL fields of study and industry are:

1.   Who are my customers?
2.   What are their needs?

After all, in order to sell something, you must first know what you're selling and to whom you're trying to sell. The problem with selling science is that science, by its very nature, is somewhat off-putting to most people. It is something that is looked upon as being too complex for most people to understand, or something that is so obscure that most people can't relate to it in a way that is meaningful for them.

We humans are emotional creatures, and we are governed, to an extent, by our relationships. Whether the relationship be positive or negative, it will affect us, just as we will affect it. Relationships are the key to selling, and the key to a healthy relationship is effective communication, so start a conversation! One of the best ways to make science more palatable is to make it entertaining in such a way that people are able to relate to it. To revisit a key point in the first sentence of this paragraph, "humans are emotional creatures." To understand something is to know something. To know something is to feel something. Creating a feeling, or an emotional connection, for your "customer" with what you are trying to "sell" is the key to helping them relate to your product. A great way to do this is through storytelling. Storytelling can be used for breaking down barriers, and for creating a way for you to connect with your audience so you are successful in gaining their trust and respect. If they trust you and respect you, they are more likely to support you.

Recommended Reading

Christensen, L. L. (2007). *The Hands-On Guide for Science Communicators: A Step-by-Step Approach to Public Outreach.* New York, NY: Springer Science Business Media.

Kouzes, J., & Posner, B. (2011). *Credibility: How Leaders Gain and Lose it, Why People Demand It* (Rev. ed.). San Francisco, CA: Jossey-Bass.

Kuchner, M. (2011). *Marketing for Scientists: How to Shine in Tough Times* (1st ed.). Washington, DC: Island Press.

GOING FORWARD WITH IIAS

Module Objectives

➢ Professional Education
➢ Citizen Science
➢ Teaching PoSSUM Science

Contributing Author:

Jason Reimuller, Ph.D

New Frontiers

As a new IIAS graduate, you have become a member of a broad international community dedicated to enabling and communicating aeronomy science. Congratulations for taking this first bold step! You are now qualified to enroll into any of IIAS's graduate programs in aeronomy, bioastronautics, flight test engineering, or space flight operations. The field of astronautics research is emerging as new technologies are enabling a greater and broader human presence in new frontiers formerly the domain only of governments. In addition to enabling novel research, IIAS graduate programs complement traditional academic programs by offering an immersive education that prepares the student for a profession in astronautics.

Immersive IIAS courses enable publishable citizen-science - professional, peer-reviewed science led by subject matter experts within IIAS and supported by its members. Though all participating members may not be experts in the field of study, IIAS seeks to demonstrate that real science may be enabled through a non-profit organization offering immersive educational services and media products of broad interest.

Professional Education Opportunities for the AST 101 Graduate

AST 101 graduates may also enroll in the Applied Astronautics Program, a credentialed curriculum designed to provide the emerging professional interested in applied astronautics research. To complete this certificate, students complete all core classes, select a concentration, and select advanced research electives. A breakdown of the curriculum is provided in Table 17. All courses are instructed by IIAS members that are subject matter experts in their respective fields.

	Class Title	Credits	Instructor(s)
Core Curriculum			
AST 101	Fundamentals of Astronautics	3 credits	J. Reimuller, E. Seedhouse
AST 102	Fundamentals of Microgravity Science	3 credits	A. Persad
EDU 101	Citizen Science Research Methods	2 credits	S. Ritter
AST 199	Thesis	6 credits	TBD
Bioastronautics (with IVA Space Suit Evaluation) Concentration			
BIO 101	Spaceflight Physiology	3 credits	E. Seedhouse
BIO 103	Microgravity Space Suit Evaluation	1 credit	A. Persad
BIO 104	Post-Landing Space Suit Evaluation	2 credits	K. Trujillo
EVA 101	Life Support Systems	2 credits	E. Seedhouse
OPS 102	Spacecraft Egress and Rescue Operations	2 credits	J. Reimuller

Bioastronautics (with EVA Space Suit Evaluation) Concentration

BIO 101	Spaceflight Physiology	3 credits	E. Seedhouse
EVA 101	Life Support Systems	3 credits	E. Seedhouse
EVA 104	Gravity-Offset EVA Space Suit Evaluation	2 credits	K. Trujillo, A. Persad
	select one of:		
EVA 102	Operational Space Medicine	3 credits	S. Pandya
EVA 103	Planetary Field Geology and EVA Tool Development	3 credits	J. Hurtado, U. Horodyskyj

Flight Test Engineering Concentration

FTE 101	Fundamentals of Flight Test Engineering	3 credits	K. Trujillo, R. Sherwood
FTE 102	Fixed Wing Performance Flight Testing	2 credits	R. Sherwood, H. Hammerstein
FTE 103	Fixed-Wing Stability and Control Flight Testing	2 credits	R. Sherwood, H. Hammerstein
	select one of:		
FTE 104	Aircraft Design	3 credits	TBD
OPS 104	Astrodynamics and Orbital Operations	3 credits	K. Ernandes

Free Electives

AER 101	Suborbital Space Environment	3 credits	A. Kleinboehl
AER 103	Remote Sensing of Noctilucent Clouds	3 credits	J. Reimuller
BIO 101	Spaceflight Physiology	3 credits	E. Seedhouse
EVA 102	Operational Space Medicine	3 credits	S. Pandya
EVA 103	Planetary Field Geology and EVA Tool Development	3 credits	J. Hurtado, U. Horodyskyj
EVA 105	Fundamentals of Underwater Analog EVA	3 credits	M. Harasymczuk
FTE 101	Fundamentals of Flight Test Engineering	3 credits	K. Trujillo, R. Sherwood
OPS 101	Systems Engineering for Human Spaceflight	3 credits	K. Trujillo
OPS 104	Astrodynamics and Orbital Operations	3 credits	K. Ernandes

Table 17. Current courses offered to AST 101 graduates (2023).

Graduate Classes in Aeronomy

AER 101: Suborbital Space Environment

AER 101 provides an understanding of the general properties and characteristics of the geospace environment and the underlying physical mechanisms. The student will understand the fundamentals of aeronomy, study of the atmospheric environment of the mesosphere and lower thermosphere (MLT) region of the atmosphere. Special emphasis is given to the environmental hazards most relevant to the operations of crewed spacecraft, including particles and radiation, impact phenomena, spacecraft charging, aerodynamic drag, and oxygen corrosion of surfaces.

AER 103: Airborne Remote Sensing of Noctilucent Clouds

AER 103 provides a foundation in flight research. Students will learn how to integrate and test imagery systems to aircraft and then organize operational field campaigns and sorties using IIAS research aircraft or balloons to study noctilucent clouds. Students will also participate in coordinated ground observation campaigns which may include lidar, incoherent scatter radar, or cameras.

Graduate Classes in Bioastronautics

BIO 101: Spaceflight Physiology

BIO 101 covers the unique aspects of health maintenance of individuals exposed to the rigors of spaceflight. An overview of the physiological changes resulting from prolonged exposures to weightlessness and the establishment of countermeasures are presented in this course as with an understanding of the methods currently in use to mitigate these changes.

BIO 103: Microgravity Space Suit Evaluation

BIO 103 provides a foundation in the microgravity environment, microgravity research campaign planning and operations, human factors and spacesuit evaluation research, biomedical monitoring systems, science communication and public outreach. Students will evaluate prototype seat concepts, suit/seat interface, the umbilical interface, and ingress and egress procedures.

BIO 104: Spacesuit Post-Landing Operational Testing

BIO 104 provides instruction on spacesuit use in nominal and off-nominal post-landing environments. Students demonstrate reliable functionality of parachute release, life preserver unit (LPU), and snorkel functionality in varying sea and lighting conditions. Students also learn the effective use of radios, beacons, signal flares, and other signaling devices in water and egress bottle use for egress operations.

Graduate Classes in Science Education

EDU 101: Citizen Science Research Methods

EDU 101 provides an overview of current citizen science research and a foundation for conducting their own research. Current citizen science research gaps in bioastronautics, extra-vehicular activity (EVA), spacecraft technology, and aeronomy, and possible ways to address these gaps are discussed. Guest instructors, including IIAS alumni, and subject matter experts within the space community, will present their citizen science work as examples.

Graduate Courses in Bioastronautics (Extravehicular Activity) Research

EVA 101: Life Support Systems

EVA 101 will familiarize the student with the essential features of life support systems required for various types of space missions and will cover the requirements and design considerations for life support systems in space. Included are an introduction to basic human physiology, a description of the space environment, a survey of historical life support systems, and a presentation of spacecraft limitations and requirements and EVA space suit operations.

EVA 102: Operational Space Medicine

EVA 102 participants will learn about space medicine, wilderness medicine, human performance, leadership and psychological resilience. The course will dedicate a special focus to extreme environment & wilderness medicine, and how the spaceflight environments may inform triage and first aid scenarios. The on-site portion of this class will focus on wilderness medicine in extreme environments, culminating with a 4-day on-site lab portion devoted to triage, scenarios and skills pertaining to wilderness medicine.

EVA 103: Planetary Field Geology and EVA Tool Development

EVA 103 covers the requirements and design considerations for EVA systems and tools for conducting planetary field geology. Included are an introduction to field science in the context of geology; an overview of the processes that shape the surface environments of Mars and Earth's moon; a survey of historical planetary surface geologic exploration by robots and humans; a survey of historical EVA systems and the design and implementation of EVA suits, tools, and procedures for effective and efficient field science operations on planetary surfaces.

EVA 104: Gravity-Offset EVA Space Suit Evaluation

EVA 104 provides an introduction to EVA space suit test and evaluation methods. Students learn fundamentals of EVA space suit operations and then use the tools and procedures developed in the EVA 102 or EVA 103 courses. IIAS's EVA space suit will be tested and validated in a gravity-offset laboratory environment that can simulate microgravity, lunar, or martian gravity environments.

EVA 105: Fundamentals of Underwater Analog EVA

EVA 105 extends upon the introductory life support system curriculum presented in EVA 101 to include specific EVA space suit systems and test and validation procedures in an underwater analog environment. The course covers a historical analysis of specific US and Russian EVA space suit development programs, EVA space suit systems, laboratory test protocols, terminology and etiquette, EVA space suit test development, and design drivers of future EVA space suit systems.

Graduate Courses in Flight Test Engineering

FTE 101: Fundamentals of Flight Test Engineering

Fundamentals of Flight Test Engineering is a class-room course which will provide basic introduction to aircraft flight test concepts, methods, and planning. Course focusses on concepts of aerodynamics, airplane performance, and stability and control. Practical exercises in aircraft performance, stability and control utilize single engine, multi-engine, and jet powered aircraft.

FTE 102: Fixed-Wing Flight Testing - Performance

Fixed-Wing Performance reviews the concepts and maneuvers for evaluating and determining the performance characteristics of fixed-wing aircraft. These concepts will be used to develop test cards and maneuvers for evaluating the performance of a Mooney M20K aircraft. Post test data analysis and final test report will be submitted by students.

FTE 103: Fixed-Wing Flight Testing - Stability and Control

Fixed Wing Stability and Control reviews the concepts and maneuvers for evaluating and determining the stability and control characteristics of single- and multi-engine aircraft. These concepts will be used to develop test cards and maneuvers for evaluating the stability and control of a Mooney M20K and a Cessna 310 aircraft. Spin test demonstrations are performed in a Super Decathlon aircraft. Post test data analysis and final test report will be submitted by students.

Graduate Courses Space Flight Operations

OPS 101: System Engineering for Human Space Flight

This course covers the roles and responsibilities of the Systems Engineer in supporting the concepts, planning, design, test, verification, operations, and disposal of aerospace systems. The course covers classical Systems Engineering processes with emphasis on spaceflight vehicles and will include assignments to introduce the students to the skills required for successful spacecraft design.

OPS 102: Spacecraft Egress and Rescue Operations

OPS 102 is the first professional education course on the landing and post-landing phase of manned spacecraft missions. this course covers nominal and contingency landing scenarios, post-landing planning, rescue and recovery architecture design, egress systems and operational procedures, de-conditioning and post-landing survivability, generalized egress skills, and emergency egress bottle use.

OPS 104: Orbital Mechanics and Mission Simulation

OPS 104 provides an overview of orbital and attitudinal dynamics. The intent is to provide a meaningful understanding of spacecraft flight dynamics with minimal mathematical emphasis. Thus, the student will gain sufficient knowledge that, when presented with mission profiles from a flight dynamics specialist, they will have a conceptual understanding of the flight profiles and the sequence of events needed to actualize the profile in a simulation environment.

Sample Questions

1. The vestibular system consists of the:
   a. The outer ear, in which are located three semicircular canals for detecting angular acceleration and the saccule and utricle which detect linear acceleration.
   b. The inner ear, in which are located three semicircular canals for detecting angular acceleration and the saccule and utricle which detect linear acceleration.
   c. The inner ear, in which are located two semicircular canals for detecting angular acceleration and the saccule and utricle which detect linear acceleration.
   d. The inner ear, in which are located three semicircular canals for detecting angular acceleration and the saccule and utricle which detect vertical acceleration.

2. Flowing through each semicircular canal is:
   a. Endolymph, which deflects small hair-like cells as the head experiences angular acceleration
   b. Endolymph, which deflects small hair-like cells as the head experiences linear acceleration
   c. Endolymph, which deflects small hair-like cells as the head experiences deceleration
   d. Endolymph, which deflects small hair-like cells as the head experiences vertical acceleration

3. Space motion sickness symptoms
   a. Always resolve within 6 hours on orbit
   b. May last more than 4 days
   c. Typically resolve within 48 to 72 hours
   d. Only affect those who suffer from terrestrial motion sickness

4. The condition in which a person is unable to maintain blood pressure on standing is termed:
   a. Orthostatic anoxia
   b. Orthostatic hypotension
   c. Orthostatic amnesia
   d. Orthostatic reflex.

5. Stroke volume is:
   a. Increased inflight by as much as 60% but compensated by a decreased heart rate.
   b. Decreased inflight by as much as 40% but compensated by a decreased heart rate.
   c. Increased inflight by as much as 10% but compensated by a increased heart rate.
   d. Increased inflight by as much as 80% but compensated by a decreased heart rate.

6. Edema may occur due to:
    a. an decrease in blood pressure, which results in a concomitant rise in capillary pressure, which in turn causes fluid to filter out of the capillary and edema of the tissues
    b. an increase in blood pressure, which results in a concomitant fall in capillary pressure, which in turn causes fluid to filter out of the capillary and edema of the tissues
    c. an increase in blood pressure, which results in a concomitant rise in capillary pressure, which in turn causes fluid to filter out of the capillary and edema of the tissues
    d. an decrease in blood pressure, which results in a concomitant fall in capillary pressure, which in turn causes fluid to filter into the capillary and edema of the tissues

7. Space Motion Sickness is:
    a. Thought to be due to a sensory conflict between vestibular and proprioceptive stimuli.
    b. Thought to be due to a sensory conflict between visual, vestibular, and proprioceptive stimuli.
    c. Only affects those who suffer from terrestrial motion sickness.
    d. Affects less than 20% of first-time astronauts.

8. +Gx direction is often referred to as:
    a. Eyeballs in acceleration.
    b. Eyeballs down acceleration
    c. Eyeballs up acceleration.
    d. Eyeballs in deceleration.

6. A properly performed anti-G straining maneuver (AGSM) can increase tolerance to +Gz by:
    a. More than 5 +Gz
    b. As much as 1 +Gz
    c. As much as 3 +Gz
    d. Less than 0.5 +Gz

7. Exposure to:
    a. 5% or lower concentrations of carbon dioxide at 1 atm can cause nausea, vomiting, chills, visual and auditory hallucinations, burning of the eyes, extreme dyspnea, and loss of consciousness.
    b. 10% or greater concentrations of carbon dioxide at 1 atm can cause nausea, vomiting, chills, visual and auditory hallucinations, burning of the eyes, extreme heat loss, and loss of consciousness.
    c. 10% or greater concentrations of carbon dioxide at 1 atm can cause nausea, vomiting, chills, visual and auditory hallucinations, burning of the eyes, extreme dyspnea, and loss of consciousness.
    d. 10% or greater concentrations of carbon dioxide at 1 atm can cause nausea, vomiting, chills, visual hallucinations, permanent blindness, extreme dyspnea, and loss of consciousness.

8. Typical signs or symptoms associated with exposure to volatile organic compounds include:
    a. Nose and throat discomfort, headache, allergic skin reaction, dyspnea, nausea, emesis, epistaxis, fatigue, dizziness
    b. Nose and throat discomfort, chills, hyperoxia, nausea, emesis, epistaxis, fatigue, dizziness
    c. Nose and throat discomfort, headache, allergic skin reaction, hyperoxia, nausea, emesis, epistaxis, fatigue,
    d. Headache, hyperoxia, nausea, emesis, epistaxis, fatigue, dizziness.

9. Very low partial pressure results in:
    a. Impaired judgment, shortness of breath, allergic skin reaction.
    b. Emesis, shortness of breath, fatigue.
    c. Impaired judgment, shortness of breath, fatigue.
    d. Impaired judgment, auditory hallucinations, fatigue.

10. The Atmosphere Revitalization Pressure Control System
    a. Ensures habitat air pressure is maintained at 147 psia and that the partial pressures of oxygen and nitrogen are maintained within nominal levels
    b. Ensures habitat air pressure is maintained at 14.7 psia and that the pressures of oxygen and nitrogen are maintained within nominal levels
    c. Ensures habitat air pressure is maintained at 4.7 psia and that the partial pressures of oxygen and nitrogen are maintained within nominal levels
    d. Ensures habitat air pressure is maintained at 14.7 psia and that the partial pressures of oxygen and nitrogen are maintained within nominal levels

11. The ARS is responsible for
    a. circulating air, ensuring humidity remains between 30% and 75%, ensuring carbon dioxide and carbon monoxide levels remain non-toxic, ensuring temperature and ventilation is regulated, and ensuring the habitat's avionics and electronics are cooled.
    b. circulating air, ensuring humidity remains between 50% and 95%, ensuring carbon dioxide and carbon monoxide levels remain non-toxic, ensuring temperature and ventilation is regulated, and ensuring the habitat's avionics and electronics are cooled.
    c. circulating air, ensuring humidity remains above 75%, ensuring carbon dioxide and carbon monoxide levels remain non-toxic, ensuring temperature and ventilation is regulated, and ensuring the habitat's avionics and electronics are cooled
    d. circulating air, ensuring humidity remains between 30% and 75%, ensuring carbon dioxide and carbon monoxide levels remain non-toxic, ensuring temperature and ventilation is regulated, and controls the fire suppression system.

12. You are hiking near Fairbanks, AK. You know that Solar Midnight at Fairbanks is at 1:55AM. You want to maintain a heading to the northeast (045 degrees). The local time is 10:00AM. Where should you maintain your shadow relative you heading in order to walk along your desired heading? What happens after you walk for 30 minutes?

13. In Barrow, AK (71.3°N, 156.8°W), what is the highest elevation the sun will reach on the summer solstice, the equinox, and the winter solstice?

14. In Barrow, AK (71.3°N, 156.8°W), how high will the sun be above the horizon at local midnight on the Summer Solstice?

15. In Barrow, AK (71.3°N, 156.8°W), for how many days in winter will there be no sunrise? For how many days in the summer will there be no sunset?

16. The mesosphere is characterized by
    a. decreasing temperatures with altitude.
    b. increasing temperatures with altitude.
    c. increasing pressure with altitude.
    d. decreasing pressure with altitude.

17. Gravity waves in the mesosphere
    a. originate often in the troposphere.
    b. display amplitude grown with altitude.
    c. propagate to outer space.
    d. slow down the mesospheric jets.

18. Ionization pressure gauges
    a. use radioactive materials to ionize the air.
    b. use an electron beam to ionize the air.
    c. do not operate above 1 mbar pressure.
    d. are standard laboratory instruments.

19. In-situ pressure measurements on sounding rockets
    a. need access to the ambient air.
    b. give often much higher results than ambient pressure.
    c. can be used to calculate a temperature profile.
    d. can only be performed in summer.

20. What is the correct order of earth's atmospheric layers from bottom to top?
    a. Stratosphere, Mesosphere, Troposphere, Thermosphere, Exosphere
    b. Stratosphere, Troposphere, Thermosphere, Mesosphere, Exosphere
    c. Troposphere, Mesosphere, Stratosphere, Thermosphere, Exosphere
    d. Troposphere, Stratosphere, Mesosphere, Thermosphere, Exosphere

21. What is the most abundant element in the earth's atmosphere?
    a.   Argon
    b.   Carbon dioxide
    c.   Nitrogen
    d.   Oxygen

22. Which layer of the atmosphere has the highest density of gas molecules?
    a.   Mesosphere
    b.   Stratosphere
    c.   Thermosphere
    d.   Troposphere

23. Which layer of the atmosphere contains the ozone layer?
    a.   Mesosphere
    b.   Stratosphere
    c.   Thermosphere
    d.   Troposphere

24. In which layer does virtually all weather phenomena take place?
    a.   Mesosphere
    b.   Stratosphere
    c.   Thermosphere
    d.   Troposphere

25. Which two atmospheric layers have temperature profiles that promote convection?
    a.   Mesosphere and Thermosphere.
    b.   Mesosphere and Troposphere.
    c.   Stratosphere and Thermosphere.
    d.   Stratosphere and Troposphere

26. Near the solstice, the wind in the mesosphere generally flows:
    a.   From summertime pole to wintertime pole
    b.   From wintertime pole to summertime pole
    c.   From east to west
    d.   From west to east

27. Why is the mesosphere coldest in the summertime?

28. Which of the following significantly drives gravity wave formation?
    a.   Atmospheric flow over mountains
    b.   Absorption of ultraviolet light in the Mesosphere
    c.   Planetary wave interactions
    d.   Variations in the Earth's gravitational constant

29. Why are NLCs observations believed to change as a result of man-made climate change?
    a. Water vapor results from an increase of CO2
    b. Methane cools the upper atmosphere
    c. Both a and b
    d. Neither a nor b

30. Why do NLCs form consistently around 83km? What factors do you think might alter the height of NLCs?

31. Mie scattering occurs around
    a. particle radii approximately 0.1 times the incident wavelength of scattering radiation.
    b. particle radii approximately 1.0 times the incident wavelength of scattering radiation.
    c. particle radii approximately 10.0 times the incident wavelength of scattering radiation.
    d. particle radii approximately 100.0 times the incident wavelength of scattering radiation.

32. If an NLC is at 83km of altitude, at what Solar Depression Angle would the column of air below be completely obscured while permitting the NLC to be directly illuminated by the Sum, assuming the Earth is spherical with a radius of 6371km?

33. Why do NLCs only form over polar regions?

34. Why do NLCs only form during the summertime?

35. Why don't we think NLCs occurred before 1885?
    a. No observations were ever reported by First Nation peoples prior to 1885.
    b. Hindcast models indicate that the atmosphere was not cold enough.
    c. Mesospheric Hindcast models indicate that there was not enough water vapor in the upper atmosphere.
    d. All of the above

37. What frequencies of electromagnetic radiation are absorbed by the earth's ozone layer?
    a. Infrared light
    b. Microwaves
    c. Radio waves
    d. Ultraviolet light

42. What is the maximum operating pressure of the Final Frontier IVA suit?

43.   In digital photography, ISO measures the:
    a.   Amount of light exposed to the sensor
    b.   The sensitivity of the sensor
    c.   The time the sensor is exposed to light
    d.   The frequency range of the light allowed to reach the sensor

44.   What is the purpose of employing High Dynamic Range imaging?

45.   Why are PoSSUM candidates trained in aerobatic aircraft
    a.   To practice Anti-G Straining Maneuver Techniques
    b.   To adapt to changing-G environments
    c.   To refine Crew Resource Management techniques
    d.   All of the above

46.   Why are PoSSUM Scientist-Astronaut Candidates exposed to hypoxic conditions?
    a.   To teach the candidate to recognize off nominal environments in a spacesuit
    b.   To condition the candidate for low oxygen environments
    c.   To evaluate the candidate's responsiveness to stress
    d.   All of the above

47.   What is the difference between the Solar Depression Angle (SDA) as perceived from an observer on the ground versus an observer in an aircraft at 25,000ft altitude?

48.   Atmospheric background scattering is approximately 1% that of sea level above
    a.   5km
    b.   15km
    c.   30km
    d.   50km

49.   From satellites, why are noctilucent clouds observed in ultraviolet frequencies?

50.   Airglow is caused by:
    a.   Refraction of sunlight off of noctilucent clouds.
    b.   Preferential scattering produced by sodium ions.
    c.   Recombination of atoms that were photo-ionized by sunlight during the day.
    d.   All of the above

APPENDIX A: SOLAR DECLINATION

	Jan	Feb	Mar	Apr	May	Jun	Jul	Aug	Sep	Oct	Nov	Dec
1	-23.08°	-17.38°	-7.97°	4.15°	14.77°	21.91°	23.18°	18.28°	8.66°	-2.79°	-14.11°	-21.65°
2	-23.00°	-17.10°	-7.59°	4.53°	15.07°	22.05°	23.11°	18.03°	8.30°	-3.18°	-14.43°	-21.81°
3	-22.91°	-16.81°	-7.21°	4.92°	15.37°	22.18°	23.04°	17.78°	7.94°	-3.57°	-14.75°	-21.96°
4	-22.82°	-16.52°	-6.83°	5.30°	15.67°	22.31°	22.96°	17.52°	7.57°	-3.95°	-15.07°	-22.10°
5	-22.72°	-16.22°	-6.44°	5.69°	15.96°	22.43°	22.88°	17.26°	7.20°	-4.34°	-15.38°	-22.24°
6	-22.61°	-15.92°	-6.06°	6.07°	16.25°	22.54°	22.79°	16.99°	6.83°	-4.73°	-15.68°	-22.37°
7	-22.49°	-15.62°	-5.67°	6.45°	16.53°	22.65°	22.69°	16.72°	6.46°	-5.11°	-15.99°	-22.49°
8	-22.37°	-15.31°	-5.28°	6.82°	16.81°	22.75°	22.59°	16.44°	6.09°	-5.50°	-16.28°	-22.61°
9	-22.24°	-14.99°	-4.89°	7.20°	17.08°	22.84°	22.48°	16.16°	5.71°	-5.88°	-16.58°	-22.72°
10	-22.10°	-14.67°	-4.50°	7.57°	17.35°	22.93°	22.36°	15.87°	5.33°	-6.26°	-16.86°	-22.82°
11	-21.95°	-14.35°	-4.11°	7.94°	17.62°	23.01°	22.24°	15.58°	4.95°	-6.64°	-17.15°	-22.91°
12	-21.80°	-14.02°	-3.72°	8.31°	17.88°	23.08°	22.11°	15.29°	4.57°	-7.02°	-17.43°	-23.00°
13	-21.64°	-13.69°	-3.33°	8.68°	18.13°	23.15°	21.98°	14.99°	4.19°	-7.39°	-17.70°	-23.08°
14	-21.47°	-13.36°	-2.93°	9.04°	18.38°	23.21°	21.84°	14.69°	3.81°	-7.77°	-17.97°	-23.15°
15	-21.30°	-13.02°	-2.54°	9.40°	18.62°	23.26°	21.69°	14.38°	3.43°	-8.14°	-18.23°	-23.21°
16	-21.12°	-12.68°	-2.14°	9.76°	18.86°	23.31°	21.53°	14.07°	3.04°	-8.51°	-18.49°	-23.27°
17	-20.93°	-12.33°	-1.75°	10.12°	19.10°	23.35°	21.37°	13.76°	2.66°	-8.88°	-18.74°	-23.32°
18	-20.74°	-11.99°	-1.35°	10.47°	19.32°	23.38°	21.21°	13.44°	2.27°	-9.25°	-18.99°	-23.36°
19	-20.54°	-11.63°	-0.96°	10.82°	19.55°	23.40°	21.04°	13.12°	1.88°	-9.62°	-19.23°	-23.39°
20	-20.33°	-11.28°	-0.56°	11.17°	19.76°	23.42°	20.86°	12.79°	1.50°	-9.98°	-19.47°	-23.41°
21	-20.12°	-10.92°	-0.17°	11.51°	19.97°	23.43°	20.67°	12.47°	1.11°	-10.34°	-19.70°	-23.43°
22	-19.90°	-10.56°	0.23°	11.85°	20.18°	23.44°	20.48°	12.13°	0.72°	-10.70°	-19.92°	-23.44°
23	-19.67°	-10.20°	0.62°	12.19°	20.38°	23.44°	20.29°	11.80°	0.33°	-11.05°	-20.14°	-23.44°
24	-19.44°	-9.83°	1.02°	12.53°	20.57°	23.43°	20.09°	11.46°	-0.06°	-11.41°	-20.35°	-23.43°
25	-19.20°	-9.47°	1.41°	12.86°	20.76°	23.41°	19.88°	11.12°	-0.45°	-11.75°	-20.55°	-23.42°
26	-18.96°	-9.09°	1.80°	13.19°	20.95°	23.39°	19.67°	10.78°	-0.84°	-12.10°	-20.75°	-23.40°
27	-18.71°	-8.72°	2.20°	13.51°	21.12°	23.36°	19.45°	10.43°	-1.23°	-12.44°	-20.94°	-23.37°
28	-18.46°	-8.35°	2.59°	13.83°	21.29°	23.32°	19.23°	10.08°	-1.62°	-12.78°	-21.13°	-23.33°
29	-18.19°		2.98°	14.15°	21.46°	23.28°	19.00°	9.73°	-2.01°	-13.12°	-21.31°	-23.28°
30	-17.93°		3.37°	14.46°	21.61°	23.23°	18.76°	9.38°	-2.40°	-13.46°	-21.48°	-23.23°
31	-17.66°		3.76°		21.77°		18.53°	9.02°		-13.79°		-23.17°

APPENDIX B: CALCULATING SOLAR ELEVATION AND AZIMUTH WITH GREATER PRECISION

The Solar Declination (δ) is the angle between the rays of the sun and the plane of the earth's equator. The value changes throughout the year as it is a function of the day of the year relative to the solstices and, to a smaller extent, the ellipticity of the Earth's orbit.

On the solstice, the angle between the rays of the sun and the plane of the earth equator is at its maximum value of 23.45°. Therefore δ = +23.45° at the northern hemisphere's summer solstice (June 21) and δ = -23.45° at the northern hemisphere's winter solstice (December 21). On each equinox, the solar declination is zero and the projection of the sun is coplanar with the Earth's equator.

Since the eccentricity of the earth orbit is quite low, it can be approximated to a circle, and δ is approximately given by the following expression:

$$\delta = -23.45° \cdot \cos\left[\frac{360°}{365} \cdot (N + 10)\right]$$

where N is the number of days of the year that have transpired since January 1.

A more accurate solution for the position of the sun relative to the Earth can be calculated from an iterative solution of Kepler's equation.

The mean anomaly (M) of the Sun is given by:

M = 0.017202791 D – 0.0656742 radians

where D is the number of days, expressed into decimals that have elapsed since epoch 1980.0. The number of whole days is displayed below for each year from January 0.0. Add the day of the year (e.g. Feb 1st = 32) and the time in UT past 00:00 into decimal days (e.g. one hour = 0.04167 days).

Year	D
2015	12783
2016	13148
2017	13514
2018	13879
2019	14244

Thus, for 23 July 2016 at 0600 UT:

D = 13148 (January 0.0 of 2106) + 205 (July 23rd on a leap year) + 0.25 (0600/2400)
D = 13353.25

The eccentric anomaly (E) is calculated by setting E_i = M and i=0. Calcualte:

$\Delta = E_i - 0.016718 \sin(E_i) - M$

Converge in a series iteration until the magnitude of Δ is less than 0.000001 radians. To perform a series iteration, solve each step with the following relation:

$E_{i+1} = E_i - \Delta/(1-0.016718 \cos(E_i)$

The true anomaly, v, is given by:

$\tan(v/2) = 1.0168601 \tan(E/2)$

The ecliptic longitude of the Sun (λ_s) is:

λ_s = v + 282.596403 degrees.

The solar right ascension (α) and declination (δ) follow from:

$\tan(\alpha) = 0.9174821 \tan \lambda_s$
$\sin(\delta) = 0.3977769 \sin \lambda_s$

The observer's latitude (Φ) and longitude (λ) are used to determine the solar azimuth (A) and elevation (h). First, UT must be converted into local sidereal time (LST) using:

S = 0.7999589 + 0.000027378508 D
R = 6.6460656 + 2400.051262 S + 0.00002581 S^2
B = 24 − R + 24 (year − 1900)
LST = 0.0657098 d − B + 1.002738UT + λ/15.0

Where d is the whole number of days elapsed since January 0.0. The solar elevation (h) and solar azimuth (A) are determined from:

$\sin(h) = \sin(\delta)\sin(\phi) + \cos(\delta)\cos(\phi)\cos(LST-\alpha)$
$\cos(A) = (\sin(\delta) - \sin(\phi)\sin(h))/(\cos(\phi)\cos(h))$

APPENDIX C: CANON EOS MARK III CAMERA OPERATIONS

The following illustrations are provided for a basic understanding of the Canon EOS 5D Mark III camera and its functions.

Familiarization with the Camera

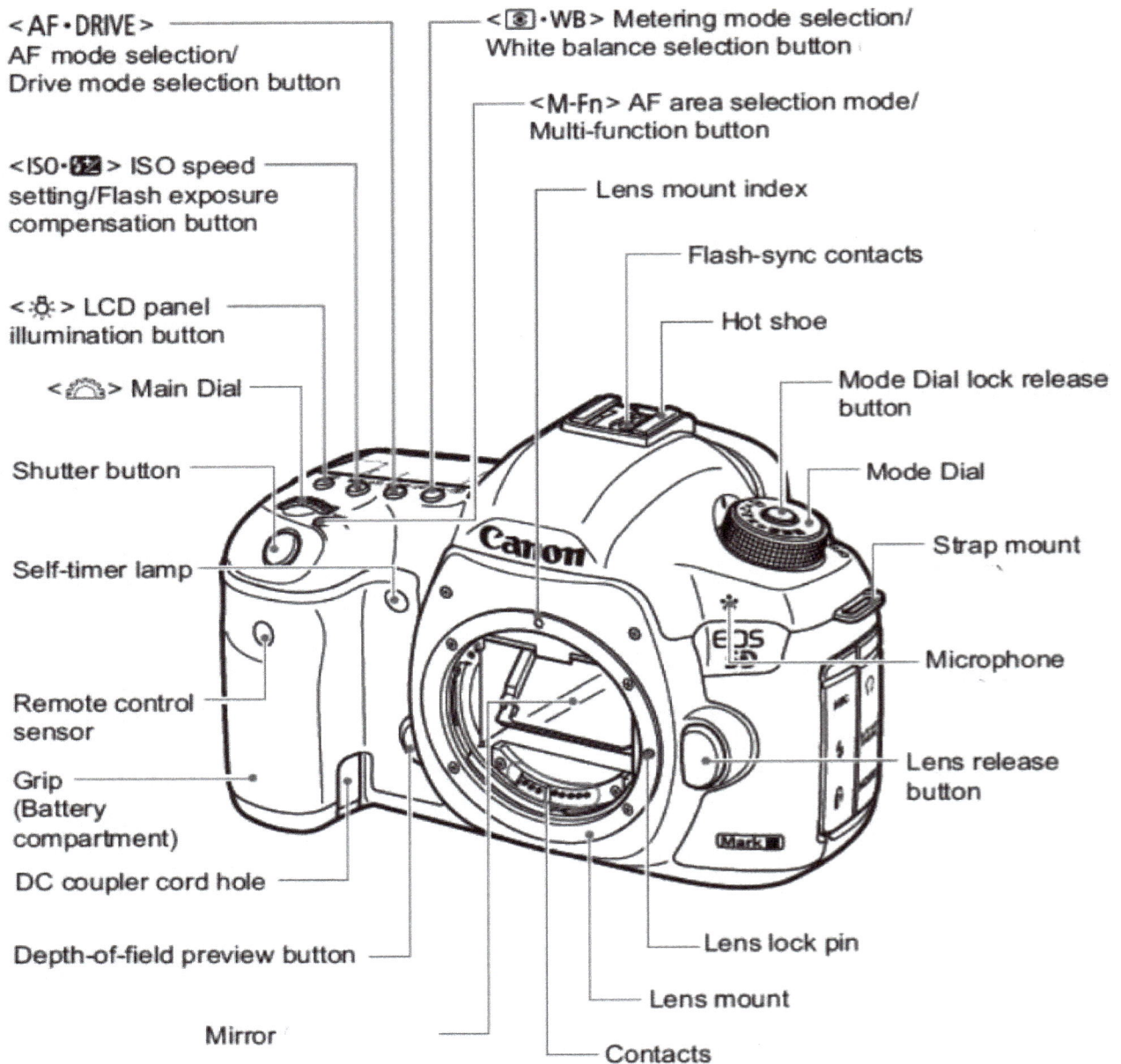

< AF•DRIVE >
AF mode selection/
Drive mode selection button

< ISO•▨ > ISO speed
setting/Flash exposure
compensation button

< ☀ > LCD panel
illumination button

< ⌂ > Main Dial

Shutter button

Self-timer lamp

Remote control
sensor

Grip
(Battery
compartment)

DC coupler cord hole

Depth-of-field preview button

Mirror

< ▣ •WB > Metering mode selection/
White balance selection button

< M-Fn > AF area selection mode/
Multi-function button

Lens mount index

Flash-sync contacts

Hot shoe

Mode Dial lock release
button

Mode Dial

Strap mount

Microphone

Lens release
button

Lens lock pin

Lens mount

Contacts

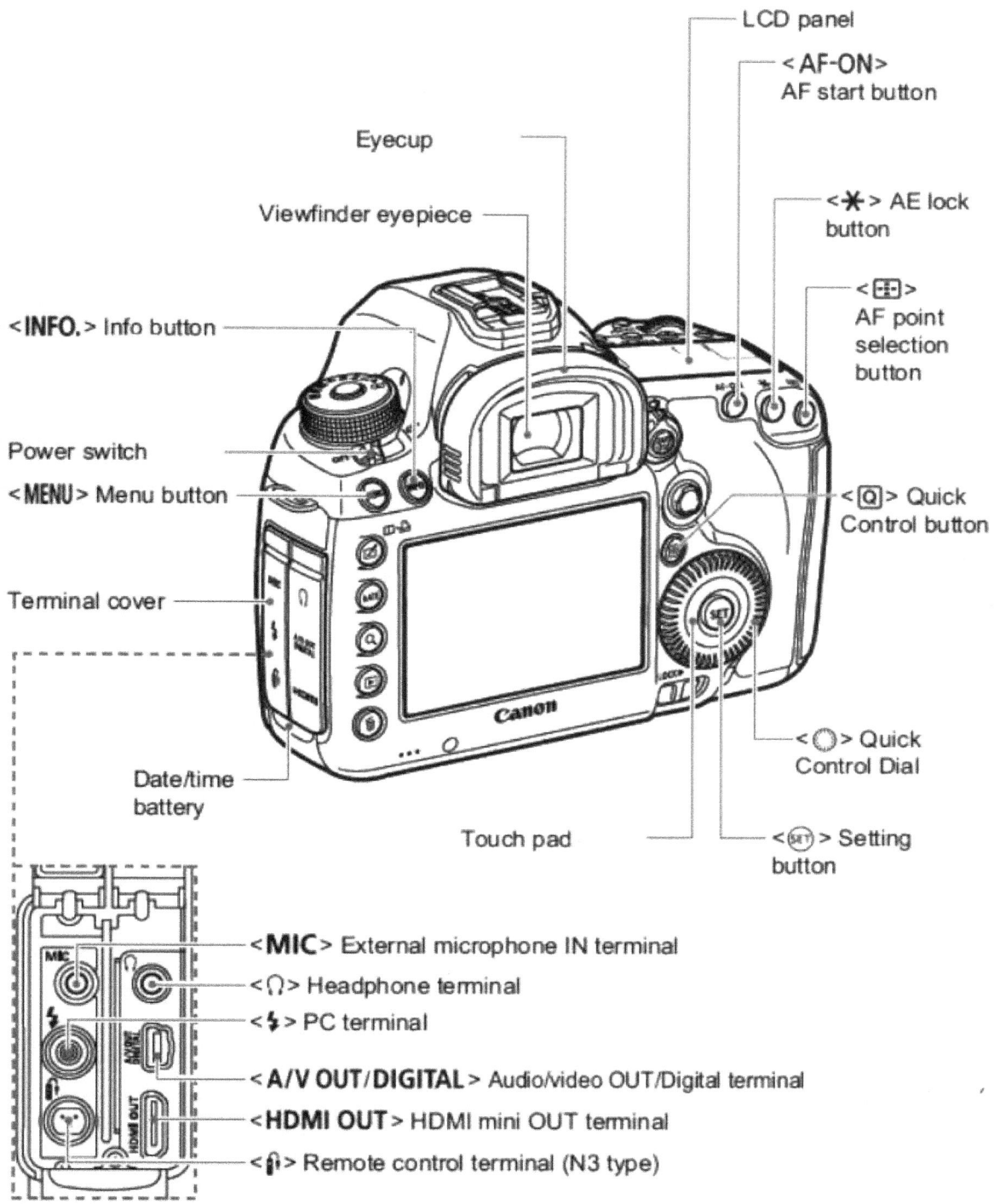

LCD panel

Eyecup

\<AF-ON\>
AF start button

Viewfinder eyepiece

\<✱\> AE lock
button

\<⊞\>
AF point
selection
button

\<**INFO.**\> Info button

Power switch

\<**MENU**\> Menu button

\<**Q**\> Quick
Control button

Terminal cover

Date/time
battery

\<◯\> Quick
Control Dial

Touch pad

\<⟨SET⟩\> Setting
button

\<**MIC**\> External microphone IN terminal

\<∩\> Headphone terminal

\<↯\> PC terminal

\<**A/V OUT/DIGITAL**\> Audio/video OUT/Digital terminal

\<**HDMI OUT**\> HDMI mini OUT terminal

\<⬡\> Remote control terminal (N3 type)

265

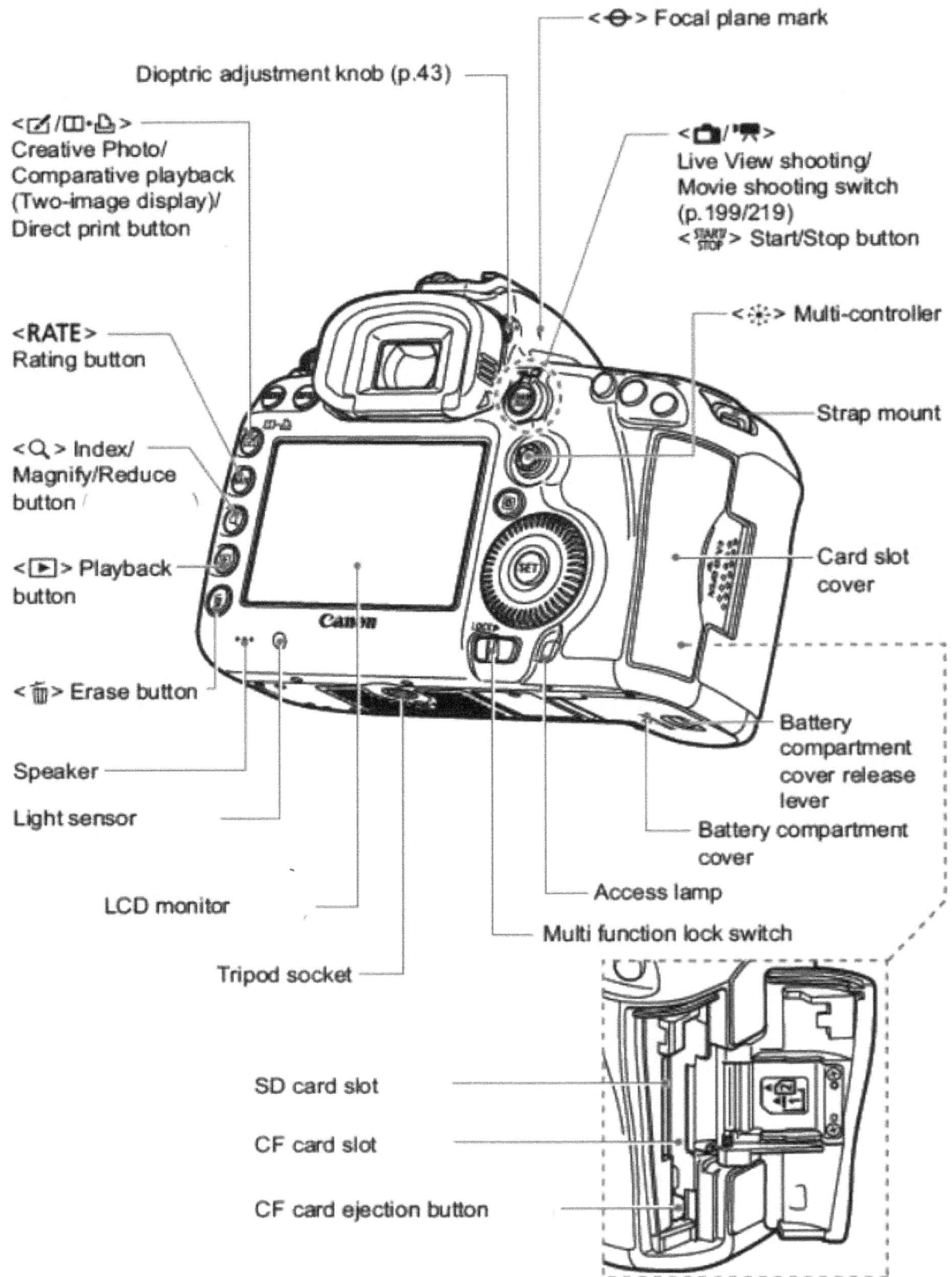

< ⊕ > Focal plane mark

Dioptric adjustment knob (p.43)

< ☑/□•⎙ > Creative Photo/ Comparative playback (Two-image display)/ Direct print button

< ⌂/'🎥 > Live View shooting/ Movie shooting switch (p.199/219)
< START/STOP > Start/Stop button

<RATE> Rating button

< ⊹ > Multi-controller

Strap mount

< Q > Index/ Magnify/Reduce button

< ▶ > Playback button

Card slot cover

< 🗑 > Erase button

Speaker

Light sensor

Battery compartment cover release lever

Battery compartment cover

Access lamp

LCD monitor

Multi function lock switch

Tripod socket

SD card slot

CF card slot

CF card ejection button

LCD Panel

Shutter speed
FE lock (FEL)
Busy (buSY)
Multi function lock warning (L)
No card warning (Card)
Error code (Err)
Cleaning image sensor (CLr)

Aperture

AF point selection
([□] AF, SEL [ ], SEL AF)
AF point registration
([□] HP, SEL [ ], SEL HP)
Card warning (Card 1/2/1.2)

Possible shots
Self-timer countdown
Bulb exposure time
Card full warning (Full)
Card error warning (Err)
Error No.
Remaining images to record

White balance
AWB Auto
☀ Daylight
🏠 Shade
☁ Cloudy
☀ Tungsten light
🔲 White fluorescent light
⚡ Flash
🔾 Custom
K Color temperature

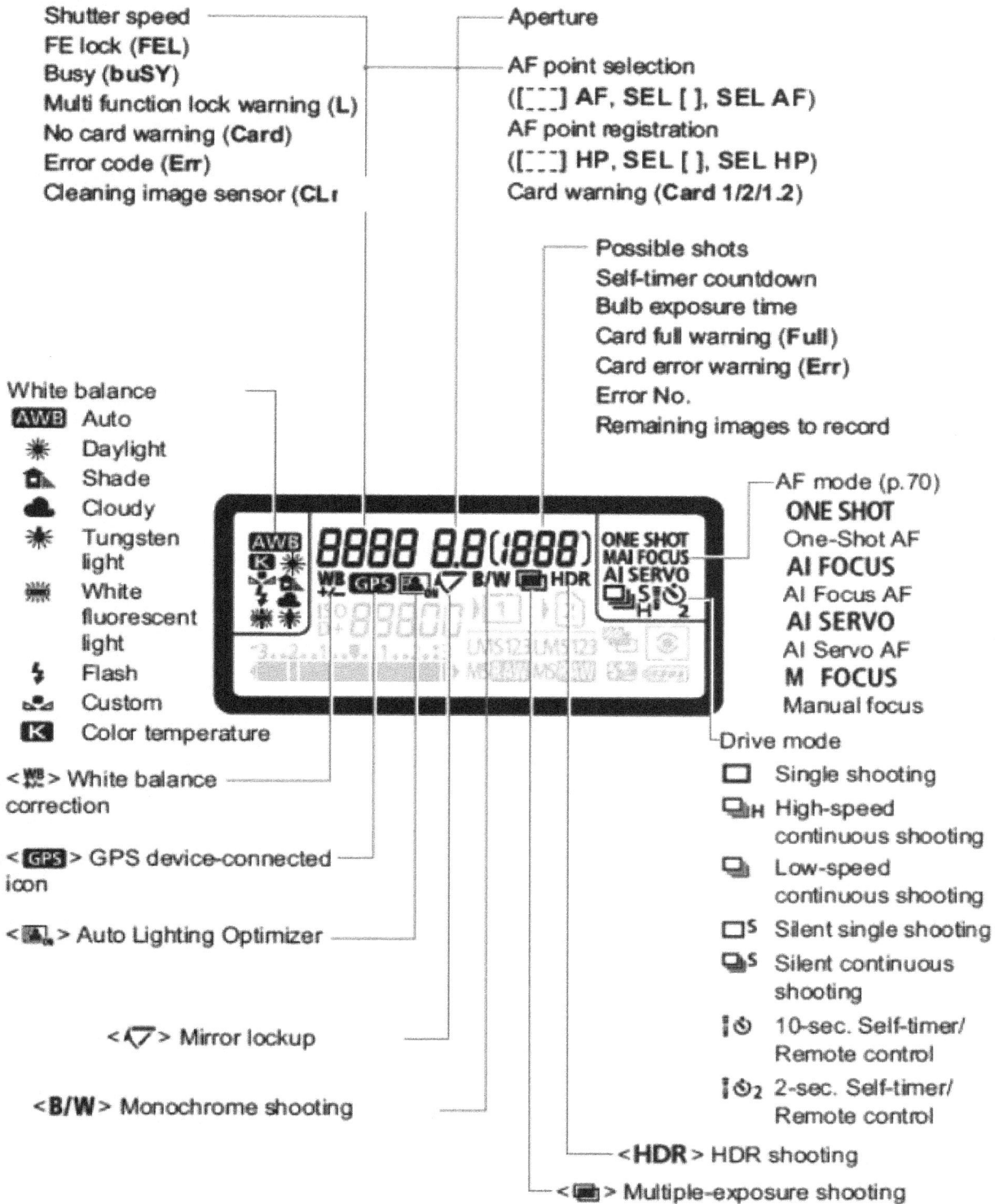

AF mode (p.70)
ONE SHOT
One-Shot AF
AI FOCUS
AI Focus AF
AI SERVO
AI Servo AF
M FOCUS
Manual focus

< > White balance correction

< GPS > GPS device-connected icon

< > Auto Lighting Optimizer

< ▽ > Mirror lockup

< B/W > Monochrome shooting

Drive mode
□ Single shooting
□H High-speed continuous shooting
□ Low-speed continuous shooting
□S Silent single shooting
□S Silent continuous shooting
⟲ 10-sec. Self-timer/ Remote control
⟲₂ 2-sec. Self-timer/ Remote control

< **HDR** > HDR shooting

< > Multiple-exposure shooting

<**1**> CF card indicator

<**▶**> CF card selection icon

ISO speed

<**ISO**> ISO speed

<**D+**> Highlight tone priority

<**▶**> SD card selection icon

<**2**> SD card indicator

<**▣**> AEB

Metering mode
- ▣ Evaluative metering
- ▢ Partial metering
- ▪ Spot metering
- ▢ Center-weighted average metering

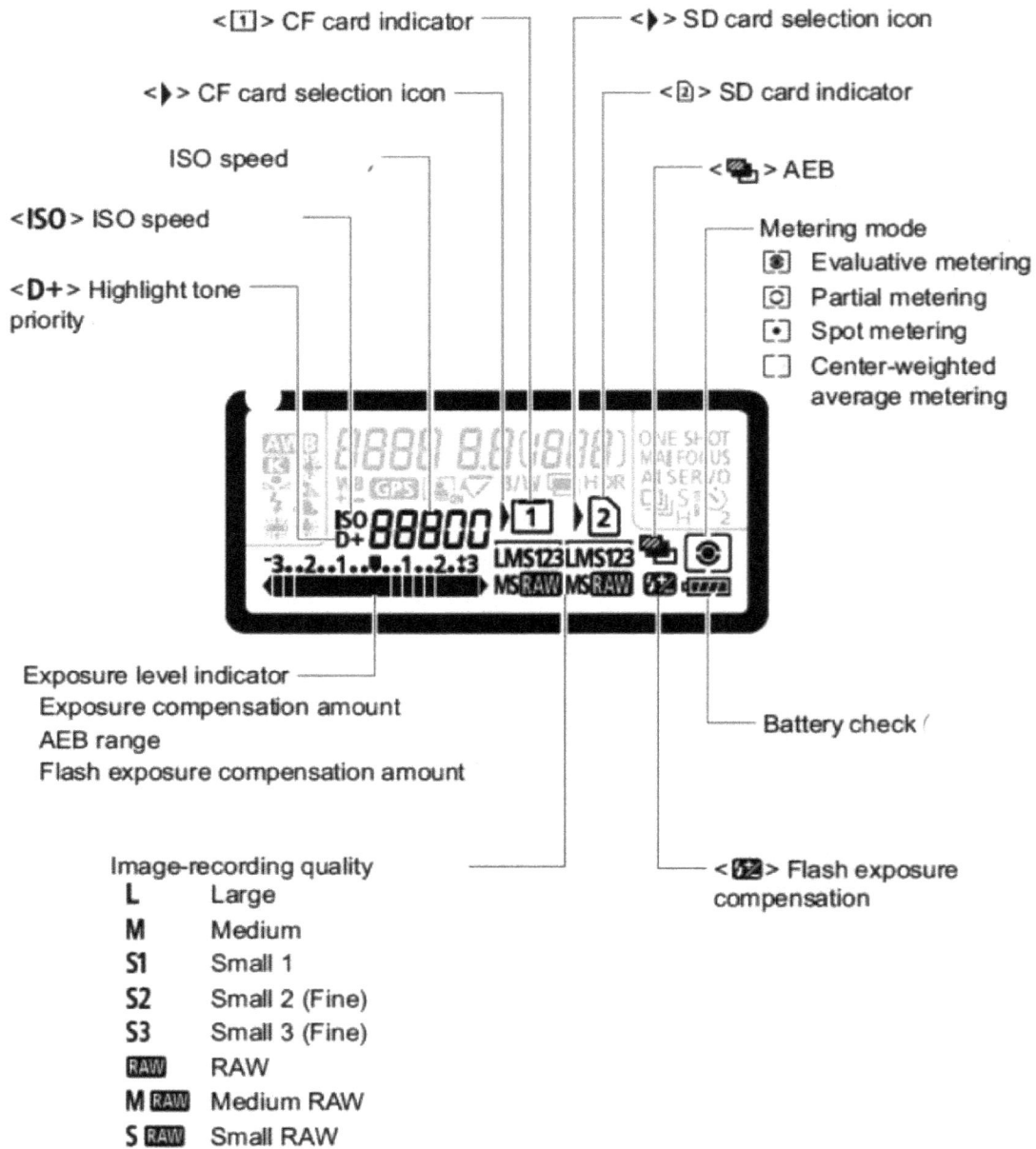

Exposure level indicator
  Exposure compensation amount
  AEB range
  Flash exposure compensation amount

Battery check

Image-recording quality

L	Large
M	Medium
S1	Small 1
S2	Small 2 (Fine)
S3	Small 3 (Fine)
RAW	RAW
M RAW	Medium RAW
S RAW	Small RAW

<**⚡±**> Flash exposure compensation